Fundamentals of Microbiology

Fundamentals of Microbiology

David Cooke

Larsen & Keller
www.larsen-keller.com

Fundamentals of Microbiology
David Cooke
ISBN: 978-1-64172-447-0 (Hardback)

Larsen & Keller
Published by Larsen and Keller Education,
5 Penn Plaza,
19th Floor,
New York, NY 10001, USA

Cataloging-in-Publication Data

Fundamentals of microbiology / David Cooke.
p. cm.
Includes bibliographical references and index.
ISBN 978-1-64172-447-0
1. Microbiology. 2. Microorganisms. 3. Biology. I. Cooke, David.
QR41.2 .F86 2020
579--dc23

The publisher's policy is to use permanent paper from mills that operate a sustainable forestry policy. Furthermore, the publisher ensures that the text paper and cover boards used have met acceptable environmental accreditation standards.

Printed and bound in China.

For more information regarding Larsen and Keller Education and its products, please visit the publisher's website www.larsen-keller.com

Table of Contents

Preface

The branch of biology which is involved in the study of unicellular, multicellular and acellular microorganisms is known as microbiology. These microorganisms are broadly divided into prokaryotic and eukaryotic microorganisms. Microbiology encompasses various sub-fields which are classified as either pure microbiology or applied microbiology. Pure microbiology includes fields such as virology, parasitology, bacteriology and mycology. Applied microbiology includes fields like medical microbiology, pharmaceutical microbiology, food microbiology and agricultural microbiology. There are widespread applications of this field such as industrial fermentation, production of amino acids and the production of biopolymers. The topics included in this textbook on microbiology are of utmost significance and bound to provide incredible insights to readers. Some of the diverse topics covered in this textbook address the varied branches that fall under this category. It will serve as a valuable source of reference for those interested in this field.

Given below is the chapter wise description of the book:

Chapter 1- Microbiology is a domain of science that deals with the study of microorganisms. The interaction between microorganisms is called microbial interaction. Two major types of interactions are positive interaction and negative interaction. The chapter examines the fundamental concepts of microbiology to provide an extensive understanding of the subject.

Chapter 2- Microorganisms are primarily classified into two types-prokaryotic and eukaryotic. A few examples of prokaryotic microorganisms are fungi, bacteria and protists while eukaryotic microorganisms include protozoa, yeast and paramecia. The topics elaborated in this chapter will help in gaining a better perspective about the diversity which exists within microorganisms.

Chapter 3- There are various subdisciplines which fall under the category of microbiology. Some of these are bacteriology, virology, parasitology and mycology. The branch of microbiology that deals with the study of morphology, biochemistry, genetics and ecology of bacteria is known as bacteriology. The chapter closely examines the different branches of microbiology to provide an extensive understanding of the subject.

Chapter 4- The study of the structure of cells along with the growth and metabolism of microorganisms within a living system is known microbial physiology. Microbial ecology is a domain that deals with the relationship of one microorganism with one another as well as with their environment. All the diverse principles of microbial physiology and ecology have been carefully analyzed in this chapter.

Chapter 5- The subject which combines microbiology with genetic engineering is known as microbial genetics. It deals with the study of microorganisms such as bacteria, virus, fungi and archaea. It has played an important role in the study of evolution, as microbes have a rapid growth rate and a short generation time. This chapter has been carefully written to provide an easy understanding of the varied facets of microbial genetics.

Chapter 6- Biotechnology is the field of biology that is concerned with living systems and organisms to develop and make products. Microbial biotechnology is a subset of biotechnology that makes use of microorganisms for industrial purposes. All the diverse principles of microbial biotechnology have been carefully analyzed in this chapter.

Chapter 7- Microbiology finds application in numerous different areas. A few of them are industrial microbiology, diagnostic microbiology, pharmaceutical microbiology and soil microbiology. This chapter closely examines these key applications of microbiology to provide an extensive understanding of the subject.

Indeed, my job was extremely crucial and challenging as I had to ensure that every chapter is informative and structured in a student-friendly manner. I am thankful for the support provided by my family and colleagues during the completion of this book.

David Cooke

1

Understanding Microbiology

Microbiology is a domain of science that deals with the study of microorganisms. The interaction between microorganisms is called microbial interaction. Two major types of interactions are positive interaction and negative interaction. The chapter examines the fundamental concepts of microbiology to provide an extensive understanding of the subject.

Microbiology is study of microorganisms or microbes, a diverse group of generally minute, simple life-forms that include bacteria, archaea, algae, fungi, protozoa, and viruses. The field is concerned with the structure, function, and classification of such organisms and with ways of both exploiting and controlling their activities.

The 17th-century discovery of living forms existing invisible to the naked eye was a significant milestone in the history of science, for from the 13th century onward it had been postulated that "invisible" entities were responsible for decay and disease. The word microbe was coined in the last quarter of the 19th century to describe these organisms, all of which were thought to be related. As microbiology eventually developed into a specialized science, it was found that microbes are a very large group of extremely diverse organisms.

Daily life is interwoven inextricably with microorganisms. In addition to populating both the inner and outer surfaces of the human body, microbes abound in the soil, in the seas, and in the air. Abundant, although usually unnoticed, microorganisms provide ample evidence of their presence—sometimes unfavourably, as when they cause decay of materials or spread diseases, and sometimes favourably, as when they ferment sugar to wine and beer, cause bread to rise, flavour cheeses, and produce valued products such as antibiotics and insulin. Microorganisms are of incalculable value to Earth's ecology, disintegrating animal and plant remains and converting them to simpler substances that can be recycled in other organisms.

Study of Microorganisms

As is the case in many sciences, the study of microorganisms can be divided into two generalized and sometimes overlapping categories. Whereas basic microbiology addresses questions regarding the biology of microorganisms, applied microbiology refers to the use of microorganisms to accomplish specific objectives.

Basic Microbiology

The study of the biology of microorganisms requires the use of many different procedures as well as special equipment. The biological characteristics of microorganisms can be summarized under the following categories: morphology, nutrition, physiology, reproduction and growth, metabolism, pathogenesis, antigenicity, and genetic properties.

Morphology

Morphology refers to the size, shape, and arrangement of cells. The observation of microbial cells requires not only the use of microscopes but also the preparation of the cells in a manner appropriate for the particular kind of microscopy. During the first decades of the 20th century, the compound light microscope was the instrument commonly used in microbiology. Light microscopes have a usual magnification factor of 1000 × and a maximum useful magnification of approximately 2000 ×. Specimens can be observed either after they have been stained by one of several techniques to highlight some morphological characteristics or in living, unstained preparations as a "wet mount."

Light Microscopy

Several modifications of light microscopy are available, such as:

- Bright Field: The specimen is usually stained and observed while illuminated; useful for observation of the gross morphological features of bacteria, fungi, algae, and protozoa.
- Dark Field: The specimen is suspended in a liquid on a special slide and can be observed in a living condition; useful for determining motility of microorganisms or some special morphological characteristic such as spiral or coiled shapes.
- Fluorescence: The specimen is stained with a fluorescent dye and then illuminated; objects that take up the fluorescent dye will "glow."
- Phase Contrast: Special condenser lenses allow observation of living cells and differentiation of cellular structures of varying density.
- Electron Microscopy: The development of the electron microscope and complimentary techniques vastly increased the resolving power beyond that attainable with light microscopy. This increase is possible because the wavelengths of the electron beams are so much shorter than the wavelengths of light. Objects as small as 0.02 nm are resolvable by electron microscopy, compared with 0.25 μm—allowing, for instance, the observation of virions and viral structures. Specimens are observed by either transmission electron microscopy or scanning electron microscopy. In TEM the electron beam passes through the specimen and registers on a screen forming the image; in SEM the electron beam moves back and forth over the surface of microorganisms coated with a thin film of metal and registers a three-dimensional picture on the screen.

Advances in microscopes and microscopic techniques continue to be introduced to study cells, molecules, and even atoms. Among these are confocal microscopy, the atomic force microscope, the scanning tunneling microscope, and immunoelectron microscopy. These are particularly significant for studies of microorganisms at the molecular level.

Nutritional and Physiological Characteristics

Microorganisms as a group exhibit great diversity in their nutritional requirements and in the environmental conditions that will support their growth. No other group of living organisms comes close to matching the versatility and diversity of microbes in this respect. Some species will grow in a solution composed only of inorganic salts (one of the salts must be a compound of nitrogen) and a source of carbon dioxide (CO_2); these are called autotrophs. Many, but not all, of these microbes are autotrophic via photosynthesis. Organisms requiring any other carbon source are called heterotrophs. These microbes commonly make use of carbohydrates, lipids, and proteins, although many microbes can metabolize other organic compounds such as hydrocarbons. Others, particularly the fungi, are decomposers. Many species of bacteria also require specific additional nutrients such as minerals, amino acids, and vitamins. Various protozoans, fungi, and bacteria are parasites, either exclusively (obligate parasites) or with the ability to live independently (facultative parasites).

If the nutritional requirements of a microorganism are known, a chemically defined medium containing only those chemicals can be prepared. More complex media are also routinely used; these generally consist of peptone (a partially digested protein), meat extract, and sometimes yeast extract. When a solid medium is desired, agar is added to the above ingredients. Agar is a complex polysaccharide extracted from marine algae. It has several properties that make it an ideal solidifying substance for microbiological media, particularly its resistance to microbial degradation.

Microorganisms vary widely in terms of the physical conditions required for growth. For example, some are aerobes (require oxygen), some are anaerobes (grow only in the absence of oxygen), and some are facultative (they grow in either condition). Eukaryotic microbes are generally aerobic. Microorganisms that grow at temperatures below 20 °C (68 °F) are called psychrophiles; those that grow best at 20–40 °C (68–104 °F) are called mesophiles; a third group, the thermophiles, require temperatures above 40 °C. Those organisms which grow under optimally under one or more physical or chemical extremes, such as temperature, pressure, pH, or salinity, are referred to as extremophiles. Bacteria exhibit the widest range of temperature requirements. Whereas bacterial (and fungal) growth is commonly observed in food that has been refrigerated for a long period, some isolated archaea (e.g., Pyrodictium occultum and Pyrococcus woesei) grow at temperatures above 100 °C (212 °F).

Other physical conditions that affect the growth of microorganisms are acidity or basicity (pH), osmotic pressure, and hydrostatic pressure. The optimal pH for most bacteria associated with the human environment is in the neutral range near pH 7, though other species grow under extremely basic or acidic conditions. Most fungi are favoured by a slightly lower pH (5–6); protozoa require a range of pH 6.7–7.7; algae are similar to bacteria in their requirements except for the fact that they are photosynthetic.

Reproduction and Growth

Bacteria reproduce primarily by binary fission, an asexual process whereby a single cell divides into two. Under ideal conditions some bacterial species may divide every 10–15 minutes—a doubling of the population at these time intervals. Eukaryotic microorganisms reproduce by a variety of processes, both asexual and sexual. Some require multiple hosts or carriers (vectors) to complete their life cycles. Viruses, on the other hand, are produced by the host cell that they infect but are not capable of self-reproduction.

The study of the growth and reproduction of microorganisms requires techniques for cultivating them in pure culture in the laboratory. Data collected on the microbial population over a period of time, under controlled laboratory conditions, allow a characteristic growth curve to be constructed for a species.

Metabolism

Collectively, microorganisms show remarkable diversity in their ability to produce complex substances from simple chemicals and to decompose complex materials to simple chemicals. An example of their synthetic ability is nitrogen fixation—the production of amino acids, proteins, and other organic nitrogen compounds from atmospheric nitrogen (N_2). Certain bacteria and blue-green algae (cyanobacteria) are the only organisms capable of this ecologically vital process. An example of microbes' ability to decompose complex materials is shown by the white and brown rot fungi that decompose wood to simple compounds, including CO_2.

Laboratory procedures are available that make it possible to determine the biochemical capability of a species qualitatively and quantitatively. Routine techniques can identify which compounds or substances are degraded by a specific microbe and which products are synthesized. Through more elaborate experimentation it is possible to determine step-by-step how the microbe performs these biochemical changes. Studies can be performed in a number of ways using growing cultures, "resting cells" (suspensions of cells), cell-free extracts, or enzyme preparations from cells.

Certain biochemical tests are routinely used to identify microbes—though more in the case of bacteria than algae, fungi, or protozoa. The adoption of routine sets of laboratory tests has allowed automated instrumentation to perform the tests. For instance, technicians often simply inoculate individual units of a "chamber" that is preloaded with a specific chemical substance (the substrate) and then place the chamber into an apparatus that serves as an incubator and analyzer. The apparatus automatically records the results and is frequently capable of calculating the degree of accuracy of the identification.

Pathogenesis

Some microorganisms cause diseases of humans, other animals, and plants. Such microbes are called pathogens. Pathogens are identified by the hosts they infect and the symptoms they cause; it is also important to identify the specific properties of the pathogen that contribute to its infectious capacity—a characteristic known as virulence. The more virulent a pathogen, the fewer the number needed to establish an infection.

Antigenic Characteristics

An antigen is a substance that, when introduced into an animal body, stimulates the production of specific substances (antibodies) that react or unite with the antigen. Microbial cells and viruses contain a variety of antigenic substances. A significant feature of antigen-antibody reactions is specificity; the antibodies formed as a result of inoculating an animal with one microbe will not react with the antibodies formed by inoculation with a different microbe. Antibodies appear in the blood serum of animals, and laboratory tests of antigen-antibody reactions are performed by using sera—hence the term serological reactions. Thus, it is possible to characterize a microorganism by its antigenic make-up as well as to identify microorganisms by using one of many different serological tests. Antigens and antibodies are important aspects of immunity, and immunology is included in the science of microbiology.

Genetic Characterization

Since the last quarter of the 20th century, researchers have accumulated a vast amount of information elucidating in precise detail the chemical composition, synthesis, and replication of the genetic material of cells. Much of this research has been done by using microorganisms, and techniques have been developed that permit experimentation at the molecular level. For instance, experiments determining the degree of similarity between different organisms' DNA and RNA have provided new insights for the classification of microorganisms. Test kits are available for the identification of microorganisms, particularly bacteria, by DNA probes.

Since the invention of recombinant DNA technology in 1973, techniques have been developed whereby genes from one cell can be transferred to an entirely different cell, as when a gene is transferred from an animal cell to a bacterium or from a bacterium to a plant cell. Recombinant DNA technology opened the door to many new medical and industrial applications of microbiology, and it plays a central role in genetic engineering.

Microbial Taxonomy

Microbial taxonomy is a means by which microorganisms can be grouped together. Organisms having similarities with respect to the criteria used are in the same group, and are separated from the other groups of microorganisms that have different characteristics.

There are a number of taxonomic criteria that can be used. For example, numerical taxonomy differentiates microorganisms, typically bacteria , on their phenotypic characteristics. Phenotypes are the appearance of the microbes or the manifestation of the genetic character of the microbes. Examples of phenotypic characteristics include the Gram stain reaction, shape of the bacterium, size of the bacterium, where or not the bacterium can propel itself along, the capability of the microbes to grow in the presence or absence of oxygen, types of nutrients used, chemistry of the surface of the bacterium, and the reaction of the immune system to the bacterium.

Numerical taxonomy typically invokes a number of these criteria at once. The reason for this is that if only one criterion was invoked at a time there would be a huge number of taxonomic groups, each consisting of only one of a few microorganisms. The purpose of grouping would be lost. By invoking several criteria at a time, fewer groups consisting of larger number of microorganisms result.

The groupings result from the similarities of the members with respect to the various criteria. A so-called similarity coefficient can be calculated. At some imposed threshold value, microorganisms are placed in the same group.

A well-known example of taxonomic characterization is the kingdom, division, class, family, genus, species and strain divisions. Such a "classical" bacterial organization, which is typified by the Bergey's Manual of Determinative Bacteriology, is based on metabolic, immunological, and structural characteristics. Strains, for example, are all descended from the same organism, but differ in an aspect such as the antigenic character of a surface molecule.

Microbial taxonomy can create much order from the plethora of microorganisms. For example, the American Type Culture Collection maintains the following, which are based on taxonomic characterization (the numbers in brackets indicate the number of individual organisms in the particular category): algae, bacteria, fungi, yeast, protozoa, animal viruses, plant viruses, and bacterial viruses. The actual number of microorganisms in each category will continue to change as new microbes are isolated and classified. The general structure, however, of this classical, so-called phenetic system will remain the same.

The separation of the microorganisms is typically represented by what is known as a dendrogram. Essentially, a dendrogram appears as a tree oriented on a horizontal axis. The dendrogram becomes increasingly specialized—that is, the similarity coefficient increases—as the dendrogram moves from the left to the right. The right hand side consists of the branches of the trees. Each branch contains a group of microorganisms.

The dendrogram depiction of relationships can also be used for another type of microbial taxonomy. In this second type of taxonomy, the criterion used is the shared evolutionary heritage. This heritage can be determined at the genetic level. This is termed molecular taxonomy.

Molecular microbial taxonomy relies upon the generation and inheritance of genetic mutations that is the replacement of a nucleotide building block of a gene by another nucleotide. Sometimes the mutation confers no advantage to the microorganism and so is not maintained in subsequent generations. Sometimes the mutation has an adverse effect, and so is actively suppressed or changed. But sometimes the mutation is advantageous for the microorganism. Such a mutation will be maintained in succeeding generations.

Because mutations occur randomly, the divergence of two initially genetically similar microorganisms will occur slowly over evolutionary time (millions of years). By sequencing a target region of genetic material, the relatedness or dissimilarity of microorganisms can be determined. When enough microorganisms have been sequenced, relationships can be established and a dendrogram constructed.

For a meaningful genetic categorization, the target of the comparative sequencing must be carefully chosen. Molecular microbial taxonomy of bacteria relies on the sequence of ribonucleic acid (RNA), dubbed 16S RNA, that is present in a subunit of prokaryotic ribosomes. Ribosomes are complexes that are involved in the manufacture of proteins using messenger RNA as the blueprint. Given the vital function of the 16S RNA, any mutation tends to have a meaningful, often deleterious, effect on the functioning of the RNA. Hence, the evolution (or change) in the 16S RNA has been very slow, making it a good molecule with which to compar microorganisms that are billions of years old.

Molecular microbial taxonomy has been possible because of the development of the technique of the polymerase chain reaction. In this technique a small amount of genetic material can be amplified to detectable quantities.

The use of the chain reaction has produced a so-called bacterial phylogenetic tree. The structure of the tree is even now evolving. But the current view has the tree consisting of three main branches. One branch consists of the bacteria. There are some 11 distinct groups within the bacterial branch. Three examples are the green non-sulfur bacteria, Gram-positive bacteria, and cyanobacteria.

The second branch of the evolutionary tree consists of the Archae, which are thought to have been very ancient bacteria that diverged from both bacteria and eukaryotic organisms billions of years ago. Evidence to date places the Archae a bit closer on the tree to bacteria than to the final branch (the Eucarya). There are three main groups in the archae: halophiles (salt-loving), methanogens, and the extreme thermophiles (heat loving).

Finally, the third branch consists of the Eucarya, or the eukaryotic organisms. Eucarya includes organisms as diverse as fungi, plants, slime molds and animals (including humans).

Single-cell Microbiology

The field of microbiology has traditionally been concerned with and focused on studies at the population level. Information on how cells respond to their environment, interact with each other, or undergo complex processes such as cellular differentiation or gene expression has been obtained mostly by inference from population-level data. New appreciation for the existence and importance of cellular heterogeneity, coupled with recent advances in technology, has driven the development of new tools and techniques for the study of individual microbial cells. As a result, scientists have been able to characterize microorganisms and their activities at unprecedented levels of detail.

Single-cell techniques have been used to more fully describe the environmental distribution and activities of microorganisms, have been a key element in revealing otherwise invisible processes such as interspecies gene transfer and chemical communication, and have been used to detail discrete physicochemical interactions between microbes and the surfaces they colonize. Single-cell methods have also been essential to our understanding of connections between cellular biochemistry and behavior and of the cellular bases of population-level phenomena. As a result, new insights into the properties of chemical signaling pathways and mechanisms behind the coordination of multicellular behaviors have been possible. Single-cell methods have also enabled direct micro- or nanoscale measurements of the mechanical properties of individual cells, including turgor pressure, elasticity, and bursting force. Microphysiological studies of metabolite, protein, or elemental localization, intracellular water dynamics, host-pathogen interactions, and surface-associated redox activity represent other areas in which single-cell techniques have been applied.

Apart from enabling fresh perspectives on issues of concern to basic science, the tools and technologies of single-cell microbiology have been brought to bear on problems of direct interest to researchers in applied science. Individual microorganisms, even those in "clonal" populations, may differ widely from each other in terms of their genetic composition, physiology, biochemistry, or behavior. This variability or heterogeneity has important practical consequences for a number of human interests, including antibiotic and biocide resistance, the productivity and stability of industrial fermentations, the efficacy of food preservatives, , and the potential of pathogens to cause disease. Additionally, methods for identification, characterization, and/or physical separation of individual microorganisms are needed for the detection of pathogens and for the identification and selection of strains with beneficial or improved properties.

Because studies made at the single-cell level are not subject to the averaging effects characteristic of bulk-phase population-scale methods, they offer a level of discrete microbial observation that is unavailable with traditional microbiological methods. Single-cell techniques have been key in probing microbial viability phenomena that are beyond the resolution of culture-based approaches, in elucidating mechanisms of pathogenesis, and in measuring the motility and the invasive forces of individual cells or hyphae.

This paper reviews some of the tools and technologies available for the study of microbes at the level of the single cell. Special interest is given to methods capable of monitoring discrete and dynamic processes occurring within living microbial cells. The limitations of traditional, population-based microbiological techniques as the motivation for the development of these single cell approaches are discussed throughout. Several of the tools and technologies discussed here have themselves been the subjects of more specialized reviews, to which the reader is referred for more detailed information. Although single-cell microbial phenomena have received attention in the past, recent advances in technology have enabled unprecedented access to processes occurring at this scale. Because our primary focus is on these recent technological advances, a historical perspective is beyond the scope of this review.

Variability is a hallmark of biological systems. Microbial cells have a remarkable capacity for displaying a multitude of genetic and nongenetic differences from each other. This inherent genetic and phenotypic plasticity forms the basis of a successful "lifestyle strategy" that enables them to adapt to and survive adverse conditions or to persist and cause disease. The central theme driving the need for methods capable of resolving the properties and activities of individual microbial cells is that of microbial heterogeneity. Bulk-scale measurements made on a heterogeneous population of cells report only average values for the population and are not capable of determining the contributions of individual cells. However, properties such as viability, protein concentration, possession of a mutant allele, or the number of flagella expressed on the cell surface are discrete and intrinsic states or properties of each individual cell. Methods capable of analyzing these properties at the level of the individual cell enable a more complete understanding of phenomena that are inaccessible to researchers using population-scale approaches.

The types of individual differences contributing to heterogeneity within a microbial population can be divided into at least four general classes: genetic differences, biochemical differences, physiological differences, and behavioral differences. The lines dividing different modes of heterogeneity are often nebulous and interactive. For example, biochemical or behavioral differences might ultimately be traced back to a genetic basis. Even physiological heterogeneity, which may be driven by forces external to the cell (e.g., nutrient limitation or the presence of antibiotics), could be viewed in terms of the organism's genetic potential to respond to these forces. However, the choice of tools used to explore cellular differences often makes it operationally clear which source of heterogeneity is the subject of investigation. For example, genetic heterogeneity is addressed using methods such as single cell PCR or fluorescence in situ hybridization (FISH), biochemical heterogeneity is measured using enzyme assays or single-cell electrophoretic separations, and behavioral heterogeneity is measured through direct observation of cellular responses to various stimuli. Examples of how individual microbial cells may vary according to their genetic, biochemical, physiological, or behavioral properties are described briefly below.

Genetic Heterogeneity

Microbial genomes can be remarkably plastic, being capable of substantial change within very short periods of time. Genetic heterogeneity in individual microorganisms can arise from a number of random, semirandom, or programmed events. Modes and mechanisms of genetic variability include spontaneous point mutations; random transcription events; phage-related phenomena (e.g., transduction and lysogeny); chromosomal duplications and gene amplification; the presence, absence, and copy number of mobile genetic elements such as plasmids and transposons; flagellar or capsular phase variation, and even intracellular genetic heterogeneity, such as that arising from transcription of multiple rRNA operons within a single cell.

Asymmetries in the distribution of genetic material between daughter cells may be important in driving processes of differentiation, as has been suggested for the strand-specific imprinting of mating-type switching in Schizosaccharomyces pombe. Processes related to cell aging, including the accumulation of DNA damage or variability in gene expression and loss of gene silencing, may also be used to describe genetic variability between individual microbial cells. Other sources of cellular heterogeneity are discussed briefly below. These can be described as nongenetic or phenotypic in nature.

Biochemical or Metabolic Heterogeneity

Biochemical or metabolic heterogeneity in a population is characterized by individual cellular differences in macromolecular composition or activity and may stem from cell cycle-related physiological processes such as turnover or from events related to aging. As the phenotypic expression of genetic phenomena, biochemical heterogeneity could also stem from mutations, programmed events associated with differentiation, or random transcription events and "noise". As with nucleic acids, proteins may also be distributed asymmetrically between mother and daughter cells. Preferential retention of oxidatively damaged proteins within the mother cell has recently been described for Saccharomyces cerevisiae, suggesting a mechanism for enhancing the fitness of newborn cells. Quantities of certain macromolecular components such as carotenoids, intracellular carbohydrate, or lipid storage polymers may also vary among individual cells, contributing to their biochemical heterogeneity.

Physiological Heterogeneity

Physiological heterogeneity stems primarily from progression through the cell cycle and describes morphological differences between individual cells, including differences in size, shape, and surface or internal characteristics. Examples of physiological heterogeneity in yeast include size differences between mother and daughter cells, bud scarring, surface wrinkling, and variation in vacuole size. Sources of physiological variation in bacteria include differences in cell volume, cell shape, buoyant density, and nucleoid morphology. More pronounced examples of cell cycle-related physiological heterogeneity occur in organisms undergoing processes of differentiation, such as sporulation or the formation of fruiting bodies. Physiological (and biochemical) heterogeneity may also be driven by microenvironmental factors acting on cells located in different strata within a colony or biofilm.

Behavioral Heterogeneity

Behavioral heterogeneity is the observable consequence of cell-to-cell variation in biochemical or physiological characteristics, such as the presence, number, state, or activity of components of chemotactic and other signaling pathways. Such variation may stem from genetic mutation or from stochastic processes affecting either gene expression or the subcellular distribution of key pathway components. Observation of individual cellular responses to chemotactic or phototactic stimuli, measurement of swimming speed or direction, and analysis of flagellar motor bias represent potential means through which behavioral heterogeneity can be explored.

Advantages of Single-cell Approaches

Plate counting and light microscopy represent the original set of tools available for single-cell analyses. As such, they have been remarkably useful for more than 100 years, and for many applications, they remain both adequate and appropriate. However, the past few decades have been marked by the introduction of a number of technological and methodological innovations, including advances in computing or imaging technologies and the development of culture-independent methods such as in situ hybridization and PCR. Progress in these areas has dramatically advanced our abilities to resolve the features and activities of individual microbial cells. Examples of some of the types of information that have been made more accessible through the use of single-cell approaches are introduced below:

1. Revealing Cryptic Processes

Microorganisms carry out a number of processes that may have substantial impact on human life. Without the proper set of tools, however, the details of these processes are inaccessible, or cryptic. Examples include gene transfer or distribution in the environment and biochemical interactions between microbial cells or between pathogens and their hosts. The nature and operation of biochemical networks occurring within individual cells and issues surrounding the gray area between cell death and viability represent other areas in which single-cell approaches have furthered our understanding of otherwise unseen microbial phenomena.

2. Observing Discrete and Dynamic Events within Living Cells

Until recently, the bacterial cell was commonly thought of as an "amorphous vessel housing a homogeneous solution of proteins". The structure of the bacterial cell, and of other microbial cells, is now recognized as being much more complex than previously imagined. Discrete subcellular domains have been observed in microbial cells in which distinct biochemical or genetic processes occur or are regulated. Additionally, certain proteins involved in control of the bacterial life cycle change their "subcellular address" over relatively short time intervals, and the activities of other proteins may be regulated according to their location within the cell. Other phenomena, such as actin polymerization in Listeria monocytogenes or protease secretion in Vibrio cholerae, occur only at the cell poles. The use of single-cell techniques allows the observation of such discrete and dynamic events occurring on or within living microbial cells with high spatial and/or temporal resolution.

3. Relating Microscopic, Mesoscopic and Macroscopic Observations

Coordinated multicellular activities such as aggregation, development of specialized structures, and colony pattern formation are visible, population-scale manifestations of individual cellular behaviors or properties. Examples of such organized phenomena include fruiting-body development in myxobacteria, mound and slug formation in Dictyostelium discoideum, chiral colony morphology in Bacillus subtilis, and coordinated movement (e.g., traveling waves, whirls, and jets) within populations of myxobacteria or B. subtilis. Methods capable of single-cell resolution enable connections to be made between these mesoscopic or macroscopic phenomena and their microscopic, cellular origins.

4. A Caveat: The "Uncertainty Principle"

Many of the methods reviewed here enable the observation of living cells under physiological or minimally invasive conditions. However, these observations may still involve the exposure of cells to potentially toxic fluorescent dyes, intense light, electric or magnetic energies, or physical manipulation using mechanical, optical, or electrokinetic forces. Alternatively, cells carrying genes for reporters such as β-galactosidase or green fluorescent protein (GFP) may experience an increased metabolic load associated with the expression of these genes. As a result, the very process of observing a cell may affect the outcome of the observation. This, in effect, is the biological equivalent of Heisenberg's "uncertainty principle". Bridson and Gould have coined the term "quantal microbiology" to describe the inherent uncertainties of microbiological phenomena at the single-cell level. An individual cell (the quantal unit here) either is exposed to a measurement or is not. Because an experiment and its control cannot be carried out on the same cell, assurances that an observation does not affect experimental results may be impossible. The inability to separate a measurement from its potential influence on an individual cell will probably be a recurrent theme in single-cell microbiology.

Tools and Technologies

A broad overview of the tools and technologies available for resolving the properties and activities of single microbial cells is provided below. Table highlights the range of studies in which these tools and technologies have been applied. Because fluorescence is of fundamental importance to many of the approaches used to investigate single-cell microbial phenomena, additional background has been included on this concept.

Fluorescence

Fluorescence is an extremely useful physicochemical property of certain molecules and compounds and, as a basic tool, has many applications in the study of single microbial cells. Fluorescence staining methods are generally rapid, are more sensitive than colorimetric techniques, and facilitate the staining of microbial cells within complex mixtures according to their individual biochemical, physiological, or taxonomic properties. Multiple fluorescent stains may be used simultaneously, allowing the collection of more than one parameter per cell, and many fluorescent stains are compatible with living cells.

Briefly, fluorescence occurs after photons from an incident light source raise electrons in a fluorophore (in many cases an organic molecule with multiple, conjugated double bonds) to a higher-energy or "excited" state. Return of the molecule to a lower-energy state is accompanied by the emission of light as fluorescence after some energy loss. Fluorescence is emitted at a lower energy (e.g., longer wavelength) than that of the original excitation light, and the difference in excitation and emission wavelengths is termed the "Stokes shift". The magnitude of the Stokes shift can be critical in ensuring spectral separation of signals from more than one fluorescent stain or when dealing with cells or sample matrices having highly autofluorescent backgrounds. Variables of practical importance to fluorescence include the intrinsic properties of the fluorophore: its excitation and emission spectra, molar absorbance coefficient, quantum yield, quantum efficiency, and photostability. The local chemical or electronic environment also plays a role, and factors such as pH, the physical proximity of other molecules in solution, and the presence of localized charge concentrations (e.g., the negatively charged backbone of DNA) can all affect the resulting fluorescence.

Fluorescent dyes and stains: Fluorescent dyes with affinities for all of the major macromolecules occurring within microbial cells are commercially available. These include stains that react with nucleic acids, proteins, or lipids or that stain polyester or polyphosphate inclusion bodies. Additionally, fluorescent enzyme or respiratory substrates, reporters of intracellular pH or ion concentration, and dye kits providing "fluorescent Gram staining" are available. The performance of these commercial kits is often validated using specific microorganisms grown under standardized conditions. However, if these assays are to be used with different microorganisms or natural populations, they must be revalidated under the new conditions, since basic physiological differences or increased biochemical heterogeneity within these populations may complicate data analysis. The difficulties in transferring multiparameter staining protocols across generic or species boundaries may be even more pronounced.

Macromolecules such as lectins, antibodies, and nucleic acid probes may be labeled with fluorescent dyes to create conjugates capable of reporting molecular recognition events. Other fluorescence-based molecular methods, such as in situ PCR or in situ reverse transcription, may result in the incorporation of fluorescently labeled deoxynucleoside triphosphates into reaction products as they are formed within the cell. Endogenous sources of fluorescence, including carotenoids, tryptophan, thiamine (after chemical derivitization), and the cell's own photopigments, may also serve as reporter molecules, for instance in industrial or environmental applications.

Dynamic microbial phenomena, including protein expression and behavior substrate uptake binding and release of individual chemoattractant molecules to cell surface receptors, selective degradation of uniparental DNA within newly formed algal zygotes bacterivory and drug efflux may also be observed or measured at the single-cell level through the use of fluorescence staining techniques. Specialized techniques such as fluorescence ratio imaging microscopy may provide insights into dynamic cellular events that are important to the outcome of microbial fermentations which highlight the physiological responses of spoilage organisms to chemical stresses or that are related to cellular inactivation resulting from treatment with antimicrobials.

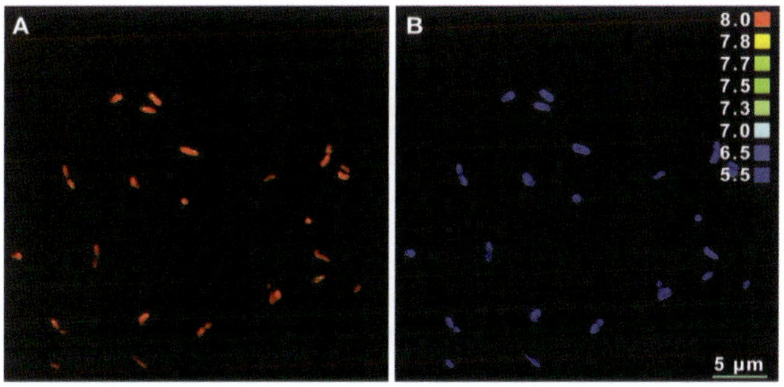

Fluorescence ratio imaging of nisin-mediated dissipation of ΔpH in Listeria monocytogenes. Fluorescence ratio imaging microscopy was used to monitor the intracellular acidification of broth-grown cells of L. monocytogenes after exposure to the membrane-permeabilizing lantibiotic nisin. The pH-dependent spectral response of carboxyfluorescein diacetate succinimidyl ester (CFSE) was used as a probe of intracellular pH (pHi). The ratio of CFSE fluorescence intensity at 490 nm to that at 435 nm was calibrated over a pH range of 5.0 to 9.0. (A) Live, intact cells of L. monocytogenes maintained pHi values between 8.0 and 8.4, even when the pH of the external medium was low (e.g. pH 5.5). (B) Nisin-mediated membrane permeabilization resulted in the equilibration of pHi with the pH of the medium after 12 min of exposure. Individual cellular responses were more heterogeneous for cells derived from colonies, suggesting the importance of microenvironmental factors in differential susceptibility to nisin (not shown). A color-coded pH scale is shown in the upper right-hand corner.

Staining with multiple fluorescent labels can yield detailed information on the identity and activities of individual microbial cells. For example, the combined use of FISH and the fluorescent respiratory substrate CTC can yield data on both genetic identity and respiratory activity. The ability to correlate single target cells with their metabolic activities could provide greater information on which to base important decisions, such as those regarding food safety or productivity in industrial fermentations.

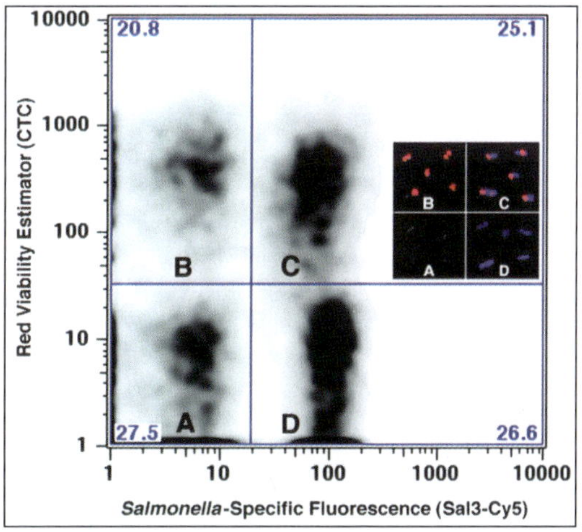

Flow cytometric analysis of a genetically and metabolically complex cell mixture. This figure illustrates the power of single-cell staining methods in combination with flow cytometric analysis for the fluorescent "dissection" of complex microbial populations. Here, Salmonella enterica serotype

Typhimurium is differentiated from a mixture of E. coli, Citrobacter freundii, Proteus vulgaris, and Shigella dysenteriae on the basis of both cytochemical activity (CTC staining) and genetic identity (Salmonella-specific FISH staining). A complex mixture containing both live and formalin-killed representatives of each cell type was incubated with CTC, fixed with 10% buffered formalin, hybridized with a Salmonella-specific DNA probe (Sal3-Cy5), and examined by flow cytometry. Four distinct populations can be seen. Clockwise from the bottom left, they are dead non-Salmonella members of the Enterobacteriaceae (A), live non-Salmonella members of the Enterobacteriaceae (B), live Salmonella (C), and dead Salmonella (D). The numbers in each quadrant represent percentages of the total population. The photographic inset provides a visual interpretation of the cytometry data.

However, multiplex fluorescence assays may be limited by the need to balance dye properties and instrument capabilities. Incompatible spectral or chemical properties and requirements for multiple excitation sources can place practical constraints on the fluorescent dye combinations that can be used. Recently, though, a new class of compounds with promise as fluorescent labels has been introduced. Fluorescent semiconductor nanocrystals, or "quantum dots," have several advantages over conventional fluorescent labels, including large extinction coefficients and reduced susceptibility to photobleaching. However, the most intriguing properties of these labels are their narrow, size-dependent (and therefore "tunable") emission spectra and the fact that differently emitting nanocrystal labels may be excited with a single UV light source. Recent work has shown that fluorescent nanocrystals can be directed to specific tissues or cell types if they are coated with antibodies or homing peptides. These fluors may also allow long-term labeling of live cells without interfering with cell growth and development. These studies highlight the potential of fluorescent nanocrystals for improving the performance of multicolor single-cell analyses while minimizing the requirements for specialized equipment.

Fluorescence in situ hybridization and immunofluorescence: In the FISH technique, fluorescently labeled nucleic acid probes are hybridized to complementary rRNA targets located on ribosomes within whole, permeabilized cells. The ribosome is a naturally amplified target molecule, especially in actively growing cells, where each cell may contain several thousand ribosomes. The aggregate signal from multiple probe-ribosome binding events leads to the sequence-specific fluorescence of target cells. Apart from rRNA, other forms of RNA (e.g., tmRNA) can serve as a target for hybridizations, especially if a signal amplification step is used. Recently, FISH-based methods have also been developed to detect low-copy-number targets on plasmid or chromosomal. This approach differs substantially from rRNA-targeted FISH in that it utilizes polynucleotide probes (~50 to 1,200 nucleotides in length), higher (1,000-fold) probe concentrations, and much longer hybridization times. The resulting fluorescent signal is also qualitatively different from that achieved with classic rRNA-targeted FISH and is characterized by the formation of a fluorescent "halo" around the periphery of target cells. The technique has thus been named RING-FISH.

Fluorescently labeled antibodies also enable the detection of diagnostic molecular binding events and can be directed against surface antigens, such as capsular, flagellar, or cell wall antigens, or against internal targets, including ribosomal proteins or cell cycle-specific cytoplasmic proteins.

FISH and immunofluorescence have substantial overlap in their applications and benefits as single-cell detection techniques. Both are whole-cell methods, and as such they can preserve a wealth of potentially valuable information that is unavailable outside the context of the intact cell. Apart from providing information on microbial identity, information about cell morphology, number, and distribution may also be collected for specific target cells. Both methods have the potential to be carried out simultaneously or in succession with other means of cell characterization, including the observation of light-scattering characteristics, staining of inclusion bodies, fluorescence-based measurements of nucleic acid or protein content, cytochemical characterization using fluorescent or colorimetric enzyme substrates, and microautoradiography. The combination of FISH or fluorescent-antibody labeling with methods for high-throughput multiparametric data collection, analysis, and sorting (e.g., flow cytometry) can be especially useful in the study of complex microbial populations.

FISH is used primarily as a means of detecting specific microbial cells, although the intensity of staining with FISH has also been used to provide an indication of physiological activity. For the most part, the use of FISH for microbial detection has involved DNA-based methods, but peptide nucleic acid probes may have substantial practical and functional advantages, especially for the detection of gram-positive bacteria.

As a means of detection, fluorescent-antibody approaches can be limited by problems with cross-reactivity, variable antigen expression under different culture conditions, or potential instability and loss of cell surface epitopes. However, unlike FISH, immunofluorescence-based detection does not require cell permeabilization and can therefore be used on living cells, potentially followed by isolation for culture. Apart from their use as taxonomic probes, fluorescent antibodies may be used for fine-structure analyses, such as the discrete localization of specific proteins within individual cells.

Neither FISH nor immunofluorescence approaches require that a cell be culturable. However, because the number of target antigens may not be as tightly coupled to the cell growth rate as is the rRNA copy number, fluorescent-antibody techniques may yield higher detection sensitivities for dormant cells than FISH does.

Green fluorescent protein and related reporters: GFP is a versatile tool for the in vivo visualization of protein expression, localization, and functionality. Because it retains its fluorescence after fixation with paraformaldehyde, GFP can be combined with fixation-dependent staining methods such as FISH. However, the true power of GFP is as a visual reporter of dynamic events occurring in living cells. For example, Raskin and de Boer used GFP fusions to probe the function of proteins associated with cell division in Escherichia coli. They observed a regular, pole-to-pole oscillation for GFP-MinD and theorized that the cell may use this protein as a "measuring device" to continuously probe the location of the center of the cell. Cluzel et al. used a cheY-gfp fusion, fluorescence correlation spectroscopy, and video microscopy to relate CheY-GFP expression levels to flagellar-rotation behavior in single cells of E. coli. These authors found that small changes in the concentration of CheY-P led to large changes in the rotational bias of the flagellar motor, suggesting that the motor itself acts as a signal amplifier and that additional cellular mechanisms exist for maintaining CheY-P concentrations within the operational range of the motor.

Other applications of GFP include the construction of whole-cell sensors for in situ monitoring of iron availability on leaf surfaces; measurement of cytoplasmic viscosity and protein diffusion rates in living cells; investigation of quorum-based interspecies communication or coordinated, multicellular behaviors measurement of the internal pH of bacterial cells and real-time reporting of fungal susceptibility to antimicrobial compounds.

GFP is especially well suited to in situ analyses of individual cells within complex consortia such as biofilms. Because its use does not require preparative steps such as dehydration, fixation, or application of exogenous probes or cofactors, GFP labeling enables the observation of microorganisms directly in these fragile structures. The range of applications of GFP has been further expanded with the introduction of fluorescence-shifted spectral variants. In a novel application of such variants, Fehr et al. created chimeric "nanosensor" proteins based on the fusion of enhanced cyan fluorescent protein (ECFP), a bacterial maltose binding periplasmic protein, and enhanced yellow fluorescent protein (EYFP). Conformational changes of these nanosensors on binding of maltose led to more efficient fluorescence resonance energy transfer (FRET) from ECFP to EYFP. When these nanosensors were expressed in S. cerevisiae, changes in ECFP/EYFP FRET ratios enabled maltose uptake and compartmentation to be monitored in individual living cells. The broad range of organic and inorganic substrates recognized by periplasmic binding proteins suggests the use of this strategy in generating fluorescent nanosensors specific for a wide variety of analytes.

Stochasticity, or noise, in gene expression can lead to substantial phenotypic variation among individual cells in an otherwise clonal population. Such noise can be either instrinsic (stemming directly from events related to the expression of a gene) or extrinsic (resulting from fluctuations in the quantities or activities of the enzymes and other cellular machinery required for gene expression). In a novel application of GFP variants, Elowitz et al. constructed strains of E. coli capable of distinguishing between these two sources of noise in gene expression. Their results indicated that both sources of noise contribute to the generation of phenotypic heterogeneity among individual cells. Their findings also suggested that any component in a cellular biochemical network that is prone to intrinsic fluctuations in concentration can serve as a source of extrinsic noise for other components in the network.

Cytometry

"Cytometry" is a general term that may apply to any technology used to measure, count, compare, or otherwise characterize biological cells. The general term has become nearly synonymous with flow cytometry, due to the popularity of this technique. However, other forms of cytometry have specialized advantages for use in single-cell microbial studies and, along with flow cytometry, are discussed below:

1. Flow cytometry: Flow cytometry is a powerful fluorescence-based diagnostic tool that enables the rapid analysis of entire cell populations on the basis of single-cell characteristics. Multiple characteristics, including cell count, cell size or content, and responses to fluorescent probes diagnostic of cell function may be collected simultaneously by this method. Cells in a liquid sample are passed individually in front of an intense light source (e.g., a laser, laser diode, or arc lamp), and data on light scattering and fluorescence are collected and saved as a data file. Detailed numerical analyses of populations and subpopulations of interest can then be carried out offline by using a number of analysis packages. Because of its capacity to collect information-rich data sets on thousands of cells, flow cytometry facilitates valuable insights into connections between single-cell and population-level processes not available with other techniques. Flow cytometers capable of sorting cells on the basis of their fluorescence characteristics or of simultaneous in-line video microscopy add to the versatility of this method. Flow cytometry has proven to be an invaluable resource in the study of apoptosis in mammalian cells. Recent work has suggested that programmed cell death is

not limited to eukaryotes but may also be active in prokaryotic systems. Therefore, flow cytometry may also be a useful tool for elucidating these processes in bacteria.

2. Laser scanning cytometry: Flow cytometry collects data on single cells in a liquid sample as they stream past the illumination source. Although multiple light scatter and fluorescence parameters may be measured, cells pass only once through the system. Because of this, flow cytometry is not suited for time-resolved studies of individual cells. An exception may be the microfluidic cell sorter described by Fu et al. in which the fluid flow may be stopped or reversed, allowing multiple observations of the same cell, but this technology is not yet widely available.

Laser scanning cytometry (LSC) is a solid-phase cytometric technology for collecting laser-induced fluorescence from cell samples on slides or on membrane filters. At their simplest, LSC instruments provide a rapid means of counting, quantifying, and recording the distribution of fluorescent events on a filter. Microscope-based LSC instruments can provide visual information on both cell morphology and the spatial distribution of fluorescence within each cell. Because LSC can be used to make multiple measurements of the same cells, this technique is well suited for the observation of cellular properties as a function of time. Examples include monitoring the kinetics of fluorescence staining in living cells (e.g., substrate uptake, enzyme activity, and dynamic changes in intracellular pH) and observing interactions between neighboring cells. Spatial "addressing" of fluorescent events may facilitate the reexamination of archived samples. The ability to concentrate cells prior to analysis gives filter-based LSC methods definite practical advantages over fluorescence microscopy or flow cytometry when working with dilute suspensions of microorganisms in filterable liquids. However, because LSC may involve exposing the sample to the excitation source for relatively long periods, photobleaching of fluorescent labels could be problematic for some applications, particularly if multiple scans are required. These effects can, in part, be minimized through the use of low-intensity (microwatt versus milliwatt) illumination sources.

3. Image cytometry: The terms "image cytometry" and "image analysis" are used interchangeably here to describe a wide range of methods by which quantitative biological information may be extracted from microscopic images. These techniques can be used to gain information on individual cell properties such as staining intensity and label specificity; cell number, size, and volume; and distribution within a field of view. Advanced image analysis techniques can be used to monitor ultradiscrete physical phenomena such as the micronewton invasive forces generated by individual fungal appressoria.

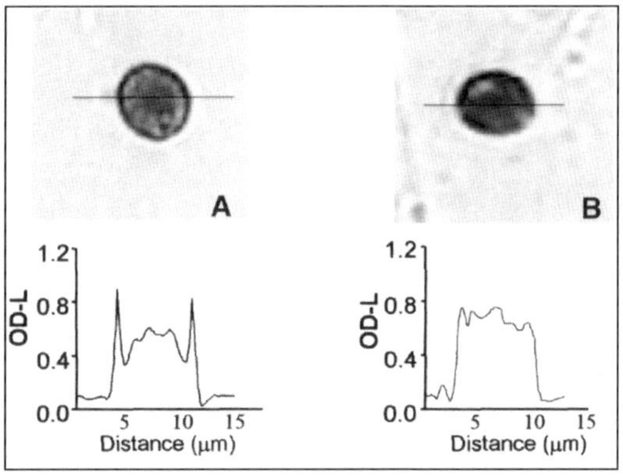

Single-cell determination of yeast glycogen content by image cytometry. The glycogen content of individual S. cerevisiae cells was determined from their optical density (OD) values after staining with Lugol's solution (I2-KI). Images were processed using a series of steps designed to extract quantitative information about cell size, shape, volume, and OD. The OD profile of cell A (glycogen poor) shows concentrated staining only on the periphery of the cell, whereas the profile of cell B (glycogen rich) shows dense staining throughout. To avoid overestimation of the mean glycogen content in glycogen-poor cells, only the central portion of each cell was used for measurement. The ability to quantitate the glycogen content in individual cells allows the determination of glycogen distribution within a population. Because the character of this distribution is related to yeast quality, image cytometry can be used as a tool for quality control. OD-L, optical density, Lugol staining.

Most image analysis methods incorporate some form of colorimetric or fluorescent cell staining (Gram staining, Lugol's solution, colorimetric enzyme substrates, 4′,6-diamidino-2-phenylindole [DAPI], FISH, etc). This provides a means of cell identification or characterization and generation of high-contrast images suitable for further processing. Alternatively, intrinsic changes in the light-scattering characteristics of a microorganism as it undergoes certain physiological processes may be sufficient to allow analysis. An example is the phase-bright to phase-dark transition of bacterial spores on germination, which has been used to investigate the variability of germination among individual spores of Clostridium botulinum (M. W. Peck, personal communication).

Image collection is often followed by a number of processing steps designed to facilitate extraction of the desired information. These include thresholding, filtering, edge detection, removal of optical artifacts (e.g. fluorescent "halos"), background subtraction, pixel averaging, and other transformations. Advantages of such thorough image processing may include the ability to differentiate target cells from background material, particularly in "difficult" sample matrices such as soil. Through the use of special algorithms, images may be automatically processed on the basis of user-defined criteria or artificial neural networks may be trained for the automatic classification of objects. Such automation can greatly aid image processing, especially where manual data extraction would be impossible, tedious, or error prone. Recently, a fully automated high-throughput microscopy system has been described that combines computer-controlled autofocusing and stage movement with advanced image segmentation, classification, and retrieval algorithms. With the ability to rapidly acquire and categorize data from large numbers of cells on slides or in microtiter plates, such high-throughput microscopy systems may eventually become competitive with flow cytometry as a method for the rapid and detailed analysis of populations on a cell-by-cell basis. However, due to the longer integration times often needed for imaging-based techniques, efforts must be made to minimize the effects of photobleaching, which are not a significant issue with flow cytometry.

In addition to single still images, multiple still images from a time series or video images may be analyzed. Video-based images, which allow the display of a time code with each frame, can provide a continuous record of a cellular measurement with high temporal resolution. Dynamic cellular properties associated with cell motility (cell speed, the number and duration of runs or tumbles, etc.) or changes in fluorescence related to some physiological characteristic can be resolved in terms of both space and time. Movements of individual cells within a larger population can also be monitored, enabling connections to be made between cell behaviors at both microscopic (individual) and mesoscopic (population) scales.

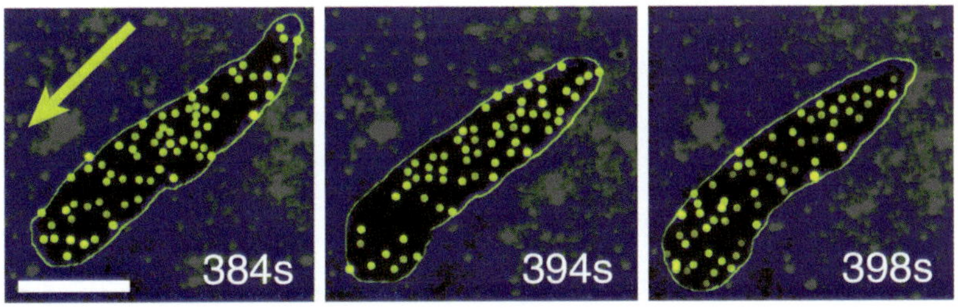

Single-molecule analysis of cyclic AMP (cAMP) receptor occupancy on the surface of a Dictyostelium discoideum cell during chemotaxis. Cells were exposed to a gradient of Cy3-labeled cAMP (Cy3-cAMP, shown to be functional as a chemoattractant) and observed for up to 10 min. Binding of Cy3-cAMP to cell surface receptors was monitored at single-molecule resolution by using total internal reflection fluorescence microscopy. Occupied cAMP receptors appear as bright yellow dots on the surface of the cell. The arrow indicates the direction of the Cy3-cAMP source. The time for each sequential image is given in seconds. Kinetic analysis showed that Cy3-cAMP receptor complexes located on anterior pseudopods dissociated faster than those on the posterior tail. This work enabled the discrete characterization of receptor dynamics in single living cells, suggesting a role for cell polarity in the chemotactic process.

Scanning Probe Microscopies

Scanning probe microscopies (SPMs) are a related group of technologies which can yield information on both the topography and the mechanical, electrochemical, electrostatic, or magnetic properties of a sample surface. In all SPM formats, samples are imaged by rastering a cantilever-mounted tip over the surface of the sample in the x-y plane. Direct (e.g., physical) and indirect (e.g., atomic force) interactions result in z-plane deviations of the cantilever. These deviations can reveal topographical details in the sample at atomic resolutions. The SPM family of tools includes scanning tunneling microscopy, atomic force microscopy (AFM), scanning electrochemical microscopy (SECM), and magnetic force microscopy. SPM technologies have found widespread use in materials science applications and are fast becoming recognized for their potential to characterize biological materials, including single living cells.

1. Atomic force microscopy: AFM is a member of the SPM family of tools, the forerunner of which was the scanning tunneling microscope (STM). The central mechanism of an AFM consists of a cantilever, or "arm," to which a very sharp probe, or "tip," is attached. An often-made and apt comparison is to the arm and needle of a (very small) phonograph. The cantilever arm may be only 100 μm long, and, ideally, the tip, or "needle," should terminate in a single atom. As the tip is scanned across a surface, tip-sample interactions cause deflections of the cantilever, which are detected and amplified by a laser. These interactions may be direct (e.g., physical), or indirect (e.g., electrostatic, electrosteric, and van der Waals' forces). Conversion of cantilever deflection data to topographical information results in both qualitative output (e.g., images) and quantitative output (e.g., measurement of interaction forces and force-distance relationships).

2. Several modes of imaging are used contact, noncontact, and tapping: Contact imaging involves "dragging" the tip across the sample and may give rise to undesirable effects such as frictional forces and sample damage. Noncontact imaging based on electrostatic deflection of the probe tip can be used to investigate charge development or distribution on biological surfaces. Tapping-mode

imaging was developed as an alternative method for measurements of "soft" biological surfaces likely to sustain damage during contact imaging. In this technique, the tip does not scrape the sample but oscillates over its surface, minimizing tip-sample frictional forces.

AFM is capable of measuring discrete interaction forces in the piconewton range. Because little sample preparation is needed and cells may be observed in liquid environments, AFM can be used for detailed ultrastructural studies of the surfaces of living microbial cells. Dynamic events, such as bacterium-mineral adhesion interactions and viral exocytosis, may be measured in real time, under native conditions of hydration and oxygen tension.

AFM tips may be chemically functionalized to study properties such as cell surface hydrophobicity. Alternatively, they may be functionalized with biomolecules such as biotin, antibodies, enzymes, or even single, intact microbial cells. AFM cantilevers with such functionalized tips can be used as "nanobiosensors" for the study of discrete receptor-ligand interactions or for characterization of cell-substrate interactions.

AFM can also be used as a method for the nanomechanical manipulation of individual microbial cells. As such, it can be used to provide quantitative data on cellular physical properties such as rigidity or elasticity. Measurements of force-distance relationships for AFM tip indentation have also provided a direct means of measuring turgor pressure in individual bacterial cells.

3. Scanning electrochemical microscopy: Electrochemical phenomena such as electron transfer and ion fluxes are associated with both energy production and intracellular signaling processes. Well-established techniques for the electrochemical characterization of single, living microbial cells include the use of microelectrodes or patch-clamping approaches. SECM is a recently introduced, SPM-based tool for mapping redox activity in living cells. In SECM, the scanning tip is an ultramicroelectrode designed for measuring charge transfer reactions. Grayscale images, or "redox maps," are generated from variations in tip current as the tip is scanned in the x-y plane above an electrochemically active cell. SECM has been used for the electrochemical visualization of oxygen production in single algal protoplasts on exposure to light, for assessment of the permeability of membrane to charged redox species, and for electrochemical studies of Rhodobacter sphaeroides cells. SECM imaging is carried out in solutions containing hydrophilic or hydrophobic redox species which function to mediate the transfer of electrons between cellular redox centers and the SECM tip. Redox mediators may differ in their abilities to penetrate various cellular permeability barriers (e.g., the outer membrane versus the cytoplasmic membrane). Therefore, carefully chosen mediators may facilitate redox studies of physiologically distinct cellular structures, such as the periplasmic space.

Microspectroscopic Methods

The term "spectroscopy" describes methods used to separate a light signal into its component wavelengths. More generally, the term can be used to describe the same process for other regions of the electromagnetic spectrum (emission or absorption) or for analogous processes such as mass spectrometry, where complex mixtures are separated on the basis of the molecular masses of their components.

In biology, as in astronomy, the spectral characteristics of an object can be used to provide information on its chemical or physical makeup. Spectroscopic methods have been used to monitor the presence and activities of natural microbial populations via remote-sensing techniques for spectral identification of bacterial suspensions in pure culture by Fourier- transform infrared, proton nuclear magnetic resonance, or mass spectroscopic methods; for noninvasive investigations of biochemical changes in P. mirabilis populations during differentiation; and for the characterization of pigmented colonies formed by various photosynthetic bacteria.

However, when they are applied at the population level, spectroscopic measurements suffer from the same drawbacks as other bulk-scale approaches, and contributions from individual microorganisms cannot be assessed. Spectroscopic methods capable of single-cell resolution (e.g., microspectroscopic methods) enable the observation of target analytes or properties within specific cells at cellular or subcellular scales. This can be especially important because information obtained within the context of a whole cell may reveal important clues to the role or function of the analyte within the cell.

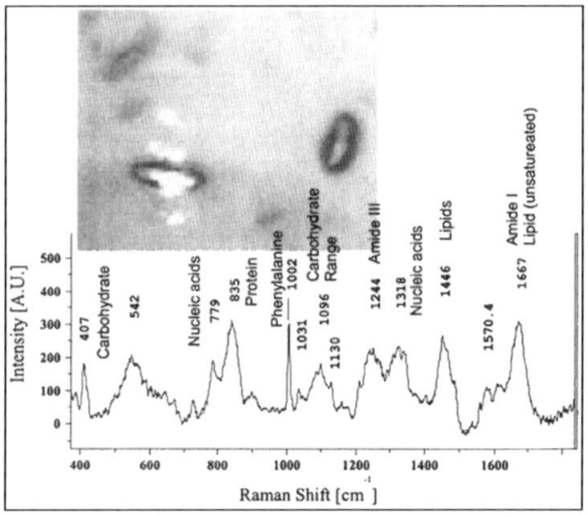

Raman spectrum of a single Clostridium beijerinckii cell. Spectral peaks ascribed to major cellular macromolecules (e.g., nucleic acids, proteins, lipids, and carbohydrates) are shown. The video inset shows a cell illuminated in the laser focus, which is of approximately the same diameter as the cell. The laser diffraction pattern, which can serve as a visual cue for achieving the proper laser focus, can also be seen. Single-cell Raman spectroscopy represents a noninvasive means of investigating the biochemical heterogeneity of microbial populations.

As an alternative to "wet-chemistry" methods, spectroscopic approaches allow target compounds to be analyzed quickly and without the need for extraction, which may be incomplete or may introduce artifacts, especially when dealing with potentially labile species or features (e.g., carotenoids or elemental inclusion bodies). Additionally, many spectroscopic approaches require only minimal sample preparation and may be used for the analysis of living cells.

Microspectroscopic methods can provide biochemical information on the overall macromolecular composition of cells or on specific analytes at either whole-cell or subcellular resolutions. Vibrational spectroscopy may also be used to generate images of individual cells by using data from the aliphatic C—H stretching within membrane lipids or from the O—H stretching of intracellular water molecules.

Not all spectroscopic methods provide direct information on the chemical composition of a cell. Methods such as electrorotation can be used to determine other characteristics of single cells, including their dielectric properties. Some of the more frequently used methods for obtaining spectroscopic data from single microbial cells are described below:

1. Raman microspectroscopy: The Raman effect is an induced emission of light resulting from the inelastic scattering of a small number of photons from a monochromatic light source. Raman spectra provide information on molecular vibrational states, which are dependent on the nature of chemical bonding within a molecule or sample. These spectra yield clues to the types and lengths of chemical bonds present and on the molecular conformation or environment. Microspectroscopic Raman probes capable of illuminating an area as small as 1 by 1 μm enable the characterization of individual cells and their subcellular components. Spectra may be collected at different points along the length of a cell or hypha or at different depths. Spectra may also be compared among different species, between mutant strains, or at different points in the cell cycle.

The range of energies used to generate Raman spectra includes UV (e.g., 257 nm), visible, and infrared excitation frequencies. Common visible sources used are argon-ion lasers (~514 nm) and helium-neon lasers (~632 nm). When the wavelength of the incident light approaches the absorption wavelength of a chromophore within a sample, scattering efficiency is greatly increased, an advantageous effect referred to as "resonance" Raman scattering. An example of an application where resonance enhancement would be expected to occur is in the UV-Raman analysis of nucleic acids.

Other applications of Raman microspectroscopy include investigations of microbial carotenoid content and subcellular distribution. Kubo et al. used Raman microspectroscopy to map the carotenoid content in Euglena and in Chlamydomonas reinhardtii. These authors also applied polarization techniques to demonstrate that the carotenoid molecules in the eyespot of C. reinhardtii are oriented parallel to the long axis of the cell. Raman microspectroscopy has also been used to investigate biochemical differences between cells in morphologically heterogeneous cultures of clostridia during solvent fermentations. Traditional, bulk-scale methods of analysis of these differentiated cultures do not facilitate connections between the morphological appearance of individual cells and their biochemistry or role in the fermentation. Analyzed by Raman microspectroscopy, morphologically distinct cells yielded spectra that differed in regions ascribed to proteins, lipids, or the storage polymer granulose. Analysis of small cell clusters also showed spectral evidence for the presence of polysaccharides, suggesting the presence of aggregation-promoting extracellular polymers. The ability to correlate morphology with biochemical characteristics may provide clues to the activities of the different cell types during solvent production. Other applications of Raman microspectroscopy include the reagentless identification of individual bacterial spores and detection of the neurotoxic amino acid domoic acid in single cells of toxigenic phytoplankton.

In related technology, coherent anti-Stokes Raman scattering microscopy allows imaging of individual microbial cells on the basis of the vibrational spectra of specific cellular components (e.g., proteins and lipids). The vibrational signatures of these molecules provide a means of generating contrast. In this way, the distribution of specific molecular components in living cells can be mapped without the need for fluorescent dyes and at relatively low power levels. Finally, time-resolved coherent anti-Stokes Raman scattering imaging can reveal dynamic processes, such as real-time changes in intracellular water concentration.

2. Microbeam analysis: Methods for microbeam analysis represent sensitive means of characterizing

single-cell elemental composition. These methods allow the measurement of the concentration, chemical state, or cellular location of biologically relevant inorganic nutrients, including phosphate, sulfur, potassium, calcium, iron, and zinc. Multiple elements can be measured in a single pass, resulting in an "elemental map" of an individual cell Available techniques include X-ray fluorescence imaging and absorption spectroscopy, as well as various ion beam-dependent methods.

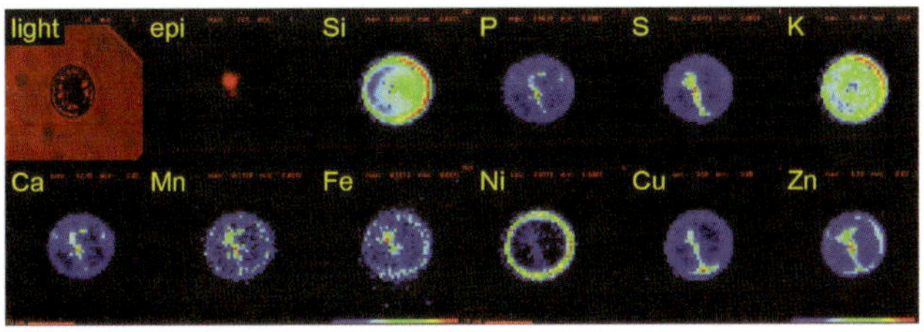

Synchrotron X-ray fluorescence mapping of the relative elemental distribution in a single diatom. An X-ray microprobe was used to focus a monochromatic X-ray beam on a diatom collected from the Southern Ocean. The sample was scanned through the focused beam in pixel steps of 0.5 µm, and the full X-ray fluorescence spectrum was collected at each step. Two-dimensional elemental maps were generated from the resulting energy spectra by using element-specific filtering. Elemental concentrations were calculated from X-ray fluorescence data using National Institute of Standards and Technology thin-film or similar standards. Comparison of light and epifluorescence (epi) micrographs with elemental maps for the same diatom enabled the discrete localization of each element within the cell. Data collected for this diatom show that Si and K map onto the cell's siliceous frustule; P, S, Ca, Mn, Fe, Cu, and Zn appear to be associated with the cytoplasm; and Ni is present only on the outer membranes or frustule. This approach to characterizing elemental distributions within individual diatoms provides biologically relevant information not available from population-scale methods of elemental analysis. The resulting data may offer unique insights into the physiological state, ambient chemical environment, and role in elemental cycling of these organisms.

X-ray microprobe techniques may not require extensive sample preparations, allowing biological materials to be examined in their natural, hydrated states. This may be essential for ensuring the stability of the chemical (oxidation) states of elements within the sample. Plant roots infected with mycorrhizal fungi have been studied by X-ray fluorescence imaging at elemental sensitivities of 500 ppb. With an X-ray beam spot of 1 by 3 µm, elemental mapping at single hyphal resolution is possible.

In contrast to the minimal preparative requirements for X-ray microprobe analysis, samples to be studied by ion beam methods may need to be dried and vacuum compatible, constraints that could hinder the analysis of many cell types. A major disadvantage of all microbeam methods described here is that they are very time-consuming. Scanning times ranging between 30 min and 4 h can be required to generate an image.

Still, single-cell microbeam analysis may provide useful information on the physiological states of individual microbial cells, as illustrated in the study by Gisselson et al. As these authors noted, sample preparation prior to traditional measurements of algal nutrient ratios may include fractionation steps designed to isolate the subset of the planktonic community to be studied. Despite such careful preparations, contributions to nutrient ratios from bacteria, protists, or particulate organic material may still skew the results. Microbeam analysis methods can be used effectively to ensure that measurements are derived from the intended cell type and to probe the nutritional heterogeneity of individual target cells within the population.

3. Electrorotation: When a cell is exposed to an electric field, a dipole is induced, whose character is dependent on the composition of the cell, the frequency of the applied electric field, and the conductivity of the medium in which the cell is suspended. In the presence of a rotating electric field, the dipole will form across the cell in synchrony with the rotation rate of the field. If the field is rotating with sufficiently high frequency, though, formation of the dipole may become asynchronous with the field's rotation rate. In this case, the cell will experience a torque and begin to rotate, either in the direction of the field ("cofield rotation") or in the opposite direction ("antifield rotation"), depending on the angular difference between the field and the induced dipole.

Precise positioning of the electrodes used to generate the rotating field is used to create a dielectrophoretic trap capable of holding individual cells in place during analysis. Electrorotation spectra are displayed as cellular rotation rate versus frequency of the applied field. Cell rotation rates can be automatically measured and documented using computer-interfaced video microscopy or interferometric methods. Because the dielectric properties of a cell are responsive to mechanical or chemical perturbation, methods for dielectric spectroscopy such as electrorotation can yield information on both the integrity and the physicochemical properties of individual cells. Compared to other methods for investigating the electrical properties of cells (e.g., the use of microelectrodes or patch clamping), electrorotation is relatively noninvasive and does not require extensive cell preparations. Applications for electrorotation include monitoring the effects of antibiotics on single yeast cells and distinguishing between nonviable and viable protozoan cysts on the basis of the direction of their rotation at specific field frequencies. The ability of electrorotation to distinguish between viable and nonviable protozoa is especially useful, since no direct culture-based methods are currently available.

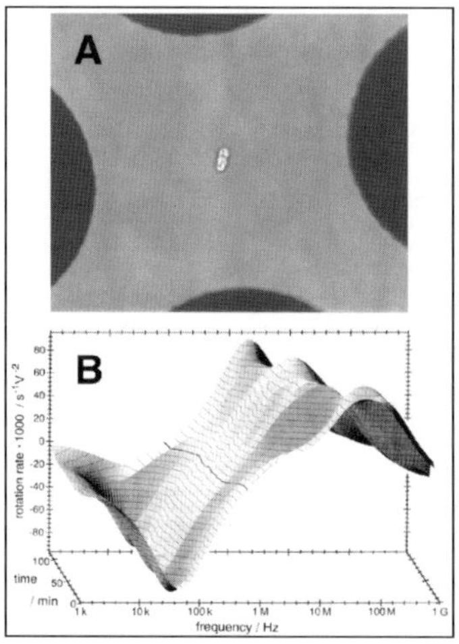

Electrorotational analyses of single yeast cells. (A) Single cell of S. pombe during analysis in a microstructured electrorotation chamber. Four circular electrodes (dark semicircles), spaced 100 μm apart, are precisely positioned to allow dielectrophoretic trapping of individual cells. (B) Cellular dielectric properties are responsive to mechanical or chemical perturbation. This panel illustrates time-resolved changes in the electrorotation spectra of a single S. cerevisiae cell treated with nystatin at t = 12 min. Nystatin-mediated leakage of intracellular ions is expected to change the dielectric properties of the cell, leading to the frequency-dependent shifts in cell rotation rates seen for both cofield and antifield rotations.

Micromanipulation

In some instances, a means of physically manipulating individual microbial cells may be needed. Examples include the isolation of cells for subsequent analyses such as single-cell PCR, the selection of cells with unique or beneficial characteristics, and the isolation of cells to obtain pure cultures of microorganisms that are difficult to purify using traditional culture-based methods. This can also be extended to include the direct isolation of dormant, stressed, or otherwise unculturable cells for further study.

Micromanipulation can be used to address a cell to a specific position in a liquid medium and hold it there in order to examine its ability to replicate. Physical segregation of daughter cells may also be used to trace the pedigree of a single cell as it undergoes multiple cycles of division or to examine adaptation processes of individual cells subjected to changes in nutrient availability.

Alternatively, a cell may be positioned in close proximity to or touched against other cells, immobilized enzyme substrates, or inorganic surfaces. In this way, discrete binding, chemical, or other interaction forces may be measured. Micromanipulative techniques also allow the stable positioning of cells for observation during single-cell assays for pharmacological or biochemical activity.

Aside from methods of physical separation or positioning, micromanipulation may permit the direct measurement of the physical or structural characteristics of an individual cell. AFM and related force transduction technologies can be used to measure turgor pressure, elasticity, bursting force, and other micro- or nanomechanical cellular properties.

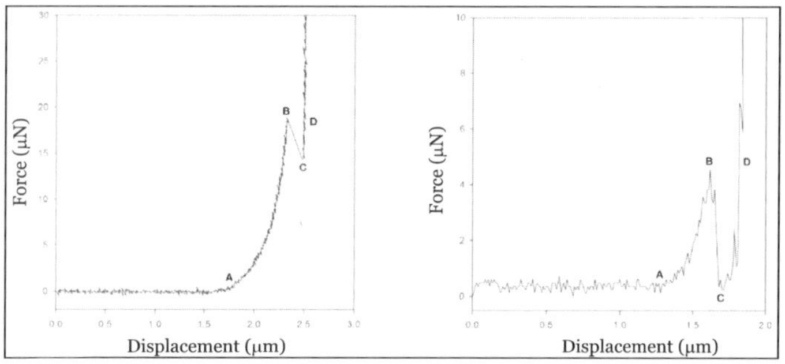

Comparison of whole-cell bursting responses from Staphylococcus epidermidis (left) and E. coli (right), as determined by micromanipulation. Force diagrams for single cells compressed between the surface of a glass slide and an optical fiber are shown. In both diagrams, datum point A indicates the first contact of the microprobe with the cell and datum point B corresponds to the point at which cell rupture takes place. The microprobe continues to advance after cell bursting, eventually compressing the cellular debris (points C and D). To correlate bursting properties with cell size, a video image was taken of each cell prior to manipulation. Data kindly provided by C. Shiu, Z. Zhang, and C. R. Thomas.

Methods of mechanical, optical, or electrokinetic micromanipulation can also be used to measure forces exerted by a microorganism on its environment, including motile power, pilus retraction forces, and the torque generated by an individual flagellar motor. Currently available methods for the manipulation of individual microbial cells are described briefly below:

1. Mechanical micromanipulation: The use of mechanical means of manipulation of single microbial cells is not a new concept. In 1951 Zelle used a microscope-mounted mechanical micromanipulator to directly monitor the pedigrees of individual E. coli cells positioned on the surface of an agar-covered slide. However, the advent of computer-assisted stage or micromanipulator movement and more sensitive methods for fluid aspiration and deposition has led to improvements in the basic technology. Together, these improvements have resulted in more accurate, more accessible, and less exacting processes for the mechanical manipulation of microbial cells.

Current technologies enable the direct isolation of individual cells of interest from within complex natural populations. For example, Frölich and König used a sterile capillary tube method to isolate individual cells of Enterococcus and Sphingomonas spp. from the diluted contents of a termite gut. Their procedure involved suspension and manipulation of cells in microvolume quantities of a cell transfer medium (phosphate-buffered saline). Dilution and microsuspension-based approaches to micromanipulation are probably less stressful to cells than are processes which result in the absorption of energy and heating (e.g., optical and electrokinetic methods).

Other mechanical methods, including AFM and methods for microprobe-based force transduction enable direct measurements of the physical or structural properties of individual cells at micro- and nanoscale resolutions. Although AFM has been used to estimate whole-cell properties such as turgor pressure, it is particularly well suited to probing local mechanical properties such as cell stiffness and elasticity. Microprobe-based force transduction methods may also yield information on cell elasticity, deformability, and bursting strength, but they provide whole-cell rather than localized measurements. Finally, mechanical force spectroscopy can be used to characterize binding interactions between cells, shedding new light on cell-cell adhesion events important for multicellular development.

2. Optical micromanipulation: Although the Sun exerts a radiation pressure on the Earth's surface, its light is diffuse and the resulting pressure is negligible. However, highly collimated light sources, such as lasers, can exert a focused radiation pressure that is substantial enough to manipulate large particles, including microbial cells. As a result, optical forces can be used to trap, move, pull, twist, or cut individual cells. Optical manipulation may also be used to measure forces exerted by a microorganism on its environment. This is accomplished by measuring the laser power needed to displace a cell or by observing the ability of a cell to escape from a known trapping force. Because optical manipulation requires no physical contact, cells can be manipulated within enclosed glass chambers under sterile conditions.

Although most applications (e.g., trapping and moving) are relatively noninvasive, cell injury and death can occur from photodamage incurred during manipulation. Variables involved in photodamage include both the wavelength and power of the light source and environmental factors such as the presence or absence of oxygen. As a means of minimizing cell damage, optical trapping is usually carried out using wavelengths in the near-infrared (NIR) region, which do not coincide with absorption or excitation maxima for most biological chromophores or fluorophores. Under certain circumstances, however, NIR optical sources can be used to elicit fluorescence, and

NIR trapping may be used in conjunction with separate excitation sources. The ability to simultaneously trap or move microorganisms while monitoring their fluorescence can simplify fluorescence-based studies of otherwise motile microorganisms and can facilitate the harvesting of specific cells on the basis of their fluorescent properties. Additional applications of optical micromanipulation methods include the photorelease of caged compounds and the manipulation of specific organelles within cells.

More invasive uses of optical micromanipulation include microsurgical applications, such as laser ablation of fungal hyphae prior to patch clamping, and insertion of individual. Agrobacterium cells into plant cells using a combination of "optical scissors" and "optical tweezers" techniques.

3. Electrokinetic micromanipulation: Although electrorotation can be thought of as a microspectroscopic tool, it also can be used as a method of micromanipulation. As such, it provides a non-contact means of holding cells in place in a liquid medium. Once a cell has been immobilized in a field of view, its response to added nutrients, antibiotics, or fluorescent enzyme substrates may be monitored visually or with dielectric measurements.

Like electrorotation, dielectrophoresis depends on the polarization of a cell exposed to an external electric field (e.g., formation of a whole-cell dipole). The interaction between a polarized cell and a nonuniform electric field leads to the generation of unequal forces on opposing sides of the dipole, resulting in net movement of the cell. Depending on differences in polarizability between the cell and the surrounding medium, net dielectrophoretic movement may be either attractive ("positive" dielectrophoresis) or repulsive ("negative" dielectrophoresis). As a micromanipulative tool, dielectrophoresis can be used to trap, move, separate, or concentrate cells based on their dielectric properties.

Other single-cell technologies may also incorporate methods of electrokinetic micromanipulation in order to separate or move individual cells. For example, in flow cytometry, cell sorting is accomplished through the electrostatic deflection of sheath fluid droplets containing cells of interest. Also, many microfluidic devices, such as the "cytometer-on-a-chip" described by Fu et al. rely on electroosmotic flow to move cells during analysis. In electroosmotic flow, ionic movement in response to an electric field results in bulk fluid movement, representing an indirect means of electrokinetic manipulation of cells. Methods for the charge-based separation of whole microbial cells, which may also fall under the rubric of "electrokinetic manipulation".

Microcapillary Electrophoresis

Charge-based microscale separations have long been a staple of analytical chemistry, and these methods are now being applied to the analysis of single microbial cells. Microcapillary methods can be used for either isoelectric focusing of whole cells or the electrophoretic separation of intracellular analytes from a single cell after lysis. Microcapillary separations are relatively rapid, and reactions can be monitored using either UV absorption or fluorescence detection methods. Fluorescence-based monitoring can be used to observe whole-cell staining with diagnostic fluorescent dyes or to characterize metabolic transformations of fluorescent enzyme substrates in single cell lysates.

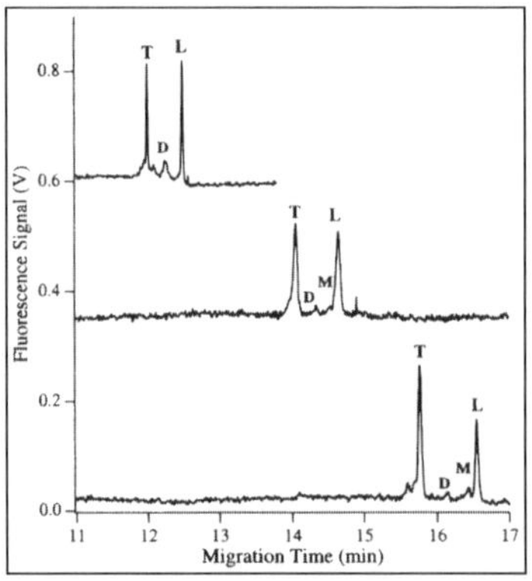

Serial microcapillary electrophoresis of three individual S. cerevisiae spheroplasts. The labeled peaks (T, D, M, and L) represent, respectively, the original substrate and three different fluorescent hydrolysis products of a tetramethylrhodamine-labeled triglucoside. The hydrolysis products accumulated in each cell due to in vivo enzymatic activity. The larger peaks (T and L) contain between 500 and 1,000 molecules of each fluorescent analyte. To ease visual comparison, the electropherograms from the first two cells have been shifted upward on the y axis. This work demonstrates the capacity of microcapillary electrophoresis to analyze sequential metabolic reactions occurring in single microbial cells (e.g., "metabolic cascades").

Whole-cell isoelectric focusing separates microbial cells on the basis of their surface properties (e.g., charge). Surface properties may vary with cell type, age, or physiological conditions and may also be altered by processes of differentiation or after exposure to antibiotics or chemical preservatives. Therefore, microcapillary methods may have wide-ranging applications in the study of single microbial cells.

By definition, separation methods are applied to cell populations, but the number of cells analyzed by some microcapillary methods may be very small (~3 to 15 cells per capillary) and single cells may be easily resolved. Technical difficulties with whole-cell microcapillary separations may include the amphoteric nature of some cell types and problems with cell clumping, adhesion to capillary walls, or cell lysis.

Microcapillary-based methods for performing biochemical separations of the contents of individual cells, including yeast, have also been described. Individual cells may be delivered into a capillary tube by suction forces or electroosmotic flow, with monitoring via microscopy. After a cell is loaded into the capillary, it is lysed to release the cell contents for analysis. The addition of surfactant may be sufficient to lyse mammalian cells, but thick-walled microorganisms such as S. cerevisiae may require spheroplasting first.

Krylov et al. found that if electrophoresis was begun immediately following cell lysis, enzymatic activity was effectively "quenched" and artifacts stemming from nonmetabolic enzymatic activity or other degradative processes were suppressed. Although this work was done on mammalian cells, it emphasizes the capacity of single-cell methods to provide high-resolution data and to avoid artifacts common to bulk-scale preparative methods. Individual cells and population-level extracts analyzed using the same method showed significantly different product distributions. In the population-level extracts, enzyme decompartmentalization stemming from the extraction process was responsible for nonmetabolic substrate degradation. The greater variability in peak heights seen between single-cell separations was attributed to the metabolic heterogeneity of individual, asynchronously grown cells.

Biological Microelectromechanical Systems

The term "biological microelectromechanical systems" (BioMEMS) describes a family of devices that combine electrical, mechanical, chemical, and/or microfluidic approaches for the microscale analysis of biological materials. BioMEMS "chips" are capable of integrating several analytical steps (e.g., cell capture, concentration, addressing, and lysis, with subsequent extraction, purification, amplification, and detection of target analytes) within a single microscale device. These devices may use pressure, acoustic energy, dielectrophoresis, or electroosmotic flow to exercise precise control over very small volumes of liquids. Steps such as the manipulation of cells, the introduction, mixing, and washing of reagents, temperature cycling, and analyte detection can be carried out sequentially within the same device. Although the amount of analytical material handled by a single BioMEMS device is small, so are the amounts of potentially expensive reagents used.

Analytical chambers fabricated at near-cellular dimensions prevent the diffusive loss of analytes expressed by individual cells, allowing their measurement at low levels. While the need for enrichment of target cells prior to analysis is a major disadvantage of macroscale detection and diagnostic methods, BioMEMS and related microscale approaches may allow the capture and analysis of individual microbial cells, which may lessen or preclude the need for such enrichment. Multiple BioMEMS devices may be operated in parallel, and they are amenable to automation, presenting the possibility for continuous, high-throughput performance of analytical processes that once were the exclusive domain of highly trained personnel. Analytical methods that have been successfully translated to the microscale and could potentially be incorporated within a BioMEMS device capable of single-cell analysis include flow cytometry and cell sorting PCR, various isothermal methods of nucleic acid amplification, and nuclear magnetic resonance.

Microfabrication techniques commonly used for the construction of BioMEMS devices include silicon micromachining and lithography, chemical etching, laser ablation, photopolymerization, micromolding, and embossing. These processes can be used to create the valves, channels, reservoirs, and other discrete microstructures critical to the function of BioMEMS devices and may also allow the incorporation of sensing or control elements such as microelectrodes or ion-selective field-effect transistors. Examples of actuators, or the "moving parts" of BioMEMS devices, include pH-responsive hydrogel valves, ferrofluidic micropumps, and even microrobotic "arms" fabricated from conducting polymer bilayers. Microrobotic devices such as these, which are capable of manipulating individual micron-scale objects within an aqueous environment, could conceivably be used for the discrete positioning or transfer of individual cells between analytical stations within a BioMEMS device.

Microbial Interaction and Types

Microorganisms interacts with each other and can be physically associated with another organisms in a variety of ways. One organism can be located on the surface of another organism as an ectobiont or located within another organism as endobiont. Microbial interaction may be positive such as mutualism, proto-cooperation, commensalism or may be negative such as parasitism, predation or competition.

Types of Microbial Interaction

- Positive interaction: mutualism, proto-cooperation, commensalism.
- Negative interaction: Ammensalism (antagonism), parasitism, predation, competition.

Mutualism

- It is defined as the relationship in which each organism in interaction gets benefits from association. It is an obligatory relationship in which mutualist and host are metabolically dependent on each other.
- Mutualistic relationship is very specific where one member of association cannot be replaced by another species.
- Mutualism require close physical contact between interacting organisms.
- Relationship of mutualism allows organisms to exist in habitat that could not occupied by either species alone.
- Mutualistic relationship between organisms allows them to act as a single organism.

Examples of Mutualism

- Lichens
    - Lichens are excellent example of mutualism.
    - They are the association of specific fungi and certain genus of algae. In lichen, fungal partner is called mycobiont.
    - Phycobiont is member of cycanobacteria ad green algae (Trabauxua).
    - Because phycobionts are photoautotrophs, the fungus get its organic carbon directly from algal partner, in turn fungi protects the phycobiont from extreme conditions and also provide water and minerals to algae.
    - Lichen grow very slowly but are able to colonies habitat that do not permit the growth of other organisms.
    - Most lichens are resistant to high temperature and drying.
- Protozoan-termite
    - Protozoan-termite relationship is the classical example of mutualism in which flagellated protozoan lives in the gut of termites.
    - Theses flagellated protozoan feeds on diet of carbohydrates acquired as cellulose or lignin by their host termites, metabolize into acetic acid which is utilized by termites.

- Paramecium-Chlorella
  - Paramecium (protozoa) can host Chlorella (algae) within its cytoplasm.
  - The algae Chlorella provide the protozoan partner with organism carbon and O_2, in turn protozoa provide protection, mortility, CO_2 and other growth factors.
  - The presence of Chlorella within Paramecium helps to survive protozoa in anaerobic condition as long as there is sufficient light.

Syntrophism

- It is an association in which the growth of one organism either depends on or improved by the substrate provided by another organism.
- In syntrophism both organism in association gets benefits.
  - Compound A
    - Utilized by population 1
  - Compound B
    - Utilized by population 2
  - Compound C
    - Utilized by both Population 1+2
  - Products
- In this theoretical example of syntrophism, population 1 is able to utilize and metabolize compound A, forming compound B but cannot metabolize beyond compound B without co-operation of population 2. Population 2 is unable to utilize compound A but it can metabolize compound B forming compound C. Then both population 1 and 2 are able to carry out metabolic reaction which leads to formation of end product that neither population could produce alone.

Examples of Syntrophism

- Methanogenic ecosystem in sludge digester:
  - Methane produced by methanogenic bacteria depends upon interspecies hydrogen transfer by other fermentative bacteria.
  - Anaerobic fermentative bacteria generate CO_2 and H_2 utilizing carbohydrates which is then utilized by methanogenic bacteria (Methanobacter) to produce methane.
- Lactobacillus arobinosus and Enterococcus faecalis:
  - In the minimal media, Lactobacillus arobinosus and Enterococcus faecalis are able to grow together but not alone.
  - The synergistic relationship between E. faecalis and L. arobinosus occurs in which E.

faecalis require folic acid which is produced by L. arobinosus and in turn lactobacillus require phenylalanine which is produced by Enterococcus faecalis.

Protocooperation

- It is a relationship in which organism in association is mutually benefited with each other.
- This interaction is similar to mutualism but the relationships between the organisms in protocooperation is not obligatory as in mutualism.

Examples of Protocooperation

- Association of Desulfovibrio and Chromatium: it is a protocooperation between carbon cycle and sulfur cycle.
- Interaction between N_2-fixing bacteria and cellulolytic bacteria such as Cellulomonas

Commensalism

It is a relationship in which one organism (commensal) in the association is benefited while other organism (host) of the association is neither benefited nor harmed.

It is an unidirectional association and if the commensal is separated from the host, it can survive.

Examples of Commensalism

- Non-pathogenic coli in intestinal tract of human:
    - E. coli is a facultative anaerobe that uses oxygen and lower the O_2 concentration in gut which creates suitable environment for obligate anaerobes such as Bacteroides. E. coli is a host which remains unaffected by Bacteroides.
- Flavobacterium (host) and Legionella pneumophila (commensal):
    - Flavobacterium excrete cystine which is used by Legionella pneumophila and survive in aquatic habitat.
    - Association of Nitrosomonas (host) and Nitrobacter (commensal) in Nitrification:
    - Nitrosomonas oxidize Ammonia into Nitrite and finally Nitrobacter uses nitrite to obtain energy and oxidize it into Nitrate.

Amensalism (Antagonism)

- When one microbial population produces substances that is inhibitory to other microbial population then this inter population relationship is known as Ammensalism or Antagonism.
- It is a negative relationship.
- The first population which produces inhibitory substances are unaffected or may gain a competition and survive in the habitat while other population get inhibited. This chemical inhibition is known as antibiosis.

Examples of Antagonism (Amensalism)

- Lactic acid produced by lactic acid bacteria in vaginal tract:
    - Lactic acid produced by many normal floras in vaginal tract is inhibitory to many pathogenic organisms such as Candida albicans.
- Skin normal flora:
    - Fatty acid produced by skin flora inhibits many pathogenic bacteria in skin.
- Thiobacillus thiooxidant:
    - Thiobacillus thioxidant produces sulfuric acid by oxidation of sulfur which is responsible to lowering of pH in the culture media which inhibits the growth of most other bacteria.

Competition

- The competition represents a negative relationship between two microbial population in which both the population are adversely affected with respect to their survival and growth.
- Competition occurs when both population uses same resources such as same space or same nutrition, so, the microbial population achieve lower maximum density or growth rate.
- Microbial population competes for any growth limiting resources such as carbon source, nitrogen source, phosphorus, vitamins, growth factors etc.
- Competition inhibits both population from occupying exactly same ecological niche because one will win the competition and the other one is eliminated.

Examples of Competition

- Competition between Paramecium cadatum and Paramecium aurelia:
    - Both species of Paramecium feeds on same bacteria population when these protozoa are placed together.
- P. aurelia grow at better rate than P. caudatum due to competition.

Parasitism

- It is a relationship in which one population (parasite) get benefited and derive its nutrition from other population (host) in the association which is harmed.
- The host-parasite relationship is characterized by a relatively a long period of contact which may be physical or metabolic.
- Some parasite lives outside host cell, known as ectoparasite while other parasite lives inside host cell, known as endoparasite.

Examples of Parasitism

- Viruses

  - Viruses are obligate intracellular parasite that exhibit great host specificity.
  - There are may viruses that are parasite to bacteria (bacteriophage), fungi, algae, protozoa etc.
- Bdellovibrio:
  - Bdellavibrio is ectoparasite to many gram negative bacteria.
  - The parasite Bdellovibrio penetrate the outer membrane of its host and enters periplasmic space but not inside host cytoplasm.

Predation

- It is a wide spread phenomenon when one organism (predator) engulf or attack other organism (prey).
- The prey can be larger or smaller than predator and this normally results in death of prey.
- Normally predator-prey interaction is of short duration.

Examples of Predation

- Protozoan-bacteria in soil:
  - Many protozoans can feed on various bacterial population which helps to maintain count of soil bacteria at optimum level.
- Bdellovibrio, Vamparococcus, Daptobacter etc are examples of predator bacteria that can feed on wide range of bacterial population.

Microbial Cooperation

Cooperation presents a fundamental challenge to customary evolutionary thinking. If only the fittest organisms survive, why would an individual ever pay a fitness cost for another organism to benefit? Traditionally, kin selection and group selection have been the most prominent explanations for the maintenance of cooperation in nature. Kin selection refers to cooperative behaviors being favored when they are preferentially directed towards relatives. In turn, group selection, also known as multilevel selection, suggests that altruistic traits can be favored because of their beneficial effect on a group, despite the individual cost of such behaviors.

While the connection between these two theories was initially unclear, recent theoretical work has brought to light many of the similarities and differences between the two concepts. In particular, it has increasingly become appreciated that the underlying theme behind all mechanisms for the evolution of altruism is the assortment of similar individuals. When similar individuals are assorted, cooperators are more likely than average to interact with other cooperators and non-cooperating defectors are, in turn, more likely to interact with defectors. This assortment is ascribed to relatedness in kin selection and between-group variance in multilevel selection models. When applied correctly and under certain assumptions, both methods are equivalent. Unfortunately, when these assumptions do not hold, both methods often resort to abstract generalities, making application difficult and prone to mistakes.

Microbes present a unique opportunity for scientists interested in the evolution of cooperation because of their well characterized and simple genetics, fast generation times, and easily manipulated and measured interactions. While these advantages are often well appreciated, other differences between organisms of the microscopic and macroscopic world are sometimes forgotten when transferring ideas and methods from the study of animals to that of microbes. Important differences include strong selection strengths, fast evolution times, high levels of diversity, and non-linear dynamics, all of which invalidate many less general techniques derived using specific assumptions, a fact too often ignored or simply unknown by non-theorists.

Four basic Classes of Cooperation

To begin, let us define cooperation as any act that increases the fitness of others. Now, if we have two strains of a microbe, one an obligate cooperator and the other a defector that never cooperates, there are still four fundamentally different classes of interactions that fit within this definition of cooperation. For example, imagine mixing the two strains at different relative fractions in a test tube, either: one strain performs better than the other at every frequency, or one strain grows faster than the other at some fractions and slower at other fractions. If one strain always does better–either the cooperator or defector–then we say that it dominates and natural selection will always favor the increase in frequency of the dominating strain. Thus, if we grow the cells in a chemostat where the conditions and total population size are kept constant by continuous dilution, then the dominated strain will eventually go extinct–barring any stochastic effects which we will ignore here.

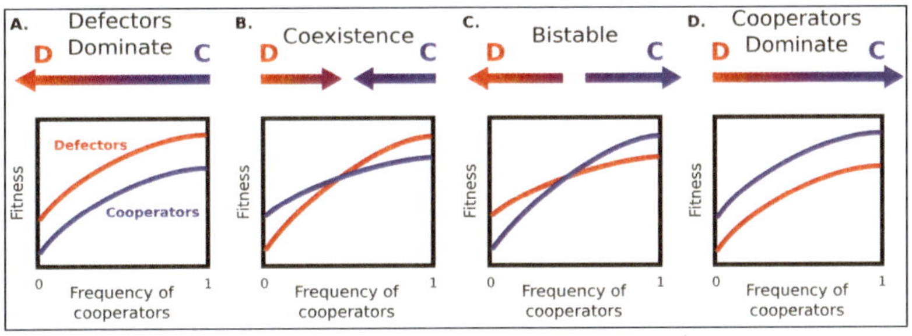

There are four basic, distinct outcomes of a cooperative interaction. (A) Defectors (red) may always have a higher fitness than cooperators (blue), which would lead to the extinction of cooperators. (B) Alternatively, cooperators may have a higher fitness than defectors when they make up a small fraction of the population and a lower fitness at high fractions, leading to coexistence of cooperators and defectors. (C) Also, cooperators may have a higher fitness only when they make up a large fraction of the population and a lower fitness otherwise. This creates an unstable fixed point, above which cooperators fix in the population and below which defectors fix, characteristic of bistability. (D) Finally, cooperators may always have a higher fitness than defectors. Note, the last case is distinct from the first because cooperators increase the fitness of the population and are therefore not arbitrarily labeled. In game theory, the popular names for these interactions (when linear) are (A) prisoner's dilemma, (B) snowdrift, (C) stag hunt, and (D) mutually beneficial or harmony game.

If, however, one strain does not always grow faster than the other, predicting the final result of such a experiment will be slightly more complicated. For instance, if cooperators grow better at low frequencies and worse at high frequencies, then the two strains will coexist at some intermediate frequency, independent of the starting frequency of the experiment. On the other hand, if cooperators grow better at high frequencies and worse at low frequencies, then cooperators will fix in the population if they start at a high enough frequency, but will go extinct if they start at a lower frequency.

For most cooperative interactions likely to be found in the lab or nature, there are thus only four possible outcomes: either (A) defectors will always fix in the population, (B) the two strains will coexist at some intermediate frequency, (C) one strain will fix depending the initial frequency, or (D) cooperators will always fix (although there may be complications,). Each situation shows drastically different dynamics and should be treated as distinct circumstances. When cooperators dominate defectors, natural selection acts to increase the mean fitness of the population, the socially optimal outcome. On the other hand, if defectors dominate cooperators, the mean fitness decreases and we are often interested in how this helpful cooperative behavior can be maintained despite this dominance, a question kin and group selection were developed to answer.

Before While it is often implicitly assumed that any costly behavior displayed by a bacteria that helps other cells is dominated by a defector cell that does not display that behavior, this is actually rarely the case. In fact, coexistence between cooperators and defectors has been found both in the laboratory and in nature for many cooperative systems, including protease and siderophore production, β-lactam antibiotic degradation, defective viruses, fruiting body formation, and sucrose metabolism in yeast. These traits are said to have negative frequency dependence because the relative fitness of a cooperator decreases as their frequency increases. In particular, the stable mixed equilibrium is where the relative fitness equals one. A common theme among these negative frequency dependent traits is that a cooperator somehow receives a more than average share of the public good that it personally produces. Despite this knowledge, simple linear models still cannot explain the coexistence of cooperators and defectors in these systems, and more accurate non-linear models are required. A prime example of this is the cooperative sucrose metabolism system in budding yeast, Saccharomyces cerevisiae.

S. cerevisiae cannot import the sugar sucrose directly into the cell very efficiently. Instead, yeast is forced to break the disaccharide sucrose into glucose and fructose outside of the cell with the extracellular enzyme invertase and then import the monosaccharides glucose and fructose. Unfortunately for yeast, because this happens in the periplasmic space, approximately 99% of the created glucose (and fructose) diffuses away to its neighbors and the cell can only capture about 1% of the products of sucrose hydrolysis (the capture efficiency, ε, is thus 0.01). This would suggest that a cheater strain that does not pay the cost of producing invertase could potentially take advantage of the invertase producing cooperators if the cost is higher than the benefit of the extra 1% of glucose captured. In a simple model, where growth is a linear function of the available glucose, cooperators would either always do better or always do worse than defectors depending on whether or not the normalized cost, c, is greater than ε. However, experimental results suggest coexistence between cooperators and defectors therefore, one of the assumptions of the simple linear model is incorrect.

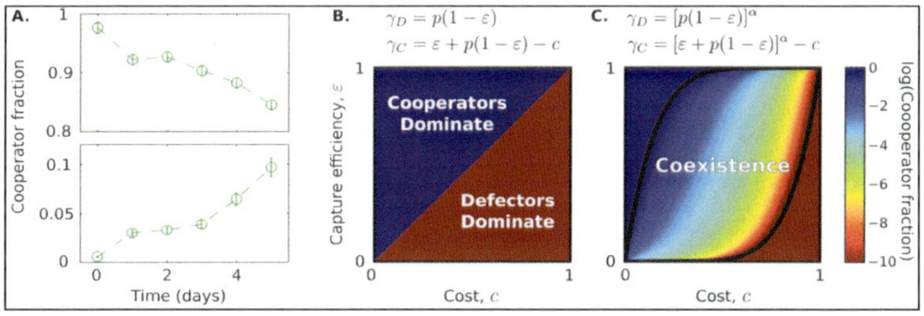

Non-linear dynamics can lead to coexistence of cooperators and defectors. (A) Cooperators and defectors coexist in the cooperative sucrose metabolism system, which can be seen by their mutual invasability, where cooperators can invade pure defector cultures (bottom) and defectors can invade pure cooperator cultures (top). (B) In a simple model of the cooperative yeast sucrose metabolism system, if the normalized growth rate, γ, is a linear function of the available glucose, then cooperators or defectors always dominate depending on the parameters. Here, we assume that cooperators pay a cost, c, and capture a small fraction, ε, of the glucose made personally, with the rest, 1 – ε, diffusing away to be used by other cells. The amount of available glucose to a cell is then approximately given by ε + p(1 – ε), where p is the proportion of cooperators in the population. Thus cooperators always grow faster than defectors if c < ε and grow slower otherwise. (C) If, however, the growth rate is a non-linear function of the available glucose, then there is a large region where coexistence between cooperators and defectors is measurable (between dark lines). The exponent, α, was experimentally measured to be 0.15 ± 0.01.

As it turns out, the growth rate is actually a concave function of the available glucose, which leads to coexistence of cooperators and defectors because there are diminishing returns to increased glucose. Thus, when the local glucose concentration is low, the increased glucose due to invertase production outweighs the cost of cooperation, but as there is more glucose in the environment from other cooperators, the additional growth rate per glucose molecule reduces below the cost of invertase production. This qualitative model has proven to be robust to experimental manipulation of the costs and benefits of cooperation and provides insight to how non-linear dynamics can lead to coexistence.

In contrast to deterministic coexistence, some systems may instead show bistability, which is characterized by mutual non-invasability: cooperators cannot invade a pure defector culture, but also defectors cannot invade a pure cooperator culture. Bistability is predicted whenever cooperators preferentially interact with themselves. Perhaps the most well-known example of this is the hypothetical "green beard" gene. Imagine a gene that simultaneously codes for a green beard–or any other arbitrary tag–and a cooperative behavior directed towards individuals with green beards. Presumably, this green beard gene would have an advantage and could spread in the population. However, if the cost of this gene is too high, then it may still be disfavored at low frequencies because it does not receive any benefit. Alternatively, two different versions of a green beard gene may compete and could also show bistability. Despite the theoretical appeal, green beard genes are rare in nature, mostly because it is difficult for a single gene to code for a "beard," a recognition mechanism, and a cooperative action. With that said, the fio1 occulation gene in yeast has recently been identified as a green beard gene by causing bearers of the gene to stick together, protecting the inside of the clump from harm, like the build up of ethanol in the environment. Also, the csaA self adhesion gene in Dictyostelium purpureum works by a similar mechanism to exclude non-bearers from fruiting bodies.

If however, cooperators always grow slower than the defectors that they are mixed with, then an additional mechanism is required to justify the existence of cooperation. This situation corresponds to cooperators paying some cost, not necessarily a fixed amount, for others to benefit in some way. While in the well mixed case, cooperators are destined for extinction, the dynamics become more complicated when structure is added to the population. Now, if cooperators on average gain a high enough proportion of the benefits from other cooperators by interacting with them more often than defectors do, then the increased benefits may outweigh the costs of cooperation and they can spread in the population. Thus, the simplest and most fundamental explanation for the maintenance of costly cooperation is that cooperators must interact more often with cooperators than defectors do. This assortment of similar individuals (i.e. cooperators with cooperators and defectors with defectors) is the ultimate outcome of every mechanism for the evolution of otherwise dominated cooperation, including kin discrimination, group selection, direct and indirect reciprocity, spatial structure and many others.

Price's Equation

The Price equation provides a general and exact mathematical description of evolution, with applications in kin and multilevel selection models. First assume each individual, i, has a genotype, Gi, that can be quantitatively described. For example, Gi is often arbitrarily set to 1 if the (haploid) individual has the cooperative gene and 0 if it doesn't, Now, if we want to know how the genotype frequency changes in the population over time, we first need to define an individual's fitness, Wi, to be the number of i individuals at the next time point divided by the number present now. For example, the two time points could be in generation times, in which case Wi would be a measure of how many offspring an individual has. The change in average genotype in the population, $\Delta\bar{G}$ is then given by the Price Equation:

$$\overline{W}\,\Delta\overline{G} = Cov(W,G) + \overline{W_i \Delta G_i}$$

Where, Cov(*W*,*G*) is the covariance between fitness and genotype in the population and $\overline{W_i \Delta G_i}$ accounts for the average change in genotype between parent and offspring (e.g. mutation). For asexual microbes, mutation and other effects that might change an offspring's genotype are often assumed to be small, in which case the Price equation becomes:

$$\overline{W}\,\Delta\overline{G} = Cov(W,G) = \beta_{WG} Var(G)$$

Where β_{WG} is the slope of the least-squares linear regression of fitness on genotype and Var(*G*) is the variance of genotypes in the population–the last equality being justified by the definition of covariance. Increasing the variance or the average effect of genotype on fitness, β_{WG}, increases the speed of evolution. In fact, when fitness is perfectly heritable, Price's equation reduces to Fisher's Fundamental theorem of natural selection which states:

$$\Delta\overline{W} = \frac{Var(W)}{\overline{W}}$$

Which is valid when G is replaced with W ($\beta_{WW} = 1$) and the last term in is neglected. Price's equation is a simple and elegant formula that is widely used in the kin and multilevel selection literature; however, some of its limitations should be noted. First, there is a subtle point about the distinction between probability theory and statistics, which has consequences for the application and interpretation of the Price equation in derivations and models. Because this point is very subtle and out of the scope of this review, we refer the interested reader to On another technical note, the Price equation lacks dynamical sufficiency by only telling the current speed and direction of selection and not the path of evolution. Finally, with generality comes a loss of specificity, and Price's equation can mask many of the important intricacies of the system. This is a point we will elaborate on later because the same criticism holds for many kin selection and multilevel selection models that use Price's equation. While powerful and general, the Price equation does not usually ease empirical calculations and can only aid in understanding a social behavior when it is coupled with an explicit model. We will now move on to a generalized form of Hamilton's rule, a simple two variable model, that uses Price's equation in its derivation.

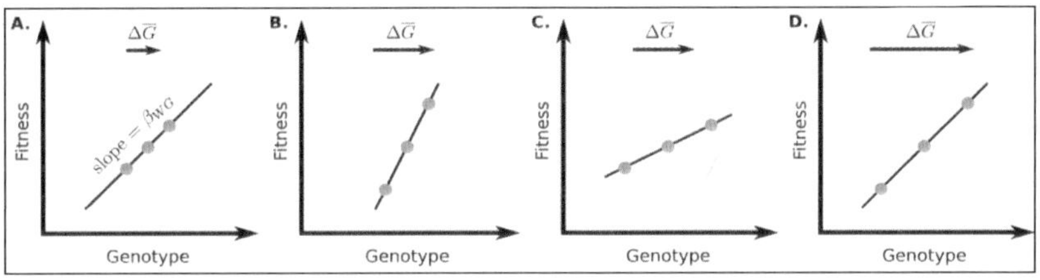

Understanding how the Price equation works can aid in understanding evolution. (A) In its simplest form, Price's equation states that the change in genotype in a population is $\Delta\bar{G} = Cov(W,G) = \beta_{WG}Var(G)$, where W is relative fitness here. (B) Doubling the slope of the linear regression, β_{WG}, while keeping everything else constant, doubles $\Delta\bar{G}$. (C) Also note that increasing the spread of genotypes by a factor of 2 reduces β_{WG} by a factor of ½, but it multiplies the variance by $2^2 = 4$ so $\Delta\bar{G}$ still doubles. (D) Effects multiply and combining (B) and (C) leads to a quadrupling of $\Delta\bar{G}$. For the purpose of illustration, the data points all sit perfectly on the regression, but that is not necessary for Price's equation to work.

Hamilton's rule:

In 1964, Hamilton stated his famous rule: a cooperative act will be favored by natural selection if the cost, *c*, of performing the cooperative act is less than the benefit, *b*, given to the other individual times the relatedness, *r*, between the two individuals, or equivalently if:

$$rb - c > 0$$

While pleasingly simple, the assumptions that went into deriving this rule almost never hold even in simple models, much less microbial systems. Therefore, a more general inequality is required for which we will use Price's equation to derive. Similar to how we regress fitness, *W*, against genotype, *G*, in Price's equation, we will first regress fitness against an individual's genotype, G_X, and the genotype of its interactants, G_Y, which results in

$$W = W_0 + \beta_{WG_X} G_X + \beta_{WG_Y} G_Y + \varepsilon$$

Where, W_0 is an individual's base fitness and the β's are the partial regression coefficients. β_{WG_X} is the average effect an individuals genotype has on its own fitness, ignoring the effect of the environment. For example, if the only difference between a cooperator, $G_X = 1$, and a defector, $G_X = 0$, is that a cooperator pays a cost, c, for others to benefit, then β_{WG_X} would be $-c$. Similarly, β_{WG_Y} is the average fitness effect the environment, G_Y, has on an individual's fitness, ignoring the effect of the individual's genotype. Again, when G_Y measures the number of cooperators in an individual's environment, and each cooperator gives its interactant a benefit, b, then β_{WG_Y} is b. The last term, ε, is the residual, which may be different for every individual and describes the difference between an individual's actual fitness and the fitness predicted by the regression. After substitution into $\overline{W}\,\Delta\overline{G} = Cov(W,G) = \beta_{WG}Var(G)$ and simplifying, we get that a cooperative act is favored when:

$$\beta_{G_YG_X}\beta_{WG_Y} + \beta_{WG_X} > 0$$

where, β_{WG_Y} and β_{WG_X} are the same as above and now $\beta_{G_Y G_X}$ is a measure of how an individual's social environment covaries with the individual's genotype and is the linear regression definition of relatedness, r. Thus, in the simplest case where every cooperator pays a fixed cost c for other individuals to gain a fixed benefit b–all measured in number of offspring–cooperators are favored when $rb - c > 0$.

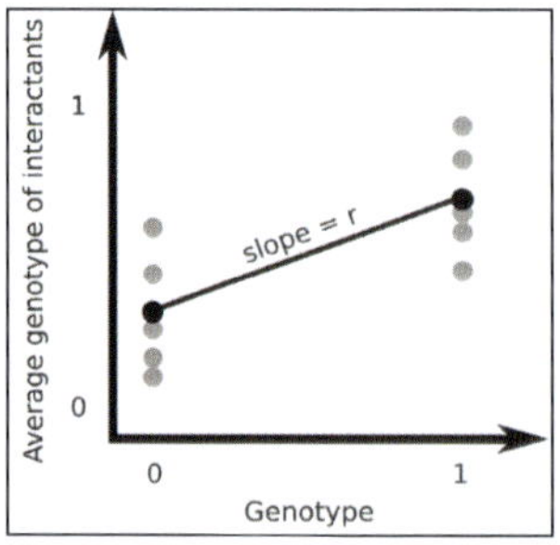

Relatedness as a regression coefficient. To apply the general form of Hamilton's rule, we need to assign a number to each genotype, and if we are working with two strains (e.g. a cooperator and defector) we can arbitrarily set the cooperator genotype to 1 and the defector genotype to 0. The relatedness, r, is then the linear regression coefficient connecting an individual's genotype with the genotype of its interactants. Note that when everyone interacts with the same number of individuals, the regression passes through the average of each's interactants (black dots). This allows us to redefine relatedness as $r = \mathbb{P}(C \mid C) - \mathbb{P}(C \mid D)$: the probability for a cooperator to interact with a cooperator minus the probability for a defector to interact with a cooperator. Also note that scaling the x axis would also scale the y axis by the same amount, leaving the slope the same, so arbitrarily setting the genotypes to 0 and 1 does not lose generality.

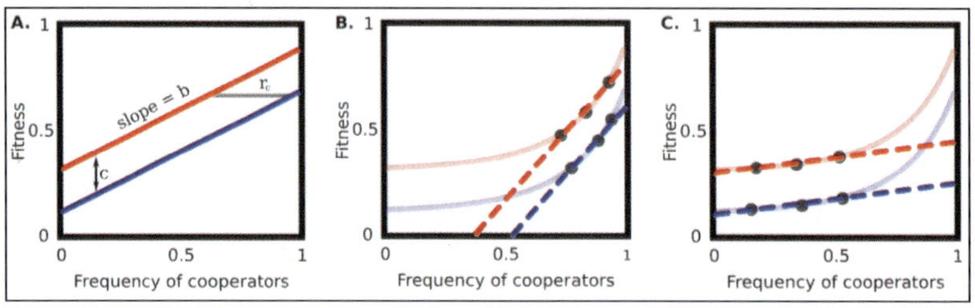

Hamilton's rule is a linear regression to generally non-linear data. (A) In the simplest case, where everything is linear, the slope of the fitness function is b and the difference between a defector and cooperator's fitness with the same frequency of cooperators is c. For a cooperator to have a higher fitness than a defector, it must interact with r more cooperators than defectors do Because of the geometry, this results in a critical relatedness, $r_c = \frac{c}{b}$, above which cooperators are favored (i.e. when $rb - c > 0$). Unfortunately, life, and particularly microbes, is rarely this simple. (*B* and *C*) Because linearity is implicitly assumed with Hamilton's rule, when the fitness is a non-linear function of the fraction of cooperators (light lines), *b* and *c* become linear regression coefficients based on the sample points (gray dots). Unfortunately this masks many of the interesting (non-linear) qualities of the system and provides no prediction power because *b* and *c* change with the population structure. Also note that this method can lead to negative fitness values (B) and situations where b < c (C), which would incorrectly suggest that cooperation can never evolve. Note that *b* and *c* determine the necessary r for cooperation to evolve, but b and c are in turn affected by the population structure and thus r, which makes disentangling fitness effects from population structure often impossible. It is important to note that the cooperator and defector fitnesses are not independently regressed, rather only one regression is done with the slope (b), vertical difference between defectors and cooperators at a fixed cooperator fraction (c), and base fitness (y-intercept) as the three parameters.

Before going on to more complicated cases, a more thorough discussion of "relatedness" is warranted. Contrary to the popular use of the word, "relatedness" describes a population of interacting individuals, where r refers to how assorted similar individuals are in the population. *r* is 1 if every individual only interacts with genetically identical individuals ($G_X = G_Y$) and *r* is 0 if interactions are random (G_Y does not change with G_X). Thus, if we have ten clonal colonies, all from a different strain of bacteria and we put each one in a separate test tube to grow, this would be full assortment and r would be 1. Any otherwise dominated cooperative strains with $b > c > 0$ would end up growing better than the defector strains and thus would increase in frequency in the ten test tube population because $1 \cdot b - c > 0$. On the other hand, if we take an equal number of cells from each colony and put a little bit in each of the ten test tubes, then there would be no assortment and r would be 0. Similarly, when two strains are competing in a well mixed environment like a single test tube, then relatedness is again 0 because each individual is no more likely than average to interact with its own type. In both these cases where *r* = 0, any cooperators dominated by defectors would decrease in frequency, again consistent with Hamilton's rule because $0 \cdot b - c < 0$.

There is, however, a measure of how related two individuals in a population are. It is important to remember that Hamilton's rule only asks if a trait will increase in *frequency* and not necessarily number, so even if we are only interested in the relatedness between two individuals, it still must take into account the population. The inclusive fitness definition of relatedness describes how similar an actor of an action is to its recipient, relative to the population, and is given by:

$$r = \frac{\mathbb{P}_{A \to R} - \mathbb{P}_{A \to P}}{\mathbb{P}_{A \to A} - \mathbb{P}_{A \to P}},$$

Where, $\mathbb{P}_{A \to R}$ is, formally, the probability that a gene drawn at random from the actor at the focal loci (e.g. the gene coding for the cooperative behavior) is identical in state to a gene drawn at random from the recipient at the focal loci. Similarly, $\mathbb{P}_{A \to P}$ compares the actor to a member of the population drawn at random and $\mathbb{P}_{A \to A}$ compares the actor to itself ($\mathbb{P}_{A \to A} = 1$ for haploids). In the case of a rare gene in a large population, $\mathbb{P}_{A \to P} \to 0$ and $r \to \frac{\mathbb{P}_{A \to R}}{\mathbb{P}_{A \to A}}$ which gives the classic result that r is one half for siblings and one eighth for cousins (in the absence of inbreeding). Equation 7 also gives $r = 1$ for clonemates, and $r = 0$ for two individuals drawn at random from the population. Note that the population "relatedness" is the average relatedness of interacting individuals, but, we will stick to the linear regression definition of relatedness in equation because it is easier to visualize and calculate.

Returning to our linear regression formula for Hamilton's rule it is important to understand how this is applied to more complicated (i.e. realistic) situations. The fitness of a microbe will very rarely be a linear function of its environment, if only because linearity is just one of the infinite number of possibilities. When an individual's fitness is a non-linear function of its genotype and environment, then the linear regression definitions of the "benefits" and "costs" depend on the data points sampled in the population (and C, data points are dark dots). Now the measured b and c are no longer inherent properties of the interaction, but rather functions of the population structure. In fact, b and c are completely dependent on the sampled data points and change every time they are measured.

The measured b and c can also give nonsensical results depending upon the population structure being sampled. For example, Hamilton's rule can predict negative fitness values for some cooperator frequencies. Alternatively, the measured b could be less than c for some population structures, which would incorrectly suggest that cooperation is in general impossible to evolve. Also, if at some point b is measured to be 2 and c to be 1, then as long as the relatedness, r, is greater than $\frac{c}{b} = \frac{1}{2}$ at that moment, cooperation will increase in frequency; however, this does not mean that ½ is the general condition for cooperation to be favored because b and c change with the population structure and thus also with r. Note that while a linear regression for the costs and benefits can always be done, it often is not very informative. Rather than being a powerful predictive tool, this method can often simply become a cumbersome way to add up fitnesses.

Also, on the practical side, G_Y is often hard to define. In the simplest case, where every individual interacts equally with the same number of individuals (e.g. in well-mixed test tubes seeded with the same number of cells), then G_Y is just the average genotype of the interactants. However, when different interactions are stronger than others, like on an agar plate where interactions weaken with distance, or when individuals interact with different numbers of individuals, then G_Y becomes more complicated. While Hamilton's rule can still be applied, the relatedness and benefit terms, $\beta_{G_Y G_X}$ and β_{WG_Y}, become less intuitive and potentially arbitrary. It should also be noted that when multiple alleles are competing, G_X, and thus also $\beta_{G_Y G_X}$, can become arbitrary. In fact, $\Delta\overline{G}$ loses much of its meaning in this case. Therefore, in cases of multiple competing alleles or nontrivial population structure, Hamilton's rule is often replaced with explicit modeling.

More complicated forms of Hamilton's rule are also possible. For example, one can take a regression of cooperator and defector's fitnesses separately, resulting in two different slopes. Importantly, this method can predict bistability and coexistence, a distinction that is traditionally out of reach for Hamilton's rule. While this can be a more accurate view of the interaction and should be favored in certain cases, it still cannot capture the full picture and has some of the same pitfalls of traditional Hamilton's rule without the advantages that come from its simplicity.

Alternatively, a highly modified form of Hamilton's rule has recently been suggested for the study of microbes. To apply this unique formula, one first fits two (non-linear) regressions to the cooperator and defector fitnesses. A Taylor series expansion is then taken of these two functions and the coefficients of these Taylor series are used in vector forms of b and a new synergy term *d*. *r* is now a vector containing the traditional definition of relatedness, but also additional terms describing higher order moments of assortment. In this formulation, the population structure is completely separated from the fitness effects–that is, *b*, *c*, and the synergy term d are now constants and do not change with each iteration of the calculation. We see Smith et al.'s reformulation as a major step forward for the application of Hamilton's rule to microbes, but it is not without its limitations. First, although it may be a more accurate view of microbial interactions, higher order relatedness and fitness terms can be much less intuitive than Hamilton's simple rule. Also, it cannot accommodate multiple strains competing or dependence on density or time, at least as it is currently written. Finally, the reformulation still is not a replacement for a mechanistic model, which can identify biologically relevant parameters and aid in predictions.

Weak Selection

Before moving on to alternative methods of conceptualizing and analyzing cooperation, it is important to understand why the simple form of Hamilton's rule–without any regressions–is usually inappropriate for the study of microbes. Hamilton thought of evolution as a very slow and gradual process where each mutant varied only slightly from the wild type and this can be seen in many of the assumptions used in deriving the simple form of Hamilton's rule, namely the assumption of weak selection. In the kin selection literature, (δ) weak selection refers to the assumption that everyone in the population plays approximately the same strategy and any mutant can only play a slightly different strategy (wild type strategy $\pm\delta$ as $\delta \to 0$;). This is analogous to every microbe in a population producing the same amount of a public good like invertase and the only mutations allowed are those that produce slightly more or slightly less invertase. Any mutants therefore also only have a vanishingly small difference in fitness, leading to dynamics dominated by random drift, and thus average frequencies of roughly 50% for competing alleles in finite populations.

In evolutionary game theory, a similar concept is also used called w weak selection. In contrast to δ weak selection, individuals in the population can play distinct, discrete strategies, but the fitness difference between any two strategies is always very small (mixed strategies are also allowed depending on the model). This would correspond to two yeast cells that produce the maximum and minimum amount of invertase having only an infinitesimal difference in fitness. While weak selection is often useful for gaining insight by simplifying the dynamics in analytical models, it is not the general case and is often invalid for cooperative microbial behaviors because of large differences in fitness.

The intuition behind weak selection partly comes from the idea that any one interaction between two animals often has only a small effect on their fitness; for example, whether or not an animal shares a berry with its group on a particular day is unlikely to have a large impact on anyone's survival.

However, many forms of microbial interactions are literally do or die. If sucrose is the only available sugar, then a microbe will starve if it cannot break it down. The difference in generation times is also important because while an animal may have many small interactions in its much longer lifetime, a microbe is likely to have only a few interactions, many of which will be important in determining if it will divide in the next hour. Additionally, while the complexity of animal behavior makes evolution by small changes more likely in animals, mutations of large effect are common in microbes.

So the concept of weak selection is inappropriate for microbes, why is this important? First of all, while the linear regression form of Hamilton's rule still works, inclusive fitness theory and many of its results are provably correct only in the limit of weak selection. For example, the commonly cited rule that natural selection acts to maximize inclusive fitness has only been proven using stringent assumptions, including weak selection. Also, assuming weak selection gives a quantitatively different answer than more accurate models. This is particularly relevant when empiricists try to apply results from a weak selection model to actual data. In addition, measuring relatedness via neutral genetic markers is only valid when the markers and the allele of interest are under weak selection. Because genes like invertase and other cooperative alleles are often not under weak selection, using a genetic marker approach to measure relatedness in microbes will likely give a wrong result. In fact, despite their seemingly simpler genetics, relatedness is often harder to estimate for microbes than for animals.

Also, even if the costs and benefits of an interaction are linear functions of ones genotype and environment, the measured b and c will not be linear unless selection is weak. This results from growth being an exponential process, so the total growth difference between a cooperator and defector grows exponentially with the costs and benefits of the interaction. This will be a linear function only when b and c $\ll 1$, which is the limit of weak selection. Another way to arrive at the same conclusion is by imagining that, in the limit of δ weak selection, every individual's phenotype would be almost the same, so all the gray dots in and C would be right next to each other. This closeness allows the linear regressions to be replaced by derivatives and turning the otherwise hard frequency dependent situation into an easy frequency independent maximization problem. While this may be alluring to the theorist, this technique is often ill-suited for the empirical study of microbial cooperation.

Isolation and Cultivation of Microorganisms

Microorganisms occur in natural environment like soil. They are mixed with several other forms of life. Many microbes are pathogenic. They cause a number of diseases with a variety of symptoms, depending on how they interact with the patient. The isolation and growth of suspected microbe in pure culture is essential for the identification and control the infectious agent.

The primary culture from natural source will normally be a mixed culture containing microbes of different kinds. But in laboratory, the various species may be isolated from one another. A culture which contains just one species of microorganism is called a pure culture. The process of obtaining a pure culture by separating one species of microbe from a mixture of other species, is known as isolation of the organisms.

Methods of Isolation

There are special techniques employed to obtain pure cultures of microorganisms. In few cases it is possible to secure pure culture by direct isolation or direct transfer. This can be done only in those situations in which pure culture occurs naturally. Kinds of specimens taken for culturing will depend on the nature and habitat of microbes.

Different pathogens can be isolated from body tissues and fluids such as blood, urine, sputum, pus, faces, spinal fluid, bile, pleural fluids, stomach fluids etc. In the blood stream of a patient suffering with typhoid fever, the bacteria Salmonella typhosa may be present.

A pure culture of this bacterium may be obtained by drawing blood sample using a sterilized hypodermic syringe and treating the blood with anticoagulant such as heparin and potassium oxalate. The presence of the anticoagulant prevents the pathogenic microbe from entrapping in fibrin clot. The sample of the blood may be inoculated into a suitable medium.

Following isolation methods are employed to isolate microbes from mixed cultures:

1. Streaking,
2. Plating,
3. Dilution,
4. Enriched procedure, and
5. Single cell technique.

Streaking

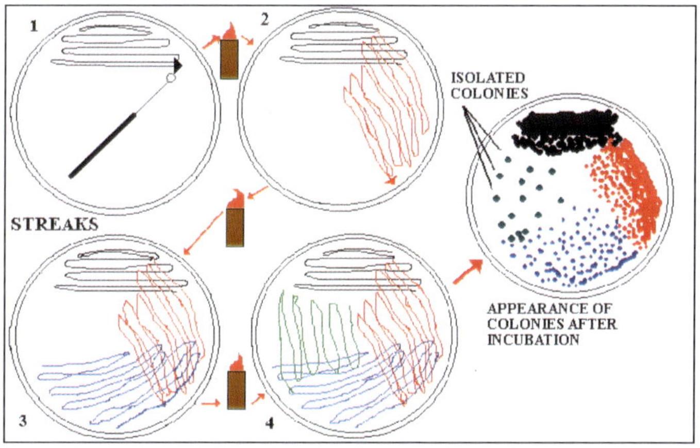

Different patterns used for obtaining pure culture by streaking.

This is most widely used method of isolation. The technique consists of pouring a suitable sterile medium into sterile petriplate and allowing the medium to solidify. By means of a sterile loope or straight needle or a sterile bent glass-rod a small amount of growth preferably from a broth culture or bacterial suspension is streaked back and forth across the surface of agar until about one third of the diameter of the plate has been covered.

The needle is then flamed and streaking in done at right angles to and across the first streak. This serves to drag bacteria out in a long line from the initial streak. When this streaking is completed the needle is again flamed and streaking is done at right angles to the second streak and parallel to the first.

Plating

It includes diluting of a mixture of microorganisms until only a few hundred bacteria are left in each millilitre of the suspension. A very small amount of the dilution is then placed in a sterile petriplateby means of a sterile loop or pipette. The melted agar medium is cooled to about 45 °C and is poured into plate. The microorganism and agar are well mixed. When the agar is solidified the individual bacterium will be held in place and will grow to a visible colony.

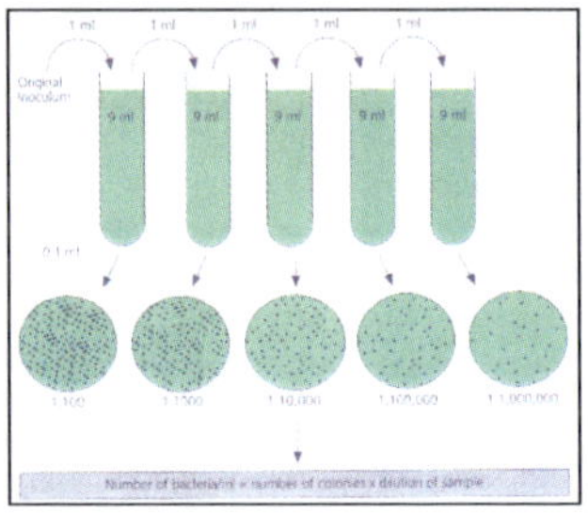

Dilution method of isolation.

Dilution

This method is used for the microorganisms which cannot be easily isolated by streaking or plating method. Sometimes when several organisms are present in a mixture, with one organism predominating, the predominating form may be isolated by this method. For example, when raw milk is allowed to sour at room temperature it will, at the time of curding, have a mixture of microorganisms with high percentage of Streptococcus lactis.

If 1 ml of the sour milk is taken into a tube containing 9 ml. of sterile milk (in which no organisms are present) then 1 ml. of this mixture is transferred with a sterile pipette into a second tube of sterile milk and the procedure is repeated i.e. from second to third tube, third to fourth tube until a series of about 10 tubes are inoculated. By this serial dilution, the chances are that a pure culture of S. lactis will be obtained.

Enrichment Procedure

This procedure involves the use of media and conditions of cultivation which favour the growth of the desired species. For example, when a man suffers with typhoid, the intestinal discharge posses small number of Salmonella typhosa when compared with E. coli and other forms.

It is almost impossible to isolate the typhoid organisms because they represent only a fraction of a per cent of the total microorganisms present. The media are therefore derived, which allow the rapid growth of the desired organisms, at the same time inhibiting the growth of other microorganisms.

Single Technique

This is one of the most ideal and difficult method of securing pure culture. In this method a suspension of the pure culture is placed on the under-side of a sterile cover-glass mounted over a moist chamber on the stage of the microscope.

While looking through the microscope, a single cell is removed with the help of sterile micropipette and transferred to a small drop of sterile medium on a sterile cover-glass and is mounted on a sterile hanging drop slide, which is then incubated at suitable temperature. If the single cell germinates in this drop, few cells are transferred into a tube containing sterile culture medium which is placed in the incubator to obtain pure culture originated from single cell.

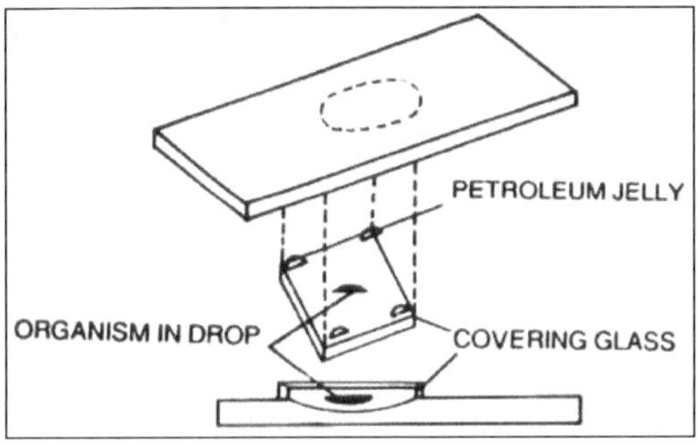

Hanging drop slide preparation.

Other Methods

The isolation of anaerobic microorganisms is very difficult. There are certain special techniques by which these organisms are isolated.

Cultivation of Microorganisms

For cultivating microbes in laboratory, we require culture media. The various mixtures of nutritive substances used for the laboratory cultivation of microorganisms are collectively known as culture media. The culture media serve as soil in which bacteria are planted for the purpose of study.

Culture Media

Culture media must contain all the essential nutrients required by the organism for its growth and reproduction. A suitable source of energy, building materials and growth factors must be supplied in adequate amounts.

So, a culture medium must contain:

Carbohydrate (Sugar)	Source of hydrogen, carbon & chemical-bond energy.
Nitrite ions or Ammonium ions	Source of nitrogen
Sodium/Magnesium suphate	Minerals
Micronutrients	Growth factores, Vitamins & certain amino acids.

Since microorganisms show a considerable variation in their nutritional requirements, no single medium is suitable for growth of all of them.

The different types of culture media employed fall into three groups:

Defined or Synthetic Media

These are the media prepared from chemical compounds. They are highly purified and specific, an investigator working in another laboratory can duplicate them.

Complex or Non-synthetic Media

Media that are prepared from ingredients that have not been precisely defined. It contains hydrolysed proteins and vitamin extracts. This type of medium cannot be duplicated by another worker in another laboratory. Peptone is usually produced by boiling beef, by the hydrolysis of its protein. Casein peptone and milk peptone are also used in complex media as the source of amino acids and nitrogen.

All liquid media, whether complex or synthetic may be converted to solid media by adding either gelat in (a protein) or agar-agar, (a complex polysaccharide) extracted from red marine algae. The use of agar has an advantage. The most bacteria are unable to hydrolyze this molecule into more simple components. Since gelatin is a liquid at room temperature, the use agar allows the medium to remain in a solid form while microbes are growing on its surface.

Living Cells

These are used for the cultivation of viruses. For example, fertilized eggs incubated for 8 to 12 days are able to support the growth of many viruses.

In another classification culture media are grouped into following four types:

Natural Media

Includes substances occurring in nature, as 1) Milk 2) Eggs 3) Blood 4) Extract of plant and animal tissues.

Derived Media

Includes known substances but the chemical composition of each is unknown. Examples are 1. Nutrient broth (prepared by boiling beef to extract nutrients and adding an amino acid-nitrogen source.) 2. Nutrient agar 3. Nutrient gelatin.

Chemically Defined Media

Exact chemical composition known.

Special Media

Include combinations of the other three types of media.

There are four categories of media used in laboratory:

1. Enrichment,
2. Selective,

3. Differential,
4. Propagation.

Enrichment Media

They are prepared with ingredients that will enhance the growth of certain microbes. Enrichment media encourage the growth of the suspected pathogen so that it will become the most pre-dominant type of microbe in the culture. Two types of enrichment media are blood agar and chocolate agar.

Selective Media

They are prepared with ingredients that inhibit the growth of unwanted microbes which might be in the specimen. The inhibitor may be an antibiotic, salt or other chemical. Mixed culture of microbes originally grown in enrichment media may be inoculated into selective media to isolate the desired microbe.

Differential Media

They are designed to differentiate among microbes. Different bacterial species may produce dissimilar colony colours when grown on differential agar. While in differential broth cultures, the media change colour. Differential media are used to confirm the identity of a microbe that has already been isolated by culturing in enrichment and selective media.

Propagation Media

They are used to propagate, or keep microbes growing for a long lime. Samples grown on these media may be taken for analysis. The most common propagation media are nutrient broth and agar.

Preparation of Media

There are three main steps in the preparation of media:

1. Preparation as solutions of chemicals and adjusting the pH.
2. Dispensing the media.
3. Sterilization.

A broth is prepared by dissolving the appropriate amount of the components in distilled water and pH is adjusted by the addition of either dilute NaOH or Hcl. The broth is dispensed into clean rimless 'Pyrex' test tubes which are plugged with non-absorbant cotton wool plugs. The test tubes are placed in wire baskets which are covered with grease proof paper.

The media are sterilized by autoclaving at a temperature of 121 °C and a pressure of 151 b/in^2 for 15 minutes. But medium containing heat- sensitive substances are sterilized either by filtering the solution at room temperature, using bacteria-proof filter or by a process called Tyndallization.

In this method, the liquids are steamed for one hour a day on three consecutive days and the liquids are incubated at 25-30 °C. During the first steaming, all the heat sensitive vegetative cells are killed, leaving only the spores. During the first incubation period, most of the spores germinate in to vegetative cells. These vegetative cells are killed by the second steam period.

In the second incubation period, the rest of the spores germinate into vegetative cells which are killed by the third steaming period. In this way, the liquids are sterilized without temperature rising above 100 °C.

Maintenance of Pure Culture

After obtaining the pure culture of a particular microbe, it may be grown and maintained as a pure culture in different ways:

1. The most common practice is to grow the culture on suitable medium until it reaches the stationary phase of growth, and then store in a refrigerator. If they are to be kept alive for a long period all culture must be transferred to a fresh sterile medium. Thus by successive transfer, a culture may be kept for an indefinite period.

2. A second method involves freezing of young culture and desiccating it under vaccum. The cells of a pure culture will remain viable for a long period of time if they are mixed with sterile blood serum, sterile skimmed milk, before freezing and drying. They dried cultures are kept in the sealed, evacuated tubes and are stored in cool places.

3. This method of maintaining pure culture is most suitable for spore forming species. The microorganisms are grown in pure culture in suitable media. A suspension of microorganisms is then transferred to cotton stoppered tubes of sterilized dry soil. The spores remain viable, though dormant, for long periods of time, in dry soil. The organism can be grown after a desired period, by transferring the soil into a suitable medium and incubating it under suitable temperature.

References

- Microbiology, science: britannica.com, Retrieved 14 February, 2019
- Microbial-taxonomy, science: encyclopedia.com, Retrieved 19 April, 2019
- Microbial-interaction-and-types, mutualism-syntropism, proto-cooperation-commensalism-antagonism-parasitism-predation-competition: onlinebiologynotes.com, Retrieved 4 January, 2019
- Isolation-and-cultivation-of-microorganisms, microorganisms: biologydiscussion.com, Retrieved 27 March, 2019

2

Microorganisms and their Diversity

Microorganisms are primarily classified into two types-prokaryotic and eukaryotic. A few examples of prokaryotic microorganisms are fungi, bacteria and protists while eukaryotic microorganisms include protozoa, yeast and paramecia. The topics elaborated in this chapter will help in gaining a better perspective about the diversity which exists within microorganisms.

Microorganisms or microbes are living forms, whose size is so small that they are usually not visible to naked human eye.

Types of Microorganisms

Microorganisms include bacteria, viruses, protozoa, microscopic fungi and algae. They are defined as follows:

Bacteria

They are prokaryotic, unicellular microorganisms, which lack chlorophyll pigments.

Viruses

Viruses are ultra-microscopic, non-cellular living particles, composed solely of a nucleic acid (DNA or RNA) core, surrounded by a protein envelope called capsid.

Protozoa

They are eukaryotic, unicellular microorganisms, which lack cell wall.

Microscopic Fungi

They are eukaryotic, heterotrophic microorganisms that fail to show any cellular differentiation into true tissues like root, stem or leaf and in which vascular system is absent.

Microscopic Algae

They are eukaryotic, autotrophic microorganisms that fail to show any cellular differentiation into true tissues like root, stem or leaf and in which vascular system is absent.

Importance of Microorganisms

Microorganisms are present in almost all places on the earth. Despite their minute size, they have tremendous importance in the maintenance of life on the earth. Due to the difference in the activities of different microbes, they influence life in different ways.

Some of them are very helpful, for example, those which fix atmospheric nitrogen into biologically useful forms, those which help in curd formation, those which recycle dead materials by degrading them into simpler substances and those which help in preparation of wine.

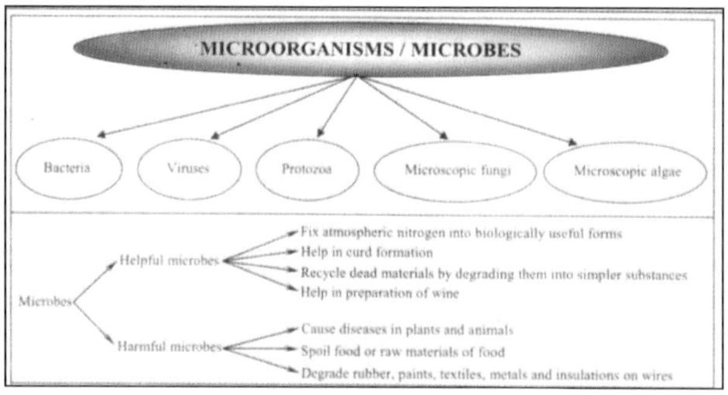

Microbes: Helpful and harmful.

Some other microbes are very harmful in that they cause diseases in plants and animals, spoil food or raw materials of food kept in normal conditions as well as they degrade rubber, paints, textiles, metals and even insulations on electric wires.

Therefore, it is essential to study these microbes in details, so that by various techniques their harmful effects can be overcome and beneficial effects be utilised for enhancing quality of life on the earth.

Viruses

Virus is an infectious agent of small size and simple composition that can multiply only in living cells of animals, plants, or bacteria. The name is from a Latin word meaning "slimy liquid" or "poison."

The earliest indications of the biological nature of viruses came from studies in 1892 by the Russian scientist Dmitry I. Ivanovsky and in 1898 by the Dutch scientist Martinus W. Beijerinck. Beijerinck first surmised that the virus under study was a new kind of infectious agent, which he designated contagium vivum fluidum, meaning that it was a live, reproducing organism that differed from other organisms. Both of these investigators found that a disease of tobacco plants could be transmitted by an agent, later called tobacco mosaic virus, passing through a minute filter that would not allow the passage of bacteria. This virus and those subsequently isolated would not grow on an artificial medium and were not visible under the light microscope. In independent studies in 1915 by the British investigator Frederick W. Twort and in 1917 by the French Canadian scientist Félix H. d'Hérelle, lesions in cultures of bacteria were discovered and attributed to an agent called bacteriophage ("eater of bacteria"), now known to be viruses that specifically infect bacteria.

The unique nature of these organisms meant that new methods and alternative models had to be developed to study and classify them. The study of viruses confined exclusively or largely to humans, however, posed the formidable problem of finding a susceptible animal host. In 1933 the British investigators Wilson Smith, Christopher H. Andrewes, and Patrick P. Laidlaw were able to transmit influenza to ferrets, and the influenza virus was subsequently adapted to mice. In 1941 the American scientist George K. Hirst found that influenza virus grown in tissues of the chicken embryo could be detected by its capacity to agglutinate (draw together) red blood cells.

A significant advance was made by the American scientists John Enders, Thomas Weller, and Frederick Robbins, who in 1949 developed the technique of culturing cells on glass surfaces; cells could then be infected with the viruses that cause polio (poliovirus) and other diseases. (Until this time, the poliovirus could be grown only in the brains of chimpanzees or the spinal cords of monkeys.) Culturing cells on glass surfaces opened the way for diseases caused by viruses to be identified by their effects on cells (cytopathogenic effect) and by the presence of antibodies to them in the blood. Cell culture then led to the development and production of vaccines (preparations used to elicit immunity against a disease) such as the poliovirus vaccine.

Scientists were soon able to detect the number of bacterial viruses in a culture vessel by measuring their ability to break apart (lyse) adjoining bacteria in an area of bacteria (lawn) overlaid with an inert gelatinous substance called agar—viral action that resulted in a clearing, or "plaque." The American scientist Renato Dulbecco in 1952 applied this technique to measuring the number of animal viruses that could produce plaques in layers of adjoining animal cells overlaid with agar. In the 1940s the development of the electron microscope permitted individual virus particles to be seen for the first time, leading to the classification of viruses and giving insight into their structure.

Advancements that have been made in chemistry, physics, and molecular biology since the 1960s have revolutionized the study of viruses. For example, electrophoresis on gel substrates gave a deeper understanding of the protein and nucleic acid composition of viruses. More-sophisticated immunologic procedures, including the use of monoclonal antibodies directed to specific antigenic sites on proteins, gave a better insight into the structure and function of viral proteins. The progress made in the physics of crystals that could be studied by X-ray diffraction provided the high resolution required to discover the basic structure of minute viruses. Applications of new knowledge about cell biology and biochemistry helped to determine how viruses use their host cells for synthesizing viral nucleic acids and proteins.

The revolution that took place in the field of molecular biology allowed the genetic information encoded in nucleic acids of viruses-which enables viruses to reproduce, synthesize unique proteins, and alter cellular functions-to be studied. In fact, the chemical and physical simplicity of viruses has made them an incisive experimental tool for probing the molecular events involved in certain life processes. Their potential ecological significance was realized in the early 21st century, following the discovery of giant viruses in aquatic environments in different parts of the world.

General Features

Viruses occupy a special taxonomic position: they are not plants, animals, or prokaryotic bacteria (single-cell organisms without defined nuclei), and they are generally placed in their own kingdom. In fact, viruses should not even be considered organisms, in the strictest sense, because they are not free-living; i.e., they cannot reproduce and carry on metabolic processes without a host cell.

All true viruses contain nucleic acid-either DNA (deoxyribonucleic acid) or RNA (ribonucleic acid)-and protein. The nucleic acid encodes the genetic information unique for each virus. The infective, extracellular (outside the cell) form of a virus is called the virion. It contains at least one unique protein synthesized by specific genes in the nucleic acid of that virus. In virtually all viruses, at least one of these proteins forms a shell (called a capsid) around the nucleic acid. Certain viruses also have other proteins internal to the capsid; some of these proteins act as enzymes, often during the synthesis of viral nucleic acids. Viroids (meaning "viruslike") are disease-causing organisms that contain only nucleic acid and have no structural proteins. Other viruslike particles called prions are composed primarily of a protein tightly complexed with a small nucleic acid molecule. Prions are very resistant to inactivation and appear to cause degenerative brain disease in mammals, including humans.

Viruses are quintessential parasites; they depend on the host cell for almost all of their life-sustaining functions. Unlike true organisms, viruses cannot synthesize proteins, because they lack ribosomes (cell organelles) for the translation of viral messenger RNA (mRNA; a complementary copy of the nucleic acid of the nucleus that associates with ribosomes and directs protein synthesis) into proteins. Viruses must use the ribosomes of their host cells to translate viral mRNA into viral proteins.

Viruses are also energy parasites; unlike cells, they cannot generate or store energy in the form of adenosine triphosphate (ATP). The virus derives energy, as well as all other metabolic functions, from the host cell. The invading virus uses the nucleotides and amino acids of the host cell to synthesize its nucleic acids and proteins, respectively. Some viruses use the lipids and sugar chains of the host cell to form their membranes and glycoproteins (proteins linked to short polymers consisting of several sugars).

The true infectious part of any virus is its nucleic acid, either DNA or RNA but never both. In many viruses, but not all, the nucleic acid alone, stripped of its capsid, can infect (transfect) cells, although considerably less efficiently than can the intact virions.

The virion capsid has three functions: (1) to protect the viral nucleic acid from digestion by certain enzymes (nucleases), (2) to furnish sites on its surface that recognize and attach (adsorb) the virion to receptors on the surface of the host cell, and, in some viruses, (3) to provide proteins that form part of a specialized component that enables the virion to penetrate through the cell surface membrane or, in special cases, to inject the infectious nucleic acid into the interior of the host cell.

Host Range and Distribution

Logic originally dictated that viruses be identified on the basis of the host they infect. This is justified in many cases but not in others, and the host range and distribution of viruses are only one criterion for their classification. It is still traditional to divide viruses into three categories: those that infect animals, plants, or bacteria.

Virtually all plant viruses are transmitted by insects or other organisms (vectors) that feed on plants. The hosts of animal viruses vary from protozoans (single-celled animal organisms) to humans. Many viruses infect either invertebrate animals or vertebrates, and some infect both. Certain viruses that cause serious diseases of animals and humans are carried by arthropods. These vector-borne viruses multiply in both the invertebrate vector and the vertebrate host.

Certain viruses are limited in their host range to the various orders of vertebrates. Some viruses appear to be adapted for growth only in ectothermic vertebrates (animals commonly referred to as cold-blooded, such as fishes and reptiles), possibly because they can reproduce only at low temperatures. Other viruses are limited in their host range to endothermic vertebrates (animals commonly referred to as warm-blooded, such as mammals).

Size and Shape

The amount and arrangement of the proteins and nucleic acid of viruses determine their size and shape. The nucleic acid and proteins of each class of viruses assemble themselves into a structure called a nucleoprotein, or nucleocapsid. Some viruses have more than one layer of protein surrounding the nucleic acid; still others have a lipoprotein membrane (called an envelope), derived from the membrane of the host cell, that surrounds the nucleocapsid core. Penetrating the membrane are additional proteins that determine the specificity of the virus to host cells. The protein and nucleic acid constituents have properties unique for each class of virus; when assembled, they determine the size and shape of the virus for that specific class. The genomes of Mimiviruses and Pandoraviruses, which are some of the largest known viruses, range from 1 to 2.5 Mb (1 Mb = 1,000,000 base pairs of DNA).

Most viruses vary in diameter from 20 nanometres (nm; 0.0000008 inch) to 250–400 nm; the largest, however, measure about 500 nm in diameter and are about 700–1,000 nm in length. Only the largest and most complex viruses can be seen under the light microscope at the highest resolution. Any determination of the size of a virus also must take into account its shape, since different classes of viruses have distinctive shapes.

Shapes of viruses are predominantly of two kinds: rods, or filaments, so called because of the linear array of the nucleic acid and the protein subunits; and spheres, which are actually 20-sided (icosahedral) polygons. Most plant viruses are small and are either filaments or polygons, as are many bacterial viruses. The larger and more-complex bacteriophages, however, contain as their genetic information double-stranded DNA and combine both filamentous and polygonal shapes. The classic T4 bacteriophage is composed of a polygonal head, which contains the DNA genome and a special-function rod-shaped tail of long fibres. Structures such as these are unique to the bacteriophages.

Animal viruses exhibit extreme variation in size and shape. The smallest animal viruses belong to the families Parvoviridae and Picornaviridae and measure about 20 nm and about 30 nm in diameter, respectively. Viruses of these two families are icosahedrons and contain nucleic acids with limited genetic information. Viruses of the family Poxviridae are about 250 to 400 nm in their longest dimension, and they are neither polygons nor filaments. Poxviruses are structurally more complex than simple bacteria, despite their close resemblance. Animal viruses that have rod-shaped (helical) nucleocapsids are those enclosed in an envelope; these viruses are found in the families Paramyxoviridae, Orthomyxoviridae, Coronaviridae, and Rhabdoviridae. Not all enveloped viruses contain helical nucleocapsids, however; those of the families Herpesviridae, Retroviridae, and Togaviridae have polygonal nucleocapsids. Most enveloped viruses appear to be spherical, although the rhabdoviruses are elongated cylinders.

The criteria used for classifying viruses into families and genera are primarily based on three structural considerations: (1) the type and size of their nucleic acid, (2) the shape and size of the capsids, and (3) the presence of a lipid envelope, derived from the host cell, surrounding the viral nucleocapsid.

Nucleic Acid

As is true in all forms of life, the nucleic acid of each virus encodes the genetic information for the synthesis of all proteins. In almost all free-living organisms, the genetic information is in the form of double-stranded DNA arranged as a spiral lattice joined at the bases along the length of the molecule (a double helix). In viruses, however, genetic information can come in a variety of forms, including single-stranded or double-stranded DNA or RNA.

The nucleic acids of virions are arranged into genomes. All double-stranded DNA viruses consist of a single large molecule, whereas most double-stranded RNA viruses have segmented genomes, with each segment usually representing a single gene that encodes the information for synthesizing a single protein. Viruses with single-stranded genomic DNA are usually small, with limited genetic information. Some single-stranded DNA viruses are composed of two populations of virions, each consisting of complementary single-stranded DNA of polarity opposite to that of the other.

The virions of most plant viruses and many animal and bacterial viruses are composed of single-stranded RNA. In most of these viruses, the genomic RNA is termed a positive strand because the genomic RNA acts as mRNA for direct synthesis (translation) of viral protein. Several large families of animal viruses, and one that includes both plant and animal viruses (the Rhabdoviridae), however, contain genomic single-stranded RNA, termed a negative strand, which is complementary to mRNA. All of these negative-strand RNA viruses have an enzyme, called an RNA-dependent RNA polymerase (transcriptase), which must first catalyze the synthesis of complementary mRNA from the virion genomic RNA before viral protein synthesis can occur. These variations in the nucleic acids of viruses form one central criterion for classification of all viruses.

A distinctive large family of single-stranded RNA viruses is called Retroviridae; the RNA of these viruses is positive, but the viruses are equipped with an enzyme, called a reverse transcriptase, that copies the single-stranded RNA to form double-stranded DNA.

Protein Capsid

The protein capsid provides the second major criterion for the classification of viruses. The capsid surrounds the virus and is composed of a finite number of protein subunits known as capsomeres, which usually associate with, or are found close to, the virion nucleic acid.

There are two major classes of viruses based on the protein capsid: (1) those in which a single (or segmented) linear nucleic acid molecule with two free ends is essentially completely extended or somewhat coiled (a helix) and (2) those in which the nucleic acid, which may or may not be a covalently closed circle, is wound tightly into a condensed configuration, like a ball of yarn. These two classes of virus assume in the first case a long, extended rodlike structure and in the second case a symmetrical polygon.

By far the best-studied example of a helical rod-shaped virus is the tobacco mosaic virus, which was crystallized by Wendell Stanley in 1935. The tobacco mosaic virus contains a genome of single-stranded RNA encased by 2,130 molecules of a single protein; there are 16 1/3 protein molecules for each turn of the RNA helix in the ratio of three nucleotides for each protein molecule.

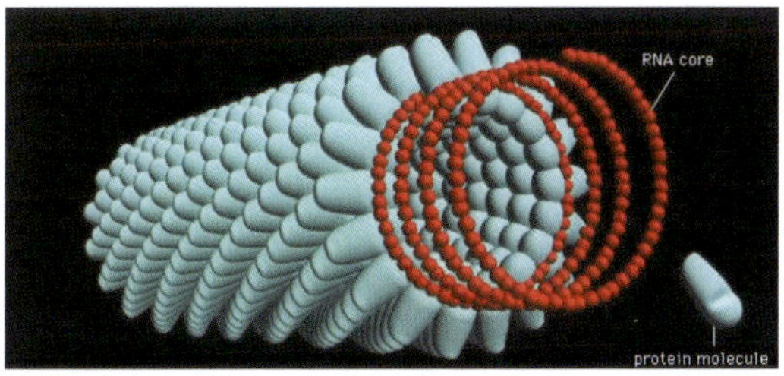

Schematic structure of the tobacco mosaic virus. The cutaway section shows the helical ribonucleic acid associated with protein molecules in a ratio of three nucleotides per protein molecule.

Under the right environmental conditions, viral RNA and protein molecules in liquid suspension will assemble themselves into a perfectly formed and fully infectious virus. The length of the helical virus capsid is determined by the length of the nucleic acid molecule, which is the framework for the assembly of the capsid protein. The various helical viruses have different lengths and widths, depending on the size of the nucleic acid as well as on the mass and shape of the protein molecule. Some of these helical viruses form rigid rods, while others form flexible rods, depending on the properties of the assembled proteins.

Polygonal viruses vary greatly in size, from 20 to 150 nm in diameter, essentially proportional to the size of the nucleic acid molecule coiled up inside the virion. Most, if not all, of the polygonal viruses are icosahedral; like a geodesic dome, they are formed by equilateral triangles, in this case 20. Each triangle is composed of protein subunits (capsomeres), often in the form of hexons (multiples of six) that are the building blocks of the capsid. There are 12 vertices (the apical junctions of these 20 triangles), each comprising a penton (five subunits). These icosahedral virions have three axes of fivefold, threefold, and twofold rotational symmetry. The number of capsomeres is a basis for taxonomic classification of these virus families. Certain icosahedral viruses, usually those that are more complex, contain internal proteins adhering to the nucleic acid that are not accessible at the surface of the virions.

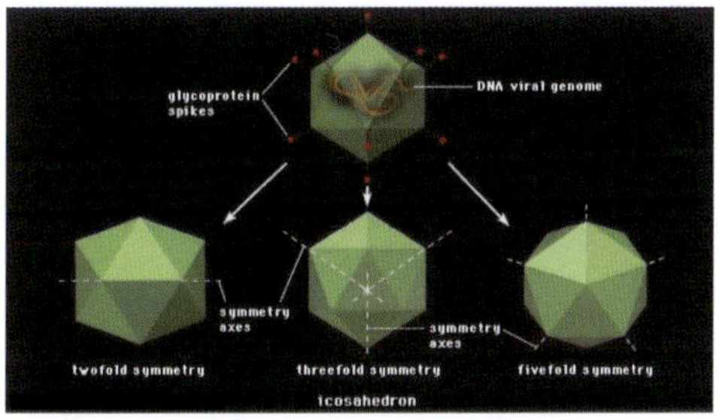

A virus icosahedron (20-sided structure) shown in the (left) twofold, (centre) threefold, and (right) fivefold axes of symmetry. Edges of the upper and lower surfaces are drawn in solid and broken lines, respectively.

Lipoprotein Envelope

Surrounding viruses of either helical or icosahedral symmetry are lipoprotein envelopes, unit membranes of two lipid layers interspersed with protein molecules (lipoprotein bilayer). These viral membranes are composed of phospholipids and neutral lipids (largely cholesterol) derived from cell membranes during the process known as budding. Virtually all proteins of the cell membrane, however, are replaced by proteins of viral origin during budding. Although all the viral envelope lipids originate from the cell, their relative proportions vary from those in the cell membrane because the viral proteins select only certain lipids during budding.

Associated with the virion membrane are “integral” glycoproteins, which completely traverse the lipid bilayer, and “peripheral” matrix proteins, which line the inner surface. The glycoproteins contain regions of amino acids that, in the first step of viral infection, recognize host-cell receptors. Matrix proteins appear to function in the selection of regions of the cell membrane to be used for the viral membrane, as well as to aid the virion in entering cells.

Classification

Distinguishing Taxonomic Features

Viruses are classified on the basis of their nucleic acid content, their size, the shape of their protein capsid, and the presence of a surrounding lipoprotein envelope.

The primary taxonomic division is into two classes based on nucleic acid content: DNA viruses or RNA viruses. The DNA viruses are subdivided into those that contain either double-stranded or single-stranded DNA. The RNA viruses also are divided into those that contain double-stranded or single-stranded RNA. Further subdivision of the RNA viruses is based on whether the RNA genome is segmented. If the viruses contain single-stranded RNA as their genetic information, they are divided into positive-strand viruses if the RNA is of messenger sense (directly translatable into proteins) or negative-strand viruses if the RNA must be transcribed by a polymerase into mRNA.

All viruses falling into one of these nucleic acid classifications are further subdivided on the basis of whether the nucleocapsid (protein coat and enclosed nucleic acid) assumes a rodlike or a polygonal (usually icosahedral) shape. The icosahedral viruses are further subdivided into families on the basis of the number of capsomeres making up the capsids. Finally, all viruses fall into two classes depending on whether the nucleocapsid is surrounded by a lipoprotein envelope.

Some virologists adhere to a division of viruses into those that infect bacteria, plants, or animals; these classifications have some validity, particularly for the unique bacterial viruses with tails, but there is otherwise so much overlap that taxonomy based on hosts seems unworkable. Classification based on diseases caused by viruses also is not tenable, because closely related viruses frequently do not cause the same disease. Eventually, it is likely that the classification of viruses will be based on their nucleotide sequences and their mode of replication rather than on structural components, as is now the case.

The basic taxonomic group is called a family, designated by the suffix -viridae. The major taxonomic disagreement among virologists is whether to segregate viruses within a family into a specific genus and further subdivide them into species names. In the first decade of the 21st century, there occurred a shift toward the use of binomial nomenclature, dividing viruses into italicized genera and species. This move was prompted in large part by the International Committee on Taxonomy of Viruses (ICTV), a member group of the International Union of Microbiological Societies. The ICTV oversees the ongoing process of devising and maintaining a universal classification scheme for viruses. In the virus classification hierarchy, the ICTV recognizes orders, families, subfamilies, genera, and species. The placement of viruses in these groups is based on information provided by study groups composed of experts on specific types of viruses.

In the ICTV system, each species of virus is generally recognized as representing a group of isolates, or viruses with distinct nucleic acid sequences. Thus, a single species of virus may sometimes contain more than one isolate. Although the isolates of a species possess unique genetic sequences, they all descend from the same replicating lineage and therefore share particular genetic traits. Furthermore, isolates of a species also share in common the ability to thrive within a specific ecological niche. As scientists identify new isolates and species, the classification of viruses is expected to become increasingly complex. The following scheme presents examples of well-characterized DNA and RNA viruses as they are classified on the basis of the ICTV system.

Protozoa

Protozoa are single-celled eukaryotes. This means that they are single celled organisms that have a nuclei as well as a number of other important organelles within the cytoplasm and enclosed by a membrane. They exist as free-living organisms or as parasites. This makes protozoa a diverse group of unicellular organisms, varying in shape and size.

Bodily Structure

Given that they are eukaryotes, protozoa are larger cells of between 10 and 100 micrometer in diameter (compared to prokaryotes) with a more complex structure. This means that they have a cell membrane which bounds the organelles, a DNA that is also bound by a membrane, nucleoli, ribosome, Golgi apparatus and multiple linear chromosomes with histones among others.

It is worth noting that organelles present in these cells will vary from one type to another. There are also a number of organelles that are exclusive to protozoa, these include:

- The Trichocysts of Paramecium.
- Certain skeletal structures.
- Contractile vacuoles.

Compared to other ciliates, the nucleus of protozoa is vesicular. As such, the chromatic is scattered resulting in a nucleus that is diffuse in appearance. However, this also varies from one to another. For instance, in the Phylum Apicomplexa, the vesicular nucleus had one or more nucleoli with DNA while the endosone of trypanosomes are lacking DNA

Protozoa also have in place locomotory structures such as pseudopodia, flagella and cilia which are used for movement. These structures are also surrounded by the plasma membrane. On the other hand, the pellicle (outer surface of some like the Giardia) is rigid enough to support and maintain a distinctive shape while at the same time allowing for twisting and bending when moving.

Classification

Because of their diversity, protozoa present several problems when it comes to classification. They are considered to be under the sub-kingdom protista with more than 50,000 species being described as free-living (these are the type that do not directly depend on others for survival).

Free-living protozoa can be found in virtually every possible habitat. Based on both light and electron microscopy morphology, they have been classified into six major phyla with a majority of disease causing protozoa falling under the phyla Sacromastigophora and Apicomplexa.

The following are some of the sub-phyla and classes within these sub-phyla based on locomotive structures:

- Plasmodroma: The locomotive structures of this sub-phylum may be flagella, pseudopodia or none at all. Classes that fall under this sub-phyla include; Mastigophora (use one or more flagella for locomotion), Sarcodina (used pseudopodia for locomotion and for capturing food) and Sporozoa which lack locomotive structures.
- Ciliophora: These are in the Sub-phyla Ciliophora use cilia or sucking tentacles in some stages or throughout their life span. Ciliata (which use cilia throughout) and Suctoria (which use cilia when young and tentacles as adults) are some of the class that fall under this sub-phylum.
- Sarcomastigopohora: The locomotive structures used in this sub-phylum include pseudopodia or flagella. Here, the nuclei is also of one kind (monomorphic). Super class Mastigophora, which falls under this sub-phyla are flagellates and thus use flagella for locomotion.

The Phytomastogophoerea also falls under this sub-phyla and use flagella in some cases. Under the Class Phytomastogophoerea is Order Chrysomonadida, which includes such organisms like Chrys amoeba, synura and ochromonas among others.

These are just a few in the classification. It is extensive and contains many more organisms.

Classification based on Mode of Existence

Of the existing protozoa, there are about 21,000 species that occur as free-living in a variety of habitats while another 11,000 species occur as parasitic microbes in both vertebrate and invertebrates hosts.

The free-living species can be found in various habitats and particularly in soil and water. These types of protozoa have little impact on human health given that they do not directly depend other organisms for their survival. However, some of the free-living can cause pathology when introduced into a human host.

Others will also affect human health by producing toxins. The following are some of the free-living amoebae that can also cause human disease:

- Naegleria fowleri - This species is mostly found in moist soil and can be located all across the world. It causes acute primary amebic meningoencephalitis.
- Acanthamoeba - Found in soil and water, acanthamoeba can cause chronic granulomatous amebic encephalitis, amebic keratitis, granulomatous skin as well as lung lesions.
- Balamuthia mandrillaris - Causes sub-acute to chronic granulomatous amebic encephalitis as well as granulomatous skin and lung lesions.
- Sappinia diploidea.

Parasitic Protozoa

Parasitic protozoa are the type that depend on the host for survival. As such, they live inside the host and even cause health problems.

The following are some of the parasitic:

- Sarcodina (e.g. Entamoeba) - Entamoeba histolytica is a type of amoeba that lives in the human alimentary canal. For most part, they are harmless and feed on various bacteria and particles that may be present in the intestine. Although they are mostly harmless, this parasite may invade the intestinal wall or the rectum where they cause ulcerations and even bleeding along with pain, vomiting, and diarrhea among other symptoms.
- Trypanosomes - This is a flagellate that lives in the blood stream. Various species of this parasite cause such diseases as:
  - Sleeping sickness,
  - Leishmaniasis,
  - Chaga's disease.
- Mastigophora (e.g. Giardia) - This is a flagellate that is mostly found in the small intestine of the host. The giardia typically attach themselves on to the intestinal lining causing inflammation, diarrhea as well as abdominal pain among other types of symptoms.
- Sporozoa (e.g. Plasmodium) - The plasmodium species is a parasite that lives in the blood stream of human beings, Once in the red blood cells, the parasite feeds on their cytoplasm. As they continue multiplying within the cells, this causes the cells to burst which in turn results in many more parasites being released into the circulatory system.

Life Cycle

Parasitic Protozoa

For the parasitic forms, the life cycle stages may occur intercellular, intracellular or in the lumen of given organs. Because of the diversity, it is not possible to describe a single or one common life cycle sequence. Here, therefore, we shall look at three of the most common broad patterns exhibited by this group of protozoa.

First Pattern

This pattern is common in the phylum Apicomplexa and involves an alteration between asexual and sexual reproductive stages.

The process starts with the cycles of asexual reproduction where the cycles of schizogony (involving mitosis and cytokinesis) in the tissues of the host results in increase population. Following this stage, some in the population start undergoing gametogony (a sexual process) to produce gametes. These gametes then unite and divide asexually to produce sporozoites through a process known as sporogeny.

It is these sporozoites that are then capable of infecting a new host and the process continues. Here, it is worth noting that transition in to a new host is through cysts, which are tough under stressful conditions. The cysts can survive external conditions (outside the body) and contain the sporozoites.

Once in a new host, the sporozoites start the reproduction cycle again. Some of the species in this phylum (Apicomplexa) require two hosts to complete their life cycle. This includes a vertebrate host where the parasite goes through schizogony and gametogony and an invertebrate where the gametes unite and sporiogony occurs in the tissues.

Second Pattern

The second pattern is common among most flagellates and involves asexual reproduction. For these, a number of morphological transformations occur during the cycle. However, they all reproduce through binary fission.

Some of the species in this group will complete this cycle in a vertebrate host as they transmit from one host to another through cysts, which can survive tough conditions better. Therefore, as in the case with the Apicomplexa phylum, some species in this group will also require two hosts to complete their life cycle.

Third Pattern

This is particularly common among amoebas and involves asexual reproduction. Unlike the others, a single host is required to complete the reproduction cycle. Here, trophozoites in the host live in the lumen of the gut and continue to multiply through binary fission. Here, under certain conditions, the trophozoites may be stimulated to encyst as they undergo nuclear division within the cyst. Once the cyst is ingested by another host, the cycle continues.

Life Cycle of Free Living Protozoa

For this group, the life cycle largely involves the growth and increase in size of the organism which is then followed by binary fission (or other forms of asexual reproduction). For the free-living, sexual reproduction only occurs under unfavorable conditions (unfavorable temperature, or reduced food supplies etc). However, these factors often vary from one species to another.

During the growth and division cycle of the free-living protozoa, there is a phase of DNA synthesis, chromosome replication as well as the growth of the cells.

The phases of the cycle include:

- First division phase.
- End of division phase and beginning of DNA synthesis.
- DNA synthesis.
- End of DNA synthesis and beginning of next division.

Classification based on Nutrition

There are three main categories based on nutrition. These include:

- Photo-autotrophs.
- Photo-heterotrophs.
- Chemoheterotrophs.

Autotrophs

Autotrophs like some of the flagellates synthesize carbohydrates from carbon dioxide and water using chlorophyll. Here, radiant energy from the sun is used.

Most of the photoautotrophic flagellates including members of Euglenida, Cryptomonadida as well as Volvocida also tend to combine autotrophy with heterotrophy. For this reason, they are often described as acetate flagellates.

Some of their source of carbon include acetates, simple fatty acids as well as alcohols. While they are autotrophs in the light, these flagellates switch to heterotrophs in the dark.

Heterotrophs

A majority of the free-living protozoa fall under this category. As such, they depend on a wide range of diet. Whereas some feed on bacteria (microbivores) others feed on algae and are described as herbivores. The carnivorous feed on both of the two trophics (herbivores and microbivores).

The free living are also divided into two groups (morphological). These include those with a mouth/cytostome and those that lack a mouth or a definite point of entry for food. For instance, whereas some flagellates and many ciliates (apart from some apostomatida) have a cytostome the Sarcodina lack a mouth.

Chemoheterotrophic - This group includes those that require energy and organic carbon sources.

Microscopy

As previously mentioned, protozoa are very diverse. As such, they are distinguished from one another based on their different structural features, means of locomotion as well as the formation of spores.

Using a light microscope, it is possible to view different types of protozoa.

Sample Collection

Protozoa can be obtained from almost any given habitat. Whereas the free-living species can be found in water as well as various moist habitats, the parasitic can be found in most metazoan (developed animals).

For students, it would prove easier to use the free-living protozoa, which can be obtained from such habitats as mud, ponds and transient bodies. Here, it is worth noting these are very fragile. For this reason, they should be handled with care.

It is also important to be careful given that even free-living protozoa can become parasitic. Protozoa can also be cultured in order to increase their numbers for observation. Some of the medium used include split pea (for Eglena) distilled water with wheat grains (for chilomonas) as well as hay (for peranema) among others.

Microscopic observation

Some of the requirements for microscopy include:

- A microscope
- Microscope slides
- Microscope clips
- Distilled water (or tap water)
- Dropper.

Wet Mounting Technique

Wet mounting technique is the technique that simply introducing the sample/specimen on to a drop of water and viewing it under the microscope. If the sample was obtained from a pond, then the following process can be used:

- Gently shake the container (to distribute the protozoa in the water).
- Using a dropper, obtain a sample of the pond water from the container.
- Place a drop of the sample onto the center of a microscope slide and cover with a cover slip (always ensure that the microscope slide and slip are clean to avoid introducing other microorganisms).
- Place the slide onto the microscope stage for viewing.

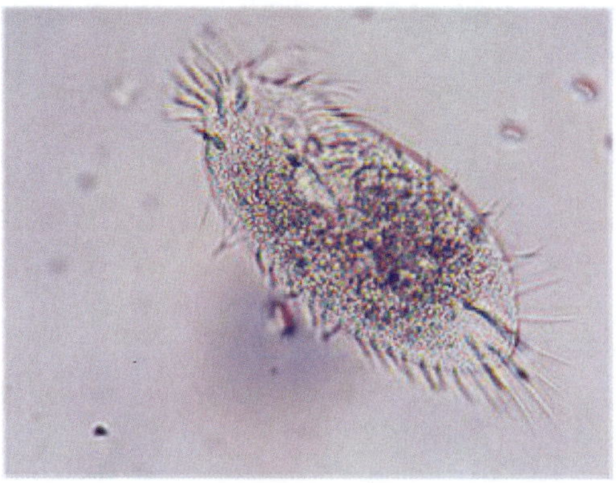

In some cases, staining may be used to increase contrast and get a clearer view. Some of the stains used here include:

- Bismarck Brown
- Brilliant Cresyl Blue

- Bromothymol Blue
- Carmine Powder
- Methylene Blue.

Bacteria

Bacteria is a singular bacterium, any of a group of microscopic single-celled organisms that live in enormous numbers in almost every environment on Earth, from deep-sea vents to deep below Earth's surface to the digestive tracts of humans.

Bacteria lack a membrane-bound nucleus and other internal structures and are therefore ranked among the unicellular life-forms called prokaryotes. Prokaryotes are the dominant living creatures on Earth, having been present for perhaps three-quarters of Earth history and having adapted to almost all available ecological habitats. As a group, they display exceedingly diverse metabolic capabilities and can use almost any organic compound, and some inorganic compounds, as a food source. Some bacteria can cause diseases in humans, animals, or plants, but most are harmless and are beneficial ecological agents whose metabolic activities sustain higher life-forms. Other bacteria are symbionts of plants and invertebrates, where they carry out important functions for the host, such as nitrogen fixation and cellulose degradation. Without prokaryotes, soil would not be fertile, and dead organic material would decay much more slowly. Some bacteria are widely used in the preparation of foods, chemicals, and antibiotics. Studies of the relationships between different groups of bacteria continue to yield new insights into the origin of life on Earth and mechanisms of evolution.

The Bacterial Cell

Bacteria as Prokaryotes

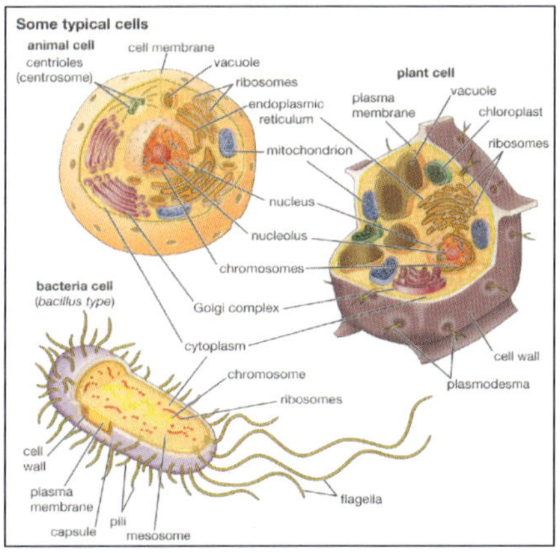

Bacterial cells differ from animal cells and plant cells in several ways. One fundamental difference is that bacterial cells lack intracellular organelles, such as mitochondria, chloroplasts, and a nucleus, which are present in both animal cells and plant cells.

All living organisms on Earth are made up of one of two basic types of cells: eukaryotic cells, in which the genetic material is enclosed within a nuclear membrane, or prokaryotic cells, in which the genetic material is not separated from the rest of the cell. Traditionally, all prokaryotic cells were called bacteria and were classified in the prokaryotic kingdom Monera. However, their classification as Monera, equivalent in taxonomy to the other kingdoms—Plantae, Animalia, Fungi, and Protista—understated the remarkable genetic and metabolic diversity exhibited by prokaryotic cells relative to eukaryotic cells. In the late 1970s American microbiologist Carl Woese pioneered a major change in classification by placing all organisms into three domains—Eukarya, Bacteria (originally called Eubacteria), and Archaea (originally called Archaebacteria)—to reflect the three ancient lines of evolution. The prokaryotic organisms that were formerly known as bacteria were then divided into two of these domains, Bacteria and Archaea. Bacteria and Archaea are superficially similar; for example, they do not have intracellular organelles, and they have circular DNA. However, they are fundamentally distinct, and their separation is based on the genetic evidence for their ancient and separate evolutionary lineages, as well as fundamental differences in their chemistry and physiology. Members of these two prokaryotic domains are as different from one another as they are from eukaryotic cells.

Prokaryotic cells (i.e., Bacteria and Archaea) are fundamentally different from the eukaryotic cells that constitute other forms of life. Prokaryotic cells are defined by a much simpler design than is found in eukaryotic cells. The most-apparent simplification is the lack of intracellular organelles, which are features characteristic of eukaryotic cells. Organelles are discrete membrane-enclosed structures that are contained in the cytoplasm and include the nucleus, where genetic information is retained, copied, and expressed; the mitochondria and chloroplasts, where chemical or light energy is converted into metabolic energy; the lysosome, where ingested proteins are digested and other nutrients are made available; and the endoplasmic reticulum and the Golgi apparatus, where the proteins that are synthesized by and released from the cell are assembled, modified, and exported. All of the activities performed by organelles also take place in bacteria, but they are not carried out by specialized structures. In addition, prokaryotic cells are usually much smaller than eukaryotic cells. The small size, simple design, and broad metabolic capabilities of bacteria allow them to grow and divide very rapidly and to inhabit and flourish in almost any environment.

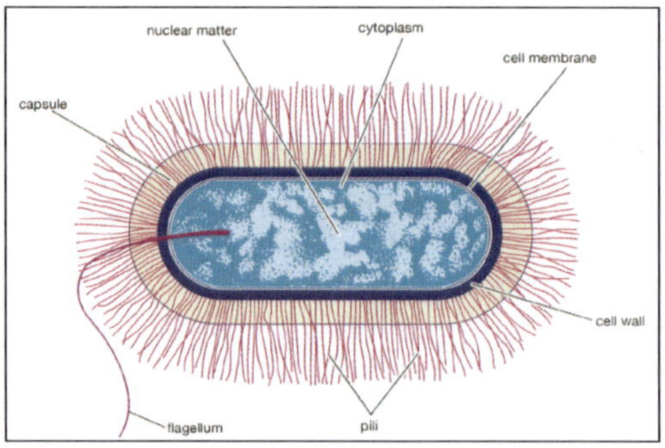

Schematic drawing of the structure of a typical bacterial cell of the bacillus type.

Prokaryotic and eukaryotic cells differ in many other ways, including lipid composition, structure of key metabolic enzymes, responses to antibiotics and toxins, and the mechanism of expression of genetic information. Eukaryotic organisms contain multiple linear chromosomes with genes that are much larger than they need to be to encode the synthesis of proteins. Substantial portions of the ribonucleic acid (RNA) copy of the genetic information (deoxyribonucleic acid, or DNA) are discarded, and the remaining messenger RNA (mRNA) is substantially modified before it is translated into protein. In contrast, bacteria have one circular chromosome that contains all of their genetic information, and their mRNAs are exact copies of their genes and are not modified.

Diversity of Structure of Bacteria

Although bacterial cells are much smaller and simpler in structure than eukaryotic cells, the bacteria are an exceedingly diverse group of organisms that differ in size, shape, habitat, and metabolism. Much of the knowledge about bacteria has come from studies of disease-causing bacteria, which are more readily isolated in pure culture and more easily investigated than are many of the free-living species of bacteria. It must be noted that many free-living bacteria are quite different from the bacteria that are adapted to live as animal parasites or symbionts. Thus, there are no absolute rules about bacterial composition or structure, and there are many exceptions to any general statement.

Individual bacteria can assume one of three basic shapes: spherical (coccus), rodlike (bacillus), or curved (vibrio, spirillum, or spirochete). Considerable variation is seen in the actual shapes of bacteria, and cells can be stretched or compressed in one dimension. Bacteria that do not separate from one another after cell division form characteristic clusters that are helpful in their identification. For example, some cocci are found mainly in pairs, including Streptococcus pneumoniae, a pneumococcus that causes bacterial lobar pneumonia, and Neisseria gonorrhoeae, a gonococcus that causes the sexually transmitted disease gonorrhea. Most streptococci resemble a long strand of beads, whereas the staphylococci form random clumps. In addition, some coccal bacteria occur as square or cubical packets. The rod-shaped bacilli usually occur singly, but some strains form long chains, such as rods of the corynebacteria, normal inhabitants of the mouth that are frequently attached to one another at random angles. Some bacilli have pointed ends, whereas others have squared ends, and some rods are bent into a comma shape. These bent rods are often called vibrios and include Vibrio cholerae, which causes cholera. Other shapes of bacteria include the spirilla, which are bent and rebent, and the spirochetes, which form a helix similar to a corkscrew, in which the cell body is wrapped around a central fibre called the axial filament.

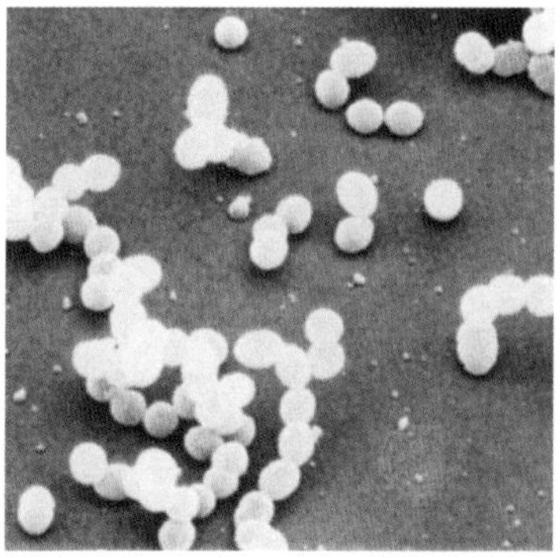

The bacterium Streptococcus mutans is an example of a spherical (coccus) bacterium. This species of bacteria commonly aggregates into pairs and short chains.

Bacteria are the smallest living entities. An average-size bacterium—such as the rod-shaped Escherichia coli, a normal inhabitant of the intestinal tract of humans and animals—is about 2 micrometres (μm; millionths of a metre) long and 0.5 μm in diameter, and the spherical cells of Staphylococcus aureus are up to 1 μm in diameter. A few bacterial types are even smaller, such as Mycoplasma pneumoniae, which is one of the smallest bacteria, ranging from about 0.1 to 0.25 ì m in width and roughly 1 to 1.5 μm in length; the rod-shaped Bordetella pertussis, which is the causative agent of whooping cough, ranging from 0.2 to 0.5 μm in diameter and 0.5 to 1 μm in length; and the corkscrew-shaped Treponema pallidum, which is the causative agent of syphilis, averaging only 0.1 to 0.2 μm in diameter but 6 to 15 μm in length. The cyanobacterium Synechococcus averages about

0.5 to 1.6 μm in diameter. Some bacteria are relatively large, such as Azotobacter, which has diameters of 2 to 5 μm or more; and Achromatium, which has a minimum width of 5 μm and a maximum length of 100 μm, depending on the species. Giant bacteria can be visible with the unaided eye, such as Thiomargarita namibiensis, which averages 750 μm in diameter, and the rod-shaped Epulopiscium fishelsoni, which ranges from 30 to more than 600 μm in length.

Bacteria are unicellular microorganisms and thus are generally not organized into tissues. Each bacterium grows and divides independently of any other bacterium, although aggregates of bacteria, sometimes containing members of different species, are frequently found. Many bacteria can form aggregated structures called biofilms. Organisms in biofilms often display substantially different properties from the same organism in the individual state or the planktonic state. Bacteria that have aggregated into biofilms can communicate information about population size and metabolic state. This type of communication is called quorum sensing and operates by the production of small molecules called autoinducers or pheromones. The concentration of quorum-sensing molecules—most commonly peptides or acylated homoserine lactones (AHLs; special signaling chemicals)—is related to the number of bacteria of the same or different species that are in the biofilm and helps coordinate the behaviour of the biofilm.

Morphological Features of Bacteria

The Gram Stain

Bacteria are so small that their presence was only first recognized in 1677, when the Dutch naturalist Antonie van Leeuwenhoek saw microscopic organisms in a variety of substances with the aid of primitive microscopes (more similar in design to modern magnifying glasses than modern microscopes), some of which were capable of more than 200-fold magnification. Now bacteria are usually examined under light microscopes capable of more than 1,000-fold magnification; however, details of their internal structure can be observed only with the aid of much more powerful transmission electron microscopes. Unless special phase-contrast microscopes are used, bacteria have to be stained with a coloured dye so that they will stand out from their background.

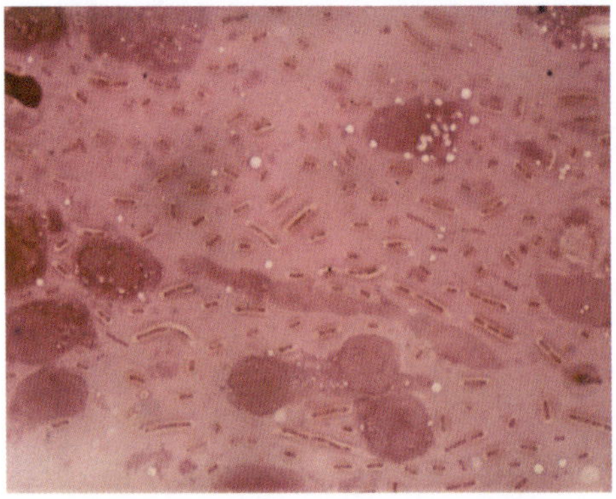

Gram-negative bacilli, Klebsiella pneumoniae, isolated from a lung abscess in a patient with pneumonia.

One of the most useful staining reactions for bacteria is called the Gram stain, developed in 1884 by the Danish physician Hans Christian Gram. Bacteria in suspension are fixed to a glass slide by brief heating and then exposed to two dyes that combine to form a large blue dye complex within each cell. When the slide is flushed with an alcohol solution, gram-positive bacteria retain the blue colour and gram-negative bacteria lose the blue colour. The slide is then stained with a weaker pink dye that causes the gram-negative bacteria to become pink, whereas the gram-positive bacteria remain blue. The Gram stain reacts to differences in the structure of the bacterial cell surface, differences that are apparent when the cells are viewed under an electron microscope.

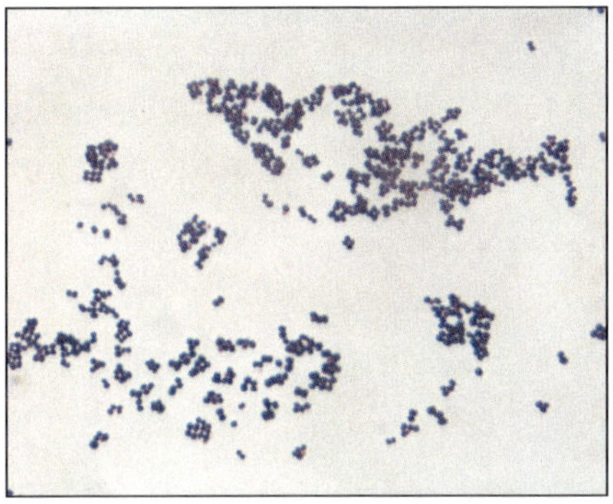

Gram-positive cocci, Staphylococcus aureus, in a laboratory culture.

The Cell Envelope

The bacterial cell surface (or envelope) can vary considerably in its structure, and it plays a central role in the properties and capabilities of the cell. The one feature present in all cells is the cytoplasmic membrane, which separates the inside of the cell from its external environment, regulates the flow of nutrients, maintains the proper intracellular milieu, and prevents the loss of the cell's contents. The cytoplasmic membrane carries out many necessary cellular functions, including energy generation, protein secretion, chromosome segregation, and efficient active transport of nutrients. It is a typical unit membrane composed of proteins and lipids, basically similar to the membrane that surrounds all eukaryotic cells. It appears in electron micrographs as a triple-layered structure of lipids and proteins that completely surround the cytoplasm.

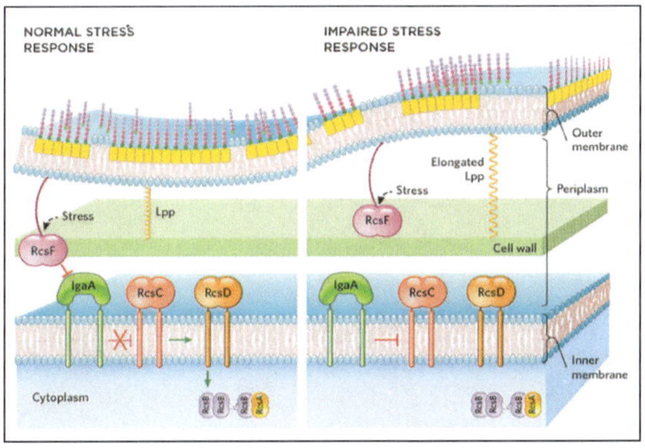

Lying outside of this membrane is a rigid wall that determines the shape of the bacterial cell. The wall is made of a huge molecule called peptidoglycan (or murein). In gram-positive bacteria the peptidoglycan forms a thick meshlike layer that retains the blue dye of the Gram stain by trapping it in the cell. In contrast, in gram-negative bacteria the peptidoglycan layer is very thin (only one or two molecules deep), and the blue dye is easily washed out of the cell.

Peptidoglycan occurs only in the Bacteria (except for those without a cell wall, such as Mycoplasma). Peptidoglycan is a long-chain polymer of two repeating sugars (N-acetylglucosamine and N-acetyl muramic acid), in which adjacent sugar chains are linked to one another by peptide bridges that confer rigid stability. The nature of the peptide bridges differs considerably between species of bacteria but in general consists of four amino acids: L-alanine linked to D-glutamic acid, linked to either diaminopimelic acid in gram-negative bacteria or L-lysine, L-ornithine, or diaminopimelic acid in gram-positive bacteria, which is finally linked to D-alanine. In gram-negative bacteria the peptide bridges connect the D-alanine on one chain to the diaminopimelic acid on another chain. In gram-positive bacteria there can be an additional peptide chain that extends the reach of the cross-link; for example, there is an additional bridge of five glycines in Staphylococcus aureus.

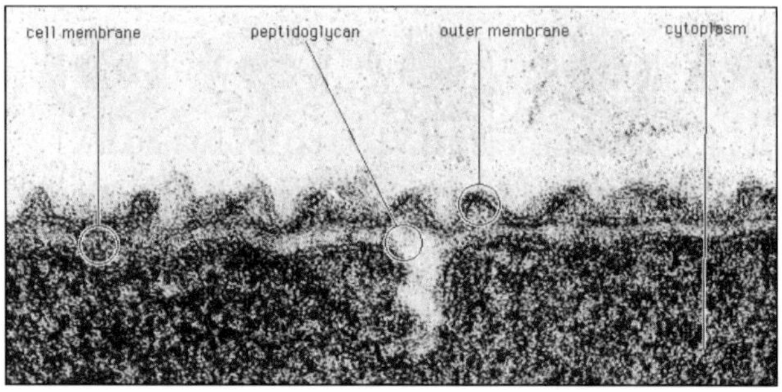

The gram-negative bacterium Aquaspirillum serpens has a thin peptidoglycan layer that lies between the cell membrane and the outer membrane.

Peptidoglycan synthesis is the target of many useful antimicrobial agents, including the β-lactam antibiotics (e.g., penicillin) that block the cross-linking of the peptide bridges. Some of the proteins that animals synthesize as natural antibacterial defense factors attack the cell walls of bacteria. For example, an enzyme called lysozyme splits the sugar chains that are the backbone of peptidoglycan molecules. The action of any of these agents weakens the cell wall and disrupts the bacterium.

In gram-positive bacteria the cell wall is composed mainly of a thick peptidoglycan meshwork interwoven with other polymers called teichoic acids and some proteins or lipids. In contrast, gram-negative bacteria have a complex cell wall that is composed of multiple layers in which an outer membrane layer lies on top of a thin peptidoglycan layer. This outer membrane is composed of phospholipids, which are complex lipids that contain molecules of phosphate, and lipopolysaccharides, which are complex lipids that are anchored in the outer membrane of cells by their lipid end and have a long chain of sugars extending away from the cell into the medium. Lipopolysaccharides, often called endotoxins, are toxic to animals and humans; their presence in the bloodstream can cause fever, shock, and even death. For most gram-negative bacteria, the outer membrane forms a barrier to the passage of many chemicals that would be harmful to the bacterium, such as dyes and detergents that normally dissolve cellular membranes. Impermeability to oil-soluble compounds is not seen in other biological membranes and results from the presence of lipopolysaccharides in the membrane and from the unusual character of the outer membrane proteins. As evidence of the ability of the outer membrane to confer resistance to harsh environmental conditions, some gram-negative bacteria grow well in oil slicks, jet fuel tanks, acid mine drainage, and even bottles of disinfectants.

The Archaea have markedly different surface structures from the Bacteria. They do not have peptidoglycan; instead, their membrane lipids are made up of branched isoprenoids linked to glycerol by ether bonds. Some archaea have a wall material that is similar to peptidoglycan, except that the specific sugar linked to the amino acid bridges is not muramic acid but talosaminuronic acid. Many other archaeal species use proteins as the basic constituent of their walls, and some lack a rigid wall.

Capsules and Slime Layers

Many bacterial cells secrete some extracellular material in the form of a capsule or a slime layer. A slime layer is loosely associated with the bacterium and can be easily washed off, whereas a capsule is attached tightly to the bacterium and has definite boundaries. Capsules can be seen under a light microscope by placing the cells in a suspension of India ink. The capsules exclude the ink and appear as clear halos surrounding the bacterial cells. Capsules are usually polymers of simple sugars (polysaccharides), although the capsule of Bacillus anthracis is made of polyglutamic acid. Most capsules are hydrophilic ("water-loving") and may help the bacterium avoid desiccation (dehydration) by preventing water loss. Capsules can protect a bacterial cell from ingestion and destruction by white blood cells (phagocytosis). While the exact mechanism for escaping phagocytosis is unclear, it may occur because capsules make bacterial surface components more slippery, helping the bacterium to escape engulfment by phagocytic cells. The presence of a capsule in Streptococcus pneumoniae is the most important factor in its ability to cause pneumonia. Mutant strains of S. pneumoniae that have lost the ability to form a capsule are readily taken up by white blood cells and do not cause disease. The association of virulence and capsule formation is also found in many other species of bacteria.

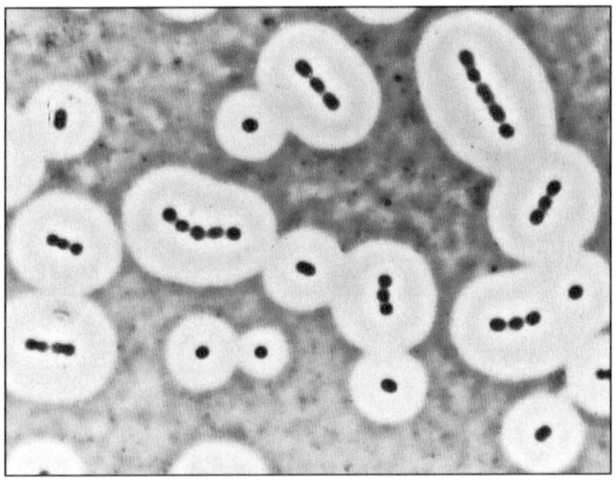

The capsular material surrounding these bacteria (Acinetobacter calcoaceticus) is revealed in a suspension of India ink and viewed through a light microscope (magnified about 2,500×).

A capsular layer of extracellular polysaccharide material can enclose many bacteria into a biofilm and serves many functions. Streptococcus mutans, which causes dental caries, splits the sucrose in food and uses one of the sugars to build its capsule, which sticks tightly to the tooth. The bacteria that are trapped in the capsule use the other sugar to fuel their metabolism and produce a strong acid (lactic acid) that attacks the tooth enamel. When Pseudomonas aeruginosa colonizes the lungs of persons with cystic fibrosis, it produces a thick capsular polymer of alginic acid that contributes to the difficulty of eradicating the bacterium. Bacteria of the genus Zoogloea secrete fibres of cellulose that enmesh the bacteria into a floc that floats on the surface of liquid and keeps the bacteria exposed to air, a requirement for the metabolism of this genus. A few rod-shaped bacteria, such as Sphaerotilus, secrete long chemically complex tubular sheaths that enclose substantial numbers of the bacteria. The sheaths of these and many other environmental bacteria can become encrusted with iron or manganese oxides.

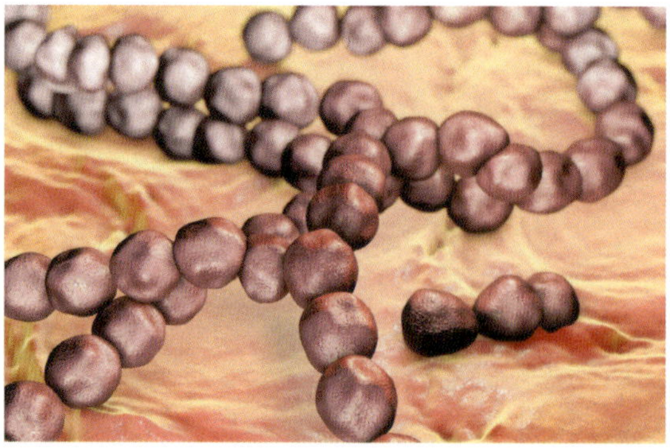

Streptococcus mutans, a bacteria found in the mouth, contributes to tooth decay.

Flagella, Fimbriae and Pili

Many bacteria are motile, able to swim through a liquid medium or glide or swarm across a solid surface. Swimming and swarming bacteria possess flagella, which are the extracellular appendages needed for motility. Flagella are long, helical filaments made of a single type of protein and located either at the ends of rod-shaped cells, as in Vibrio cholerae or Pseudomonas aeruginosa, or all over the cell surface, as in Escherichia coli. Flagella can be found on both gram-positive and gram-negative rods but are rare on cocci and are trapped in the axial filament in the spirochetes. The flagellum is attached at its base to a basal body in the cell membrane. The protomotive force generated at the membrane is used to turn the flagellar filament, in the manner of a turbine driven by the flow of hydrogen ions through the basal body into the cell. When the flagella are rotating in a counterclockwise direction, the bacterial cell swims in a straight line; clockwise rotation results in swimming in the opposite direction or, if there is more than one flagellum per cell, in random tumbling. Chemotaxis allows a bacterium to adjust its swimming behaviour so that it can sense and migrate toward increasing levels of an attractant chemical or away from a repellent one.

Not only are bacteria able to swim or glide toward more favourable environments, but they also have appendages that allow them to adhere to surfaces and keep from being washed away by flowing fluids. Some bacteria, such as E. coli and Neisseria gonorrhoeae, produce straight, rigid, spikelike projections called fimbriae or pili, which extend from the surface of the bacterium and attach to specific sugars on other cells—for these strains, intestinal or urinary-tract epithelial cells, respectively. Fimbriae are present only in gram-negative bacteria. Certain pili (called sex pili) are used to allow one bacterium to recognize and adhere to another in a process of sexual mating called conjugation. Many aquatic bacteria produce an acidic mucopolysaccharide holdfast, which allows them to adhere tightly to rocks or other surfaces.

The Cytoplasm

Although bacteria differ substantially in their surface structures, their interior contents are quite similar and display relatively few structural features.

Genetic Content

The genetic information of all cells resides in the sequence of nitrogenous bases in the extremely long molecules of DNA. Unlike the DNA in eukaryotic cells, which resides in the nucleus, DNA in bacterial cells is not sequestered in a membrane-bound organelle but appears as a long coil distributed through the cytoplasm. In many bacteria the DNA is present as a single circular chromosome, although some bacteria may contain two chromosomes, and in some cases the DNA is linear rather than circular. A variable number of smaller, usually circular (though sometimes linear) DNA molecules, called plasmids, can carry auxiliary information.

The sequence of bases in the DNA has been determined for hundreds of bacteria. The amount of DNA in bacterial chromosomes ranges from 580,000 base pairs in Mycoplasma genitalium to 4,700,000 base pairs in E. coli to roughly 9,450,000 base pairs in Myxococcus xanthus. Sorangium cellulosum, a myxobacterium, has one of the largest bacterial genomes, containing in excess of 13 million base pairs. The length of the E. coli chromosome, if removed from the cell and stretched to its fullest extent, is about 1.2 mm, which is striking in view of the fact that the length of the cell is about 0.001 mm.

As in all organisms, bacterial DNA contains the four nitrogenous bases adenine (A), cytosine (C), guanine (G), and thymine (T). The rules of base pairing for double-stranded DNA molecules require that the number of adenine and thymine bases be equal and that the number of cytosine and guanine bases also be equal. The relationship between the number of pairs of G and C bases and the number of pairs of A and T bases is an important indicator of evolutionary and adaptive genetic changes within an organism. The proportion, or molar ratio, of $G+C$ can be measured as $G+C$ divided by the sum of all the bases $\left(A+T+G+C\right)$ multiplied by 100 percent. The extent to which $G+C$ ratios vary between organisms may be considerable. In plants and animals, the proportion of $G+C$ is about 50 percent. A far wider range in the proportion of $G+C$ is seen in prokaryotes, extending from about 25 percent in most Mycoplasma to about 50 percent in E. coli to nearly 75 percent in Micrococcus, actinomycetes, and fruiting myxobacteria. The $G+C$ content within a species in a single genus, however, is very similar.

Cytoplasmic Structures

The cytoplasm of bacteria contains high concentrations of enzymes, metabolites, and salts. In addition, the proteins of the cell are made on ribosomes that are scattered throughout the cytoplasm. Bacterial ribosomes are different from ribosomes in eukaryotic cells in that they are smaller, have fewer constituents (consist of three types of ribosomal RNA [rRNA] and 55 proteins, as opposed to four types of rRNA and 78 proteins in eukaryotes), and are inhibited by different antibiotics than those that act on eukaryotic ribosomes.

There are numerous inclusion bodies, or granules, in the bacterial cytoplasm. These bodies are never enclosed by a membrane and serve as storage vessels. Glycogen, which is a polymer of glucose, is stored as a reserve of carbohydrate and energy. Volutin, or metachromatic granules, contains polymerized phosphate and represents a storage form for inorganic phosphate and energy. Many bacteria possess lipid droplets that contain polymeric esters of poly-β-hydroxybutyric acid or related compounds. This is in contrast to eukaryotes, which use lipid droplets to store triglycerides. In bacteria, storage granules are produced under favourable growth conditions and are consumed after the nutrients have been depleted from the medium. Many aquatic bacteria produce gas vacuoles, which are protein-bound structures that contain air and allow the bacteria to adjust their buoyancy. Bacteria can also have internal membranous structures that form as outgrowths of the cytoplasmic membrane.

Biotypes of Bacteria

The fact that pathogenic bacteria are constantly battling their host's immune system might account for the bewildering number of different strains, or types, of bacteria that belong to the same species but are distinguishable by serological tests. Microbiologists often identify bacteria by the presence of specific molecules on their cell surfaces, which are detected with specific antibodies. Antibodies are serum proteins that bind very tightly to foreign molecules (antigens) in an immune reaction aimed at removing or destroying the antigens. Antibodies have remarkable specificity, and the substitution of even one amino acid in a protein might prevent that protein from being recognized by an antibody.

For many bacterial species there are thousands of different strains (called serovars, for serological variants), which differ from one another mainly or solely in the antigenic identity of their lipopolysaccharide, flagella, or capsule. Different serovars of enteric bacteria—such as E. coli and Salmonella enterica, for example—are often found to be associated with the ability to inhabit different host animals or to cause different diseases. Formation of these numerous serovars reflects the ability of bacteria to respond effectively to the intense defensive actions of the immune system.

Classification of Bacteria

Taxonomic Rankings

The classification of bacteria has long presented unique challenges in biological systematics. In the 17th century, when bacteria were first observed under a microscope, only two categories of life were recognized in biological systematics: plants and animals. Lacking any obvious relation to animals, bacteria initially were classified in the plant kingdom. In the latter part of the 19th century, however, German zoologist Ernst Haeckel, recognizing the basic morphological characteristics of single-celled life—particularly the lack of a clearly defined nucleus among many of those organisms—proposed a third kingdom of "lower" life, Protista, and within it the class Monera, which would contain the structureless (nuclei-lacking) microorganisms. About the same time, German naturalist and botanist Ferdinand Cohn began to systematically organize bacteria into genera and species. Although Cohn's arrangement of the bacteria was based on morphology, he recognized that bacteria could not be adequately classified by morphology alone. In 1938, seeking to further distinguish the bacteria from other forms of life, American biologist Herbert F. Copeland elevated Monera to the level of kingdom.

Although kingdom Monera was readily adopted and later accepted as part of a five-kingdom system (whereby American biologist Robert H. Whittaker separated the fungi into their own kingdom), in the purview of some researchers, Monera remained an unsatisfactory division. The differences between prokaryotic organisms, namely between bacteria and archaea, are great; those groups of organisms are as different from one another as they are from plants and animals. Hence, in 1990 Carl R. Woese and colleagues proposed the three-domain system, dividing life into the Bacteria, the Archaea, and the Eukarya. Still other researchers disagreed with the domain system, however, and in 1998 British zoologist Thomas Cavalier-Smith presented yet another classification scheme, the six-kingdom system, which contained kingdom Bacteria with two subdivisions, Eubacteria and Archaebacteria. In 2015 Cavalier-Smith and others revised the system to include seven kingdoms, whereby kingdom Bacteria was split into two separate kingdoms—Bacteria (containing the eubacteria) and Archaea (containing the archaebacteria).

According to the rules of nomenclature under the International Code of Nomenclature of Bacteria—the body that governs the naming of prokaryotes—valid taxa for bacteria extend from subspecies to class; taxonomic categories above class (e.g., phylum) are not considered valid.

The Bacterial Species Problem

As with the general taxonomy of bacteria, the basic concept of species in bacterial systematics is problematic. Unlike in higher organisms, in which a species is defined by the ability of organisms with common characteristics to interbreed and give rise to fertile offspring, bacteria generally reproduce asexually. When genetic material is exchanged between bacterial cells, it can occur between individuals of different species through the process of horizontal gene transfer, thereby confusing the limits of interbreeding.

Speciation among bacteria is suspected of occurring at the level of subspecies, or ecotype, whereby genetically distinct populations survive within the same ecological niche until, through adaptation and natural selection, one type outcompetes the others, clearing the niche of its diversity. The process of divergence within the niche then begins anew. Hence, a single named species of bacteria can have numerous ecotypes. For that reason, some researchers consider the species level in bacterial classification to be nearly comparable to the genus level in the classification systems of other organisms.

Genetic Approaches

Genetic approaches to the classification of bacteria are aimed at identifying a degree of relatedness between organisms to obtain a more-fundamental measure of the time elapsed since two organisms diverged from a common ancestor.

DNA-based Methods

DNA-based approaches used in the identification and classification of species of bacteria include DNA-DNA hybridization, DNA fingerprinting, and DNA sequencing. DNA-DNA hybridization, initially developed in the 1980s, is used to determine the similarity of DNA sequences from different organisms. The degree of similarity is reflected in the degree to which a strand of DNA from the organism of interest passively hybridizes with (attaches to) a single strand of DNA from a known organism. The less stable the hybridization is, the more quickly the DNA strands will dissociate when heated; hence, low DNA melting temperatures typically suggest low degrees of sequence similarity. DNA-DNA hybridization is most valuable for determining genetic relatedness at the genus and species levels.

DNA fingerprinting methods for bacterial identification centre primarily on the use of the polymerase chain reaction (PCR). Repetitive element-PCR, for example, targets specific DNA segments that are repeated at random in the bacterial genome. The identification of repetitive elements is powerful, capable of resolving bacteria at intraspecies levels. DNA sequencing methods, including whole genome sequencing and multilocus sequence analysis (MLSA), likewise have proved useful in the identification of bacteria. MLSA, which entails DNA sequencing of subsets of so-called housekeeping (or conserved) genes, has been shown to provide resolution down to intraspecies levels.

16S rRNA Analysis

Evolutionary relatedness can also be assessed through the sequencing of 16S rRNA, the gene that encodes the RNA component of the smaller subunit of the bacterial ribosome (16S refers to the rate of sedimentation, in Svedberg units, of the RNA molecule in a centrifugal field). The 16S rRNA gene is present in all bacteria, and a related form occurs in all cells. The 16S rRNA gene of E. coli is 1,542 nucleotides long, and some of its regions are double-stranded, while other regions are single-stranded. Single-stranded regions often form loops, because there is a lack of complementary bases on the opposing strand. Since 16S rRNA makes very specific contacts with many different ribosomal proteins and with other parts of itself, the pace at which spontaneous random mutation can change the sequence of the bases in the rRNA is slow. Any change in sequence at one site must be compensated for by another change elsewhere within the rRNA or in a ribosomal protein, lest the ribosome fail to assemble properly or to function in protein synthesis and the cell die.

Analysis of the 16S rRNA sequences from many organisms has revealed that some portions of the molecule undergo rapid genetic changes, thereby distinguishing between different species within the same genus. Other positions change very slowly, allowing much broader taxonomic levels to be distinguished. The comparison of 16S rRNA sequences between organisms is quantitative and is based on a defined set of assumptions. The assumption that the rate at which base changes occur and are established within a species is constant is unlikely to be true. Changes in Earth's environment are expected to alter the ecological niches or selective pressures that affect the rate of mutation and the rate at which various species are able to evolve.

The radical differences between Archaea and Bacteria, which are evident in the composition of their lipids and cell walls and in the utilization of different metabolic pathways, enzymes, and enzyme cofactors, are also reflected in the rRNA sequences. The rRNAs of Bacteria and Archaea are as different from each other as they are from eukaryotic rRNA. That suggests that the bacterial and archaeal lines diverged from a common precursor somewhat before eukaryotic cells developed. That proposal also implies that the eukaryotic line is quite ancient and probably did not arise from any currently known bacteria. It had been previously believed that eukaryotic cells arose when some bacterial cells engulfed another type of bacterium. Those bacteria might have formed a symbiotic relationship in which the engulfed cell continued to survive but gradually lost its independence and took on the properties of an organelle. Although the original eukaryotic cell may or may not be derived from bacteria, it remains likely, if not certain, that eukaryotic organelles (e.g., mitochondria and chloroplasts) are descendants of bacteria that were acquired by eukaryotic cells in an example of symbiotic parasitism.

Early hypotheses about the origins of life suggested that the first cells obtained their energy from the breakdown of nutrients in a rich organic liquid environment proposed to have formed in the early oceans by the action of light and intense solar radiation on the early, anaerobic atmosphere. The process of photosynthesis might have evolved much later in response to the gradual depletion of those rich nutrient sources. On the other hand, rRNA sequence analysis places photosynthetic capability in almost all of the major bacterial divisions and shows that photosynthetic genera are closely related to nonphotosynthetic genera. Since photosynthesis is such a highly conserved, mechanistically complex process, it is unlikely that the ability to carry out photosynthesis could have evolved at different times in so many different organisms. Even more widely distributed among prokaryotes is lithotrophy, the ability to obtain energy by the transfer of electrons from hydrogen gas to inorganic acceptors. It has been proposed that the earliest forms of life on Earth used lithotrophic metabolism and that photosynthesis was a later addition to the early bacterial progenitors. The nonlithotrophic and nonphotosynthetic forms found today arose from the earliest forms of Bacteria, although they have lost their capacities for lithotrophy and photosynthesis.

The proposal that lithotrophy was widely distributed among bacterial organisms before photosynthesis developed suggests that the Archaea came from a different line of descent from that of Bacteria. The only photosynthetic archaeon, Halobacterium, has a completely different type of photosynthesis that does not use chlorophyll in large protein complexes to activate an electron, as in plants and bacteria. Rather, it uses a single protein, bacteriorhodopsin, in which light energy is absorbed by retinal, a form of vitamin A, to activate a proton (hydrogen ion).

The analysis of rRNA sequences from bacteria that are closely related to one another has revealed several surprising relationships between those organisms. For example, Mycoplasma, which appear to be different from other bacteria—in that they are very small, lack a cell wall, have a very small genome, and have sterols in their cell membranes—actually are related to some gram-positive clostridia on the basis of their nucleic acid sequences. That circumstance underscores the hazard of relying on phenotypic traits (observable characteristics such as the absence of a cell wall) for the assignment of evolutionary or genetic relationships. In fact, there are many groupings of bacteria that are not supported by RNA sequence analysis.

A limitation of 16S rRNA sequence analysis, however, is its poor resolution below the genus level. Populations of organisms in a given genus that reside within the same habitats can have unique 16S rRNA genotypes. However, whether those unique genotypes are indicative of distinct species typically cannot be determined from rRNA information alone.

Classification by Morphology, Biochemistry and other Features

Although genetic divergence highlights the evolutionary relationships of bacteria, morphological and biochemical features of bacteria remain important in the identification and classification of those organisms. Indeed, bacteria are classified on the basis of many characteristics. Cell shape, nature of multicell aggregates, motility, formation of spores, and reaction to the Gram stain are important. Those morphological features, including the shape and colour of bacterial colonies, are not always constant and can be influenced by environmental conditions. Important in the identification of a genus and species of bacteria are biochemical tests, including the determination of the kinds of nutrients a cell can use, the products of its metabolism, the response to specific chemicals, and the presence of particular characteristic enzymes. Other criteria used for the identification of some types of bacteria might be their antigenic composition, habitat, disease production, and requirement for specific nutrients. Some tests are based on the ultrastructure of the bacteria revealed under the electron microscope by negative staining and preparation of thin sections.

Microalgae

Microalgae are small-sized organisms found in fresh and saline waters, in both benthic and littoral habitats, and also throughout the ocean waters as phytoplankton, while the larger macroalgae (seaweeds) occupy the littoral zone. Microalgae are unicellular to filamentous in form. They lack roots, vascular systems, leaves and stems, and are autotrophic and photosynthetic. Microalgae are generally eukaryotic organisms, although cyanobacteria, such as spirulina, which are prokaryotes, are included under microalgae due to their photosynthetic and reproductive properties. Microalgae range in size from about 5 μm (Chlorella) to more than 100 μm (spirulina) (Becker, 2013b). The commercial cultivation of microalgae began in Japan with the cultivation of Chlorella in the 1960s, followed by the cultivation of spirulina in Mexico and the USA in the 1970s. Since then, the industrial biotechnology of microalgae has grown tremendously. The immense chemical diversity of microalgae provides numerous applications in the food, feed and pharmaceutical industries. Microalgae are cultivated for the production of whole biomass and valuable substances such as nutraceuticals, carotenoids, phycocyanin and poly-unsaturated fatty acids (PUFAs), which are utilised in the food and feed (notably aquaculture) industry. The production of biofuel from lipid- or carbohydrates-rich microalgae is under way.

Microalgae for Feed

Microalgae species appeared in the FAO statistics in 2003 with the Chinese production of several species of spirulina, which was first recorded at 16,483 t and rose sharply to 97,104 t in 2010. These microalgae have gained popularity as a nutraceutical all over the world. They are increasingly used as a dietary ingredient for poultry, and as a protein and vitamin supplement for aquaculture. Microalgae are required for the nutrition of the larvae of fish and crustaceans, either for direct consumption in the case of molluscs and penaeid shrimp, or indirectly as feed for the live prey fed to small-larvae fish. In all these cases, the post-larvae are hatched, bred, and raised in specialized hatcheries. As of 1999, aquacultural hatcheries used approximately one-fifth of the 5,000 t of yearly global microalgae production: 62% were molluscs, 21% were shrimp, and 16% were fish. It was estimated that of the 63 million t of aquacultural production in 2005, 35 million t were fed only on pond-grown green water. In 2011, the total ingestion of green water microalgae by cultured fish and crustaceans via planktivory was estimated at 240 million t.

In intensive monoculture systems, microalgae are used to feed directly, or indirectly, larval stages of bivalves, shrimp, and some species of fish. Species of algae favoured for this purpose include Chaetoceros, Thalassiosira, Tetraselmis, Isochrysis, Nannochloropsis, Pavlova, and Skeletonema. They are either consumed directly by the target animals (at the larval stage or beyond), or by other organisms such as Artemia, rotifers, and Daphnia, which are, in turn, fed to the target larval organisms. Nutriments such as fatty acids and vitamins are consumed by intermediary zooplankton and transferred to higher trophic levels (the larvae and other stages of fish, shrimp and molluscs.

In extensive polyculture systems, microalgae are used for the growth of bivalves, carp, and shrimp. This type of aquaculture is becoming advanced in developing countries, and is contributing significantly to the protein supply of the most populous countries, especially India and China. Species with different feeding requirements are reared together. Faeces from carnivorous species support phytoplankton and zooplankton blooms, which in turn are utilized by filter feeders such as carp and shrimp. The production of algal biomass was estimated at 70 million t (DM) in 2011.

Over the last decades, several hundred microalgal species have been tested as feed, but only about 40 species are grown in intensive cultivation systems. These species include diatoms, flagellated and chlorococcalean green algae as well as filamentous blue-green algae, notably spirulina.

Spirulina (Arthrospira)

Among all the algae employed in commercial aquaculture, the cyanobacterium spirulina (Arthrospira sp.) has the broadest range of applications. Arthrospira is a microscopic blue-green alga in the shape of a spiral coil, living both in sea and fresh water as a free-floating filamentous organism characterized by cylindrical, multicellular trichomes in an open left-hand helix. Spirulina is the common name for human and animal food supplements produced primarily from two Arthrospira species: Arthrospira platensis, which occurs in Africa, Asia and South America, and Arthrospira maxima, which is confined to Central America. They occur naturally in tropical and subtropical lakes with high pH and high concentrations of carbonate and bicarbonate. Arthrospira is cultivated around the world. Its main applications are human dietary supplements and whole food, and feed supplements in the aquaculture, aquarium, and poultry industries. Spirulina has a long history of being used for food in Chad, as far back as the 9th century Kanem Empire. It is still in daily use today, dried into cakes called dihé, harvested from small lakes and ponds around Lake Chad.

Processes

Preservation

As marine invertebrates depend on microalgae for their whole life cycle, mollusc or fish hatcheries need to include a microalgae production system in parallel to their fish production. However, handling live microalgae is not without problems and various attempts have been made to replace living algae by processed algae-based diets with a longer shelf life.

Spray-drying

Spray-drying is probably the most common technique used to preserve microalgae as a sole feed or supplement to the diet in aquaculture. The shelf life is up to 2 years if stored at a low temperature. Spray-dried Chlorella vulgaris and Nannochloropsis sp. are available as intact cells or broken cell walls.

Freeze-drying

Freeze-drying is a good method for producing an algal feed with a shelf life similar to that obtained with spray-drying, resulting in diets that are almost identical to fresh diets regarding size, shape, and biochemical composition.

Low Temperatures

Refrigerated (0-4 °C) concentrated pastes of 90-300 g/L (DM) have a shelf life ranging from 4 weeks to 1 year. All algal pastes are nonviable cells except for Chlorella vulgaris (130 g/L). Frozen paste kept at -20 °C may have a shelf life of up to 2 years. Cultures of Nannochloropsis oculata have been kept alive at 8 °C for over 100 days, or at 5 °C for 18 months when provided with light and nitrogen.

Microencapsulation

The encapsulation of microalgae in digestible microcapsules allows the delivery to suspension feeders without loss of nutrients to the aqueous medium. The stabilization of the capsule was best when microalgae were cultivated in 2% $CaCl_2$ concentration.

Cell Walls Disruption

Most microalgae contain cell walls that are detrimental to digestion in monogastric species. Treatment is necessary to disrupt the cell walls: physical methods include boiling and various types of high temperature drying, if necessary simply sun drying; chemical methods include autolysis or breaking of hydrogen bonds by phenol, formic acid, or urea which requires later detoxification.

Live vs. Dried Microalgae

Dried microalgae are often less efficient than live microalgae for feeding marine larvae. The oxidation of microalgae that occurs during drying causes a loss of highly unsaturated fatty acids essential for larval growth. Dried cells tend to disintegrate when kept in suspension. Due to the broken cell walls, water-solubles components are lost in the culture media and are no longer available to the host species. Dried microalgae are also more susceptible to pathogenic bacterial contamination. Algae paste exhibit similar issues. These problems can be alleviated by proper preparation (centrifugation, flocculation, or filtration) and/or preservation techniques (additives and freezing) that preserve cell wall integrity.

Forage Management

Open Cultivation Systems

Most algae for commercial use are grown in the open air. The two most common open cultivation systems are circular and raceway ponds. These systems can be developed using natural water bodies such as lagoons and ponds, or artificial ponds such as raceways. Open systems, such as coastal shallow brackish-water ponds, are extensively used for feed production in aquaculture and for other industrial applications. The open cultivation pond is cheap and they do not compete for agricultural land as they can be built on non-productive (marginal) land, or coastal regions for marine algal cultivation. Open systems require minimal investment in terms of light source and operations. However, open systems provide less control over abiotic conditions (rainfall, temperature, dust, etc.) and have a greater risk of contamination by other algae species, or by microorganisms such as rotifers. In some countries, notably in Asia, microalgae are used in integrated livestock management in manure ponds for growing fish, thereby providing feed and avoiding use of fish meal.

Closed Cultivation Systems

Commercial algae production for aquaculture, particularly for hatcheries, often relies on closed cultivation systems where culture parameters can be closely controlled. Such systems have a high biomass productivity compared with open ponds since culture parameters such as light, turbulence and air exchange can be carefully regulated. The two main closed systems are:

- Photobioreactors, which provide environments with the greatest level of control. They include tubular systems, flat-plate vessels, vertical cylinders and vertical or horizontal bag cultures.

- Fermenters, which are used for large-scale production of freshwater (Chlorella vulgaris, in South-East Asia) and marine algae (Schizochytrium sp. and Crypthecodinium cohnii).

Environmental Impact

Water systems containing organic materials, such as swine waste, fishery waste, etc. can be used to grow microalgae. Using algae in this way has been proposed as a means of reducing biological oxidation demand in water systems and improving water quality, while at the same time generating a supplemental protein source that can be used in feeding livestock.

Archaea

Archaea is any of a group of single-celled prokaryotic organisms (that is, organisms whose cells lack a defined nucleus) that have distinct molecular characteristics separating them from bacteria (the other, more prominent group of prokaryotes) as well as from eukaryotes (organisms, including plants and animals, whose cells contain a defined nucleus). Members of the archaea include: Pyrolobus fumarii, which holds the upper temperature limit for life at 113 °C (235 °F) and was found living in hydrothermal vents; species of Picrophilus, which were isolated from acidic soils in Japan and are the most acid-tolerant organisms known—capable of growth at around pH 0; and the methanogens, which produce methane gas as a metabolic by-product and are found in anaerobic environments, such as in marshes, hot springs, and the guts of animals, including humans.

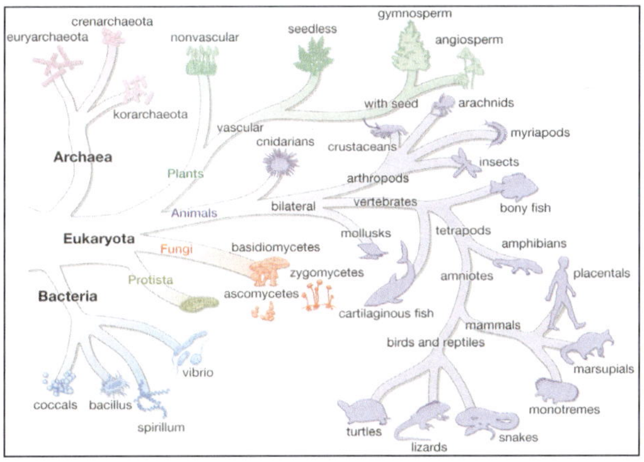

Life: three-domain classification The tree of life according to the three-domain system.

In some systems for classifying all of life, the archaea constitute one of three great domains of living creatures. In 1977 American microbiologist Carl Woese, on the basis of analyses of ribosomal RNA, proposed that the prokaryotes, long considered to be a single group of organisms (essentially, the bacteria), actually consist of two separate lineages. Woese called these two lineages the eubacteria and the archaebacteria. These names were subsequently changed to bacteria and archaea (the archaea being distinctly different from bacteria), but Woese's splitting of the prokaryotes into two groups has remained, and all living organisms are now considered by many biologists to fall into one of three great domains: Archaea, Bacteria, and Eukarya. Further molecular analysis has shown that domain Archaea consists of two major subdivisions, the Crenarchaeota and the Euryarchaeota, and one minor ancient lineage, the Korarchaeota. Other subdivisions have been proposed, including Nanoarchaeota and Thaumarchaeota.

Habitats of the Archaea

Archaea are microorganisms that define the limits of life on Earth. They were originally discovered and described in extreme environments, such as hydrothermal vents and terrestrial hot springs. They were also found in a diverse range of highly saline, acidic, and anaerobic environments.

Archaea at Midway Geyser Basin, Yellowstone National Park, Wyoming.

Although many of the cultured archaea are extremophiles, these organisms in their respective extreme habitats represent only a minority of the total diversity of the Archaea domain. The majority of archaea cannot be cultured within the laboratory setting, and their ubiquitous presence in global habitats has been realized through the use of culture-independent techniques. One commonly used culture-independent technique is the isolation and analysis of nucleic acids (i.e., DNA and RNA) directly from an environment, rather than the analysis of cultured samples isolated from the same environment. Culture-independent studies have shown that archaea are abundant and fulfill important ecological roles in cold and temperate ecosystems. Uncultivated organisms in the subdivision Crenarchaeota are postulated to be the most abundant ammonia-oxidizing organisms in soils and to account for a large proportion (roughly 20 percent) of the microorganisms present in the picoplankton in the world's oceans. In the subdivision Euryarchaeota, uncultivated organisms in deep-sea marine sediments are responsible for the removal of methane, a potent greenhouse gas, via anaerobic oxidation of methane stored in these sediments. In contrast, uncultivated methanogenic (methane-producing) euryarchaea from terrestrial anaerobic environments, such as rice fields, are estimated to generate approximately 10–25 percent of global methane emissions.

The cultured representatives of the Crenarchaeota are from high-temperature environments, such as hot springs and submarine hydrothermal vents. Likewise, cultured members of the Euryarchaeota include organisms isolated from hot environments, organisms that are methanogenic, and organisms that grow vigorously in high-salt environments (halophiles). Organisms in the Korarchaeota lineage and the proposed Nanoarchaeota lineage also inhabit high-temperature environments; however, the nanoarchaea are highly unusual because they grow and divide on the surface of another archaea, Ignicoccus. Nanoarchaea, which were discovered in 2002, contain both the smallest known living cell (1/100th the size of Escherichia coli) and the smallest known genome (490 kilobases [1 kilobase = 1,000 base pairs of DNA]; for comparison, the human genome contains 3,000,000 kilobases). Members of the Korarchaeota and Nanoarchaeota have not been detected in pure culture; rather, they have been detected only in mixed laboratory cultures.

Archaea are also found living in association with eukaryotes. For example, methanogenic archaea are present in the digestive systems of some animals, including humans. Some archaea form symbiotic relationships with sponges. In fact, Cenarchaeum symbiosum was grown in the laboratory with its host sponge and was the first nonthermophilic Crenarchaeota to be cultured and described. It was the first organism considered for classification in the proposed Thaumarchaeota lineage.

Characteristics of the Archaea

Although the domains Bacteria, Archaea, and Eukarya were founded on genetic criteria, biochemical properties also indicate that the archaea form an independent group within the prokaryotes and that they share traits with both the bacteria and the eukaryotes. Major examples of these traits include:

1. Cell walls: Virtually all bacteria contain peptidoglycan in their cell walls; however, archaea and eukaryotes lack peptidoglycan. Various types of cell walls exist in the archaea. Therefore, the absence or presence of peptidoglycan is a distinguishing feature between the archaea and bacteria.

2. Fatty acids: Bacteria and eukaryotes produce membrane lipids consisting of fatty acids linked by ester bonds to a molecule of glycerol. In contrast, the archaea have ether bonds connecting fatty acids to molecules of glycerol. Although a few bacteria also contain ether-linked lipids, no archaea have been discovered that contain ester-linked lipids.

3. Complexity of RNA polymerase: Transcription within all types of organisms is performed by an enzyme called RNA polymerase, which copies a DNA template into an RNA product. Bacteria contain a simple RNA polymerase consisting of four polypeptides. However, both archaea and eukaryotes have multiple RNA polymerases that contain multiple polypeptides. For example, the RNA polymerases of archaea contain more than eight polypeptides. The RNA polymerases of eukaryotes also consist of a high number of polypeptides (10–12), with the relative sizes of the polypeptides being similar to that of hyperthermophilic archaeal RNA polymerase. Therefore, the archaeal RNA polymerases more closely resemble RNA polymerases of eukaryotes rather than those of bacteria.

4. Protein synthesis: Various features of protein synthesis in the archaea are similar to those of eukaryotes but not of bacteria. A prominent difference is that bacteria have an initiator tRNA (transfer RNA) that has a modified methionine, whereas eukaryotes and archaea have an initiator tRNA with an unmodified methionine.

5. Metabolism: Various types of metabolism exist in both archaea and bacteria that do not exist in eukaryotes, including nitrogen fixation, denitrification, chemolithotrophy, and hyperthermophilic growth. Methanogenesis (the production of methane as a metabolic by-product) occurs only in the domain Archaea, specifically in the subdivision Euryarchaeota. Classical photosynthesis using chlorophyll has not been found in any archaea.

The metabolic strategies utilized by the archaea are thought to be extraordinarily diverse in nature. For example, halophilic archaea appear to be able to thrive in high-salt environments because they house a special set of genes encoding enzymes for a metabolic pathway that limits osmosis. That metabolic pathway, known as the methylaspartate pathway, represents a unique type of anaplerosis (the process of replenishing supplies of metabolic intermediates; in this instance the intermediate is methylaspartate). Halophilic archaeans, which include Haloarcula marismortui, a model organism used in scientific research, are thought to have acquired the unique set of genes for the methylaspartate pathway via a process known as horizontal gene transfer, in which genes are passed from one species to another.

Microscopic Fungi

Microscopic fungi are eukaryotic, heterotrophic microorganisms that fail to show any cellular differentiation into true tissues like root, stem or leaf and in which vascular system is absent.

Characteristics of Fungi

The major distinguishing characteristics of fungi are given below:

1. Absence of Chlorophyll: Due to absence of chlorophyll, they cannot synthesize their own food and depend upon food synthesised by others. Therefore, they have heterotrophic mode of nutrition.

Depending on the source of food, they are of the following types:

Saprophytic

They use dead organic matter, such as wood and hay. Examples: Mucor, yeast, mushroom.

Parasitic

They live on other living plants or animals. Examples: Rusts, smuts.

2. Thalloid Body: The body of fungi consists of a thallus. A thallus is a plant body, which is not differentiated into true tissues like root, stem or leaf. The thallus of fungi is either a single cell or consists of branching filaments called 'hyphae'. The fungal hyphae are interwoven loosely or compactly to form the 'mycelium'. Thus, a fungus can be either unicellular or filamentous.

Classification of Fungi

The classification of fungi depends on their sexual mode of reproduction. They are subdivided into the following four classes.

Phycomycetes (Alga-like Fungi)

Reproductive spores are external and uncovered. Examples: Water mold, terrestrial mold, bread mold (Rhizopus stolonifer, Mucor spp.), downy mildew (Albugo Candida), Phytopthora infestans (causative agent of 'late blight disease' in potato).

Ascomycetes (Sac Fungi)

Sexual spores, called ascospores, are produced in a sac-like structure called ascus. Examples: Yeasts (Torula spp., Candida albicans, Saccharomyces cerevisiae), molds (Aspergillus flavus, Penicillium notatum, Cladosporium fulvum, Cephalosporium acremonium).

Basidiomycetes (Club Fungi)

Reproductive spores, called basidiospores, separate from specialised stalks called basidia. Examples: Fleshy fungi, toadstools, puffballs, bracket fungi, mushrooms (edible mushroom: Agaricus campestris; poisonous mushroom: Amanita muscaria).

Deuteromycetes (Fungi Imperfecti)

No sexual reproductive phase exists. Therefore, it is also called fungi imperfecti. Examples: Alternaria solani (causative agent of 'early blight disease' in potato), Fusarium solani, Colletotrichum falcatum (causative agent of 'red rot disease' in sugarcane).

Types of Fungi

Based on structure, fungi are of the following two types:

1. Unicellular fungi: Example: Yeasts.
2. Filamentous fungi: Example: Molds.

Protists

Protists are unicellular organisms that don't possess enough characteristics to be defined as purely plant or animal. The organisms within the Kingdom Protista contain a nucleus, like all Eukaryotes, and are categorized as plant-like, animal-like or fungus-like.

Protists are abundant in the world around us, usually thriving in aqueous environments; they survive in bodies of water as well as the human body.

A sample of pond water or its moist surrounding area placed on a slide under a compound microscope yields images of living organisms such as paramecium and amoeba – inexpensive and easy, this is often a student's introduction to microscopy.

Samples are live, often moving and differ with each drop of water; these seemingly simple images provide the basis for identifying structures within a cell – an invaluable foundation for the study of prokaryote and eukaryote specimens.

Kingdom Protista

Protista posses characteristics that make them "like" multi-cellular organisms, yet they lack certain properties to be classified as animal, plant or fungus. The presence of a nucleus in all protist organisms means they are all eukaryotic.

The three primary classifications in the Protista Kingdom and subsequent phylum include:

1. Animal-like or Protozoan
   - Ciliates (including Planktonic subgroups)
   - Flagettes (Zooflagellates)
   - Sarcodines (amoebas belong to this group)
   - Sporozoans
2. Plant-like or Diatoms
   - Chlorophyta (green algae, mostly single cell)
   - Rhodophyta (red algae or seaweed, multi-cellular)
3. Fungus-like
   - Decomposers
   - Molds (slime molds, mildew)

Many protists overlap in classification, such as lichen – considered a type of algae and fungus and some, like red algae, are amongst the few multi-cellular protista organisms.

Literally defined as "the very first," Protista are believed to be the first known organisms. However, it is important to note the miscellany nature of this Kingdom, specifically that groups of protista are not related to one another in the same manner as Eukaryotes and Prokaryotes.

Containing over 80 groups and over 115,000 species, the Kingdom Protista represents a wide range of Eukaryotic organisms bound only by the fact that they act like animals, plants or fungi.

The Nucleus

Eukaryotic organisms possess a nucleus in each cell. Whether an organism contains one or multiple cells, the nucleus – the first and largest organelle discovered – is essential to cell life.

Protected in a double envelope, cytoplasm crosses through pores in the outer membrane; this is unlike prokaryotes, where cytoplasm directly crosses the cell wall.

Appearing like a darkened area inside the nucleus under a light microscope, the nucleolus is surrounded by chromatin, which contains DNA and RNA – necessary for cell division.

Serving as a conduit between the pores in the nuclear envelope, the endoplasmic reticulum is responsible for the in/out transport of compounds.

Animal-like Protista

Animal-like protists or protozoan are primarily identified by method of movement such as:

- Pseudopods or "false feet" – Amoeba and organisms belonging to Sarcondine Phylum have no true shape, moving via projections of cytoplasm.
- Cilia – Paramecium and plankton from Ciliate Phylum use tiny hairs that line the outside of the cell.
- Flagella – From the Phylum Mastigophorans; Euglenoids whip a sole flagellum, Dinoflagettes use two flagella.

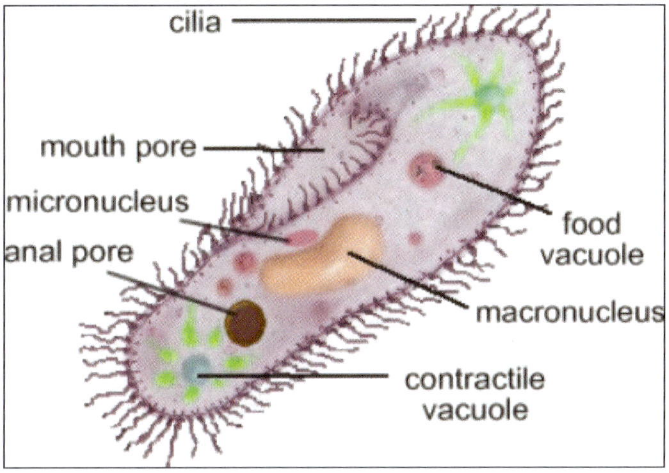

Ciliates Paramecium

In addition, parasitic protists such as sporozoans also fall under animal-like, but are categorized by the way they survive and diseases brought about in hosts.

Animal-like protozoan are also heterotrophic and contain organelles such as a cell membrane and food vacuole.

Plant-like Protista

Considered the foundation of many aquatic food chains, responsible for over forty percent of photosynthesis that occurs in salt and fresh bodies of water, as well as essential to the production of oxygen in the atmosphere, plant-like protista are classified into three phylum:

- Euglenophytes – One-celled Euglena are located in fresh water and contain chloroplasts; they possess plant and animal traits, functioning autotrophic when light and heterotrophic when dark.
- Chrysophytes – Contain chlorophyll and are autotrophic through photosynthesis; examples include:
    - Diatoms.
    - Green Algae – Can live in fresh or salt water and sometimes moist land; many, like Volvox, form colonies.

  - Red Algae (seaweed) – Multi-cellular, live in deep salt water; in uncontrolled spurts, this algae has been responsible for ecological damages.
  - Brown Algae – Type of seaweed with large leafs called "blades," contain root and air sac structures; thrive in salt water; can grow to heights in excess of 100 feet; appear most plant-like of the algae.
- Dinoflagelles – Contains chlorophyll and uses two flagella to move; creates a porous glass (silicon) shell; sometimes glows in the dark ocean floor.

Although each contains chlorophyll, organisms such as Dinoflagelles have properties that are both animal and plant like.

This overlap is exclusive to Eukaryotes in the Protista Kingdom.

Fungus-Like

Fungus-like protists have cells walls similar to plants, which contain chitin, but possess the animal-like function of heterotrophy. They release spores into the air to reproduce and have the ability to move, although this might only happen once during a lifespan.

Requiring a moist environment to survive, the three types of fungus-like protists are:

- Slime molds – often seen on decaying plant life or trees, these protists sustain on bacteria and other microorganisms that appear on rotting plants under wet conditions; the two types, plasmodial and cellular, can appear in a range of colors.
- Water molds – live in shallow or damp places; can exist as decomposer or parasite; as a mold, it can be harmful if found in gardens and farms, detrimental to potatoes, corn and cabbage and can harm a host as a parasite; looks like a combination of fuzz and threads.
- Downey mildews – similar to water molds, downey mildews are harmful to certain vegetable life.

Fungus-like organisms also have instances of overlapping. For example, certain slime molds are the result of stressed amoebas merging into a pseudoplasomodium (slug); this fungal "body" is able to relocate and reproduce by releasing spores.

Microscopy

All types of protista organisms can be studied under a simple light microscope and some, like fungus, can be seen with the naked eye.

Microscopy studies can be as easy as using a pipette to drop pond water onto a slide and viewing live paramecium as they move in their natural environment.

Advanced techniques such as dark field illumination or phase contrast are used to view structures in greater detail. Comparing images from each technique also yields valuable information.

Unicellular Eukaryotes, protists provide a foundation for viewing multi-cellular plants and animals.

Often used in introductory microscope experiments, the first image many students see through a lens is an amoeba or paramecium.

Although members of the Kingdom Protista do not contain all the organelles found in plants and animals, they do contain a well-defined nucleus, providing a foundation for more advanced microscope study.

Uses of Microorganisms

Microorganisms help in the production of many food items, making medicines, keeping the environment clean, in manufacturing and in research.

Production of Dairy Products

Bacteria are the key players here. Bacteria help in fermentation which helps in making different forms of dairy products from milk like curd, buttermilk, butter, cheese. Streptococcus is the most common genus of bacteria that are used in the commercial production of this product.

Bread Baking

A species of Streptococcus is added to the dough before making bread to bring about the required fermentation.

Alcoholic Drinks

Alcoholic drinks are prepared or manufactured by the process of fermentation. Each drink is derived from a different starting product such as potato and grapes. Then it is fermented, distilled and alcohol is prepared. The commonly used microorganism here is different types of fungus like yeast. Some even use bacteria and fungus. Alcoholic drinks include wine, rum, vodka etc.

Organic Acids

Organic acids are commercially prepared using fungi. Acetobacter, Rhizopus, Penicillium are a few fungi that are used to ferment substrates such as fruits and sugar-containing syrups. Examples of acids that are derived and manufactured on a large scale using fungi are acetic acid, citric acid, gluconic acid, fumaric acid and lactic acid.

Enzymes

Many microbes are used in the derivation of enzymes such as lipase, lactase, protease, peptidase to name a few.

Steroid Production

Some bacterial and fungal species are used in the preparation of steroids that are then injected into the human body for different purposes.

Help in Sewage Treatment

Not only are microorganisms helpful to our body, they are also helpful to the environment. They help in the secondary treatment stage of sewage treatment.

Used as Insecticides

Certain bacterial and fungal species are used to keep certain insects and pests away from crops.

Fertility of Soil

Microorganisms play a very important role in maintaining the fertility of the soil. They help in the composting process which forms manure. Also, microorganisms present in the soil help aerate it and enrich the soil with nitrates and other nutrients. These nutrients are needed by the crops for an abundant harvest.

Production of vitamins

An essential vitamin that people need for proper digestion is Vitamin B 12. Fungi are responsible for manufacturing B12.

Production of Antibiotics and Antivirals

Bacteria and viruses are isolated and their antigens and enzymes are extracted. These antigens help in the development of antibiotics and antivirals.

Biotechnology and Research

So many labs use bacteria, fungi and especially viruses for research studies. Non- virulent forms of these microorganisms are injected into subjects going through clinical trials. This in future helps in the development of medicines, vaccinations and cure for diseases. And DNA and RNA studies also make use of them.

References

- Microorganisms-definition-types-and-importance: yourarticlelibrary.com, Retrieved 7 May, 2019
- Virus, science: britannica.com, Retrieved 11 August, 2019
- Protozoa: microscopemaster.com, Retrieved 1 February, 2019
- Bacteria, science: britannica.com, Retrieved 14 June, 2019
- Archaea, science: britannica.com, Retrieved 23 March, 2019
- Microscopic-fungi-definition-characteristics-classification-and-types: yourarticlelibrary.com, Retrieved 29 July, 2019
- Microorganisms-and-its-uses, microorganisms, biology: toppr.com, Retrieved 3 January, 2019

3

Branches of Microbiology

There are various subdisciplines which fall under the category of microbiology. Some of these are bacteriology, virology, parasitology and mycology. The branch of microbiology that deals with the study of morphology, biochemistry, genetics and ecology of bacteria is known as bacteriology. The chapter closely examines the different branches of microbiology to provide an extensive understanding of the subject.

Bacteriology

Bacteriology is the study of bacteria, unique life-forms that maintain the basic physiological and genetic processes of all other types of cellular life but that have the unusual characteristic of chemicophysiological diversity.

Fundamentals of Bacterial Biology

Bacteria are one-celled prokaryotic organisms that lack a true nucleus (i.e., a nucleus defined by a membrane). Bacteria maintain their genetic material, deoxyribonucleic acid (DNA), in a single, circular chain. Bacteria also contain DNA in small circular molecules termed plasmids.

In addition to not being contained in a membrane bound nucleus, the DNA of prokaryotes is not associated with the special chromosome proteins called histones, which are found in higher organisms. In addition, prokaryotic cells lack other membrane-bounded organelles, such as mitochondria.

Although all bacteria share certain structural, genetic, and metabolic characteristics, important biochemical differences exist among the many species of bacteria. The cytoplasm of all bacteria is enclosed within a cell membrane surrounded by a rigid cell wall whose polymers, with few exceptions, include peptidoglycans—large, structural molecules made of protein carbohydrate. Bacteria also secrete a viscous, gelatinous polymer (called the glycocalyx) on their cell surfaces. This polymer, composed either of polysaccharide, polypeptide, or both, is called a capsule when it occurs as an organized layer firmly attached to the cell wall. Capsules increase the disease-causing ability (virulence) of bacteria by inhibiting immune system cells called phagocytes from engulfing them.

The shape of bacterial cells are classified as spherical (coccus), rodlike (bacillus), spiral (spirochete), helical (spirilla) and comma-shaped (vibrio). Many bacilli and vibrio bacteria have whiplike appendages (called flagella) protruding from the cell surface. Flagella are composed of tight, helical rotors made of chains of globular protein called flagellin, and act as tiny propellers, making the bacteria very mobile. On the surface of some bacteria are short, hairlike, proteinaceous projections that may arise at the ends of the cell or over the entire surface. These projections, called fimbriae, facilitate bacteria adherence to surfaces.

Other proteinaceous projections, called pili, occur singly or in pairs, and join pairs of bacteria together, facilitating transfer of DNA between them.

During periods of harsh environmental conditions some bacteria can produce within themselves a dehydrated, thick-walled endospore. These endospores can survive extreme temperatures, dryness, and exposure to many toxic chemicals and to radiation. Endospores can remain dormant for long periods (hundreds of years in some cases) before being reactivated by the return of favorable conditions.

Identifying and Classifying Bacteria

The identification schemes of Bergey's Manual are based on morphology (e.g., coccus, bacillus), staining (gram-positive or negative), cell wall composition (e.g., presence or absence of peptidoglycan), oxygen requirements (e.g., aerobic, facultatively anaerobic) and biochemical tests (e.g., in which sugars are aerobically metabolized or fermented).

Another important identification technique is based on the principles of antigenicity—the ability to stimulate the formation of antibodies by the immune system. Commercially available solutions of antibodies against specific bacteria (antisera) are used to identify unknown organisms in a procedure called a slide agglutination test. A sample of unknown bacteria in a drop of saline is mixed with antisera that has been raised against a known species of bacteria. If the antisera causes the unknown bacteria to clump (agglutinate), then the test positively identifies the bacteria as being identical to that against which the anti-sera was raised. The test can also be used to distinguish between strains, slightly different bacteria belonging to the same species.

Pathogens are disease-causing bacteria that release toxins or poisons that interfere with some function of the host's body.

Aerobic and anaerobic bacteria: Oxygen may or may not be a requirement for a particular species of bacteria, depending on the type of metabolism used to extract energy from food (aerobic or anaerobic). Obligate aerobes must have oxygen in order to live. Facultative aerobes can exist in the absence of oxygen by using fermentation or anaerobic respiration. Anaerobic respiration and fermentation occur in the absence of oxygen, and produce substantially less ATP than aerobic respiration.

During the 1860s, the French microbiologist Louis Pasteur studied fermenting bacteria. He demonstrated that fermenting bacteria could contaminate wine and beer during manufacturing, turning the alcohol produced by yeast into acetic acid (vinegar). Pasteur also showed that heating the beer and wine to kill the bacteria preserved the flavor of these beverages. The process of heating, now called pasteurization in his honor, is still used to kill bacteria in some alcoholic beverages, as well as milk.

Pasteur described the spoilage by bacteria of alcohol during fermentation as being a "disease" of wine and beer. His work was thus vital to the later idea that human diseases could also be caused by microorganisms and that heating can destroy them.

Bacterial Growth and Division

A population of bacteria in a liquid medium is referred to as a culture. In the laboratory, where growth conditions of temperature, light intensity, and nutrients can be made ideal for the bacteria, measurements of the number of living bacteria typically reveals four stages, or phases, of growth, with respect to time. Initially, the number of bacteria in the population is low. Often the bacteria are also adapting to the environment. This represents the lag phase of growth. Depending on the health of the bacteria, the lag phase may be short or long. The latter occurs if the bacteria are damaged or have just been recovered from deep-freeze storage.

After the lag phase, the numbers of living bacteria rapidly increases. Typically, the increase is exponential. That is, the population keeps doubling in number at the same rate. This is called the log or logarithmic phase of culture growth, and is the time when the bacteria are growing and dividing at their maximum speed.

The explosive growth of bacteria cannot continue forever in the closed conditions of a flask of growth medium. Nutrients begin to become depleted, the amount of oxygen becomes reduced, and the pH changes, and toxic waste products of metabolic activity begin to accumulate. The bacteria respond to these changes in a variety of ways to do with their structure and activity of genes. With respect to bacteria numbers, the increase in the population stops and the number of living bacteria plateaus. This plateau period is called the stationary phase. Here, the number of bacteria growing and dividing is equaled by the number of bacteria that are dying.

Finally, as conditions in the culture continue to deteriorate, the proportion of the population that is dying becomes dominant. The number of living bacteria declines sharply over time in what is called the death or decline phase.

Bacteria growing as colonies on a solid growth medium also exhibit these growth phases in different regions of a colony. For example, the bacteria buried in the oldest part of the colony are often in the stationary or death phase, while the bacteria at the periphery of the colony are in the actively-dividing lo phase of growth.

Culturing of bacteria is possible such that fresh growth medium can be added at a rate equal to the rate at which culture is removed. The rate at which the bacteria grow is dependent on the rate of addition of the fresh medium. Bacteria can be tailored to grow relatively slow or fast and, if the set-up is carefully maintained, can be maintained for a long time.

Bacterial growth requires the presence of environmental factors. For example, if a bacterium uses organic carbon for energy and structure (chemoheterotrophic bacteria) then sources of carbon are needed. Such sources include simple sugars (glucose and fructose are two examples). Nitrogen is needed to make amino acids, proteins, lipids and other components. Sulphur and phosphorus are also needed for the manufacture of bacterial components. Other elements, such as potassium, calcium, magnesium, iron, manganese, cobalt and zinc are necessary for the functioning of enzymes and other processes.

Bacterial growth is also often sensitive to temperature. Depending on the species, bacteria exhibit a usually limited range in temperatures in which they can growth and reproduce. For example, bacteria known as mesophiles prefer temperatures from 20°–50 °C (68°–122 °F). Outside this range, growth and even survival is limited. Other factors, which vary depending on species, required for growth include oxygen level, pH, osmotic pressure, light and moisture.

The events of growth and division that are apparent from measurement of the numbers of living bacteria are the manifestation of a number of molecular events. At the level of the individual bacterium, the process of growth and replication is known as binary division. Binary division occurs in stages. First, the parent bacterium grows and becomes larger. Next, the genetic material inside the bacterium uncoils from the normal helical configuration and replicates. The two copies of the genetic material migrate to either end of the bacterium. Then a cross-wall known as a septum is initiated almost precisely at the middle of the bacterium. The septum grows inward as a ring from the inner surface of the membrane. When the septum is complete, an inner wall has been formed, which divides the parent bacterium into two so-called daughter bacteria. This whole process represents the generation time.

Bacterial Genetics

Bacteria can exchange genetic material via conjugation. Genetic recombination between bacteria (or protists) occurs via a cytoplasmic bridge between the organisms. A primitive form of exchange of genetic material between bacteria involving plasmids also can occur. Plasmids are small, circular, extrachromosomal DNA molecules that are capable of replication and are known to be capable of transferring genes among bacteria. For example, resistance plasmids carry genes for resistance to antibiotics from one bacterium to another, while other plasmids carry genes that confer pathogenicity. In addition, the transfer of genes via bacteriophages—viruses that specifically parasitize bacteria—also serves as a means of genetic recombination.

Bioengineering uses sophisticated techniques to purposely transfer DNA from one organism to another in order to give the second organism new characteristics. For example, in a process called transformation, antibiotic susceptible bacteria that are induced to absorb manipulated plasmids placed in their environment can acquire resistance to that antibiotic substance due to the new genes they have incorporated. Similarly, in a process called transfection, specially constructed viruses are used to artificially inject bioengineered DNA into bacteria, giving infected cells some new characteristic.

Bacterial adaptation and resistance: Evolution has driven both bacterial diversity and bacterial adaptation. Some alterations are reversible, disappearing when the particular pressure is lifted. Other alterations are maintained and can even be passed on to succeeding generations of bacteria.

The first antibiotic was discovered in 1929. Since then, a myriad of naturally occurring and chemically synthesized antibiotics have been used to control bacteria. Introduction of an antibiotic is frequently followed by the development of resistance to the agent. Resistance is an example of the adaptation of the bacteria to the antibacterial agent.

Antibiotic resistance can develop swiftly. For example, resistance to penicillin (the first antibiotic discovered) was recognized almost immediately after introduction of the drug. As of the mid 1990s, almost 80% of all strains of Staphylococcus aureus were resistant to penicillin. Meanwhile, other bacteria remain susceptible to penicillin. An example is provided by Group A Streptococcus pyogenes, another Gram-positive bacteria.

The adaptation of bacteria to an antibacterial agent such as an antibiotic can occur in two ways. The first method is known as inherent (or natural) resistance. Gramnegative bacteria are often naturally resistant to penicillin, for example. This is because these bacteria have another outer membrane, which makes the penetration of penicillin to its target more difficult. Sometimes when bacteria acquire resistance to an antibacterial agent, the cause is a membrane alteration that has made the passage of the molecule into the cell more difficult. This is adaptation.

The second category of adaptive resistance is called acquired resistance. This resistance is almost always due to a change in the genetic make-up of the bacterial genome. Acquired resistance can occur because of mutation or as a response by the bacteria to the selective pressure imposed by the antibacterial agent. Once the genetic alteration that confers resistance is present, it can be passed on to subsequent generations. Acquired adaptation and resistance of bacteria to some clinically important antibiotics became a great problem in the last decade of the twentieth century.

Bacteria adapt to other environmental conditions as well. These include adaptations to changes in temperature, pH, concentrations of ions such as sodium, and the nature of the surrounding support. This adaptation is under tight genetic control, involving the expression of multiple genes.

Bacteria react to a sudden change in their environment by expressing or repressing the expression of a whole lost of genes. This response changes the properties of both the interior of the organism and its surface chemistry.

Another adaptation exhibited by a great many bacteria is the formation of adherent populations on solid surfaces. This mode of growth is called a biofilm; bacteria within a biofilm and bacteria found in other niches, such as in a wound where oxygen is limited, grow and divide at a far slower speed than the bacteria found in the test tube in the laboratory. Such bacteria are able to adapt to the slower growth rate, once again by changing their chemistry and gene expression pattern. When presented with more nutrients, the bacteria can often very quickly resume the rapid growth and division rate of their test tube counterparts.

A further example of adaptation is the phenomenon of chemotaxis, whereby a bacterium can sense the chemical composition of the environment and either moves toward an attractive compound or shifts direction and moves away from a compound sensed as being detrimental. Chemotaxis is controlled by more than 40 genes that code for the production of components of the flagella that propel the bacterium along, for sensory receptor proteins in the membrane, and for components that are involved in signaling a bacterium to move toward or away from a compound.

Bacteriocidal and bacteriostatic treatment of bacteria: Bacteriocidal is a term that refers to the treatment of a bacterium such that the organism is killed. Bacteriostatic refers to a treatment that restricts the ability of the bacterium to grow.

Bacteriocidal methods include heat, filtration, radiation, and the exposure to chemicals. The use of heat is a very popular method of sterilization in a microbiology laboratory. The dry heat of an open flame incinerates microorganisms like bacteria, fungi and yeast. The moist heat of a device like an autoclave can cause deformation of the protein constituents of the microbe, as well as causing the microbial membranes to liquefy. The effect of heat depends on the time of exposure in addition to form of heat that is supplied. For example, in an autoclave that supplies a temperature of 121 °F (49.4 °C), an exposure time of 15 minutes is sufficient to kill the so-called vegetative form of bacteria. However, a bacterial spore can survive this heat treatment. More prolonged exposure to the heat is necessary to ensure that the spore will not germinate into a living bacteria after autoclaving. The relationship between the temperature and the time of exposure can be computed mathematically.

A specialized form of bacteriocidal heat treatment is called pasteurization after Louis Pasteur, the inventor of the process. Pasteurization achieves total killing of the bacterial population in fluids such as milk and fruit juices without changing the taste or visual appearance of the product.

Another bacteriocidal process, albeit an indirect one, is filtration. Filtration is the physical removal of bacteria from a fluid by the passage of the fluid through the filter. The filter contains holes of a certain diameter. If the diameter is less than the smallest dimension of a bacterium, the bacterium will be retained on the surface of the filter it contacts. The filtered fluid is sterile with respect to bacteria. Filtration is indirectly bactericidal since the bacteria that are retained on the filter will, for a time, be alive. However, because they are also removed from their source of nutrients, the bacteria will eventually die.

Exposure to electromagnetic radiation such as ultraviolet radiation is a direct means of killing bacteria. The energy of the radiation severs the strands of deoxyribonucleic acid in many locations throughout the bacterial genome. With only one exception, the damage is so severe that repair is impossible. The exception is the radiation resistant bacterial genus called Deinococcus. This genus has the ability to piece together the fragments of DNA in their original order and enzymatic stitch the pieces into a functional whole.

Exposure to chemicals can be bacteriocidal. For example, the gas ethylene oxide can sterilize objects. Solutions containing alcohol can also kill bacteria by dissolving the membrane (s) that surround the contents of the cell. Laboratory benches are routinely "swabbed" with an ethanol solution to kill bacteria that might be adhering to the bench top. Care must be taken to ensure that the alcohol is left in contact with the bacteria for a suitable time (e.g., minutes). Otherwise, bacteria might survive and can even develop resistance to the bactericidal agent. Other chemical means of achieving bacterial death involve the alteration of the pH, salt or sugar concentrations, and oxygen level.

Antibiotics are designed to be bacteriocidal. Penicillin and its derivatives are bactericidal because they act on the peptidoglycan layer of Gram-positive and Gram-negative bacteria. By preventing the assembly of the peptidoglycan, penicillin antibiotics destroy the ability of the peptidoglycan to bear the stress of osmotic pressure that acts on a bacterium. The bacterium ultimately explodes. Other antibiotics are lethal because they prevent the manufacture of DNA or protein. Unlike bacteriocidal methods such as the use of heat, bacteria are able to acquire resistance to antibiotics. Indeed, such resistance by clinically-important bacteria is a major problem in hospitals.

Bacteriostatic agents prevent the growth of bacteria. Refrigeration can be bacteriostatic for those bacteria that cannot reproduce at such low temperatures. Sometimes a bacteriostatic state is advantageous as it allows for the long-term storage of bacteria. Ultra-low temperature freezing and lyophilization (the controlled removal of water from a sample) are means of preserving bacteria. Another bacteriocidal technique is the storage of bacteria in a solution that lacks nutrients, but which can keep the bacteria alive. Various buffers kept at refrigeration temperatures can keep bacteria alive for weeks.

Bacteria and Disease

Bacteria can multiply and cause an infection in the bloodstream. The invasion of the bloodstream by the particular type of bacteria is referred to as a bacteremia. If the invading bacteria also release toxins into the bloodstream, the malady can also be called blood poisoning or septicemia. Staphylococcus and Streptococcus are typically associated with septicemia.

The bloodstream is susceptible to invasion by bacteria that gain entry via a wound or abrasion in the protective skin overlay of the body, or as a result of another infection elsewhere in the body, or following the introduction of bacteria during a surgical procedure or via a needle during injection of a drug.

Depending on the identity of the infecting bacterium and on the physical state of the human host (primarily with respect to the efficiency of the immune system), bacteremic infections may not produce any symptoms. However, some infections do produce symptoms, ranging from an elevated temperature, as the immune system copes with the infection, to a spread of the infection to the heart (endocarditis or pericarditis) or the covering of nerve cells (meningitis). In more rare instances, a bacteremic infection can produce a condition known as septic shock. The latter occurs when the infection overwhelms the ability of the body's defense mechanisms to cope. Septic shock can be lethal.

Septicemic infections usually result from the spread of an established infection. Bacteremic (and septicemic) infections often arise from bacteria that are normal resident on the surface of the skin or internal surfaces, such as the intestinal tract epithelial cells. In their normal environments the bacteria are harmless and even can be beneficial. However, if they gain entry to other parts of the body, these so-called commensal bacteria can pose a health threat. The entry of these commensal bacteria into the bloodstream is a normal occurrence for most people. In the majority of people, however, the immune system is more than able to deal with the invaders. If the immune system is not functioning efficiently then the invading bacteria may be able to multiply and establish an infection. Examples of conditions that compromise the immune system are another illness (such as acquired immunodeficiency syndrome and certain types of cancer), certain medical treatments such as irradiation, and the abuse of drugs or alcohol.

Examples of bacteria that are most commonly associated with bacteremic infections are Staphylococcus, Streptococcus, Pseudomonas, Haemophilus, and Escherichia coli.

The generalized location of bacteremia produces generalized symptoms. These symptoms can include a fever, chills, pain in the abdomen, nausea with vomiting, and a general feeling of ill health. Not all these symptoms are present at the same time. The nonspecific nature of the symptoms may prevent a physician from suspecting bacteremia until the infection is more firmly established. Septic shock produces more drastic symptoms, including elevated rates of breathing and heartbeat, loss of consciousness and failure of organs throughout the body. The onset of septic shock can be rapid, so prompt medical attention is critical.

As with many other infections, bacteremic infections can be prevented by observance of proper hygienic procedures including hand washing, cleaning of wounds, and cleaning sites of injections to temporarily free the surface of living bacteria. The rate of bacteremic infections due to surgery is much less now than in the past, due to the advent of sterile surgical procedures, but is still a serious concern.

Bacterial infection does not always result in disease—even if a pathogen is virulent (able to cause disease). The steps of pathogenesis (the process of causing actual disease) can depend on a number of genetic and environmental factors. In some cases, pathogenic bacteria produce toxins released extracellularly (exotoxins) that migrate from the actual site of infection to cause damage to cells in other parts of the body.

Virology

Virology is the study of viruses and virus-like agents, including (but not limited to) their taxonomy, disease-producing properties, cultivation and genetics. It is often considered a part of microbiology or pathology. In the early years this discipline was dependent upon advances in the chemical and physical sciences, but viruses soon became tools for probing basic biochemical processes of cells.

Viruses have traditionally been viewed in a rather negative context as agents responsible for disease that must be controlled or eliminated. However, viruses also have certain beneficial properties that can be exploited for useful purposes (for example in gene therapy or vaccinology).

Characteristics and Classification of Viruses

Following the initial operational definition of a virus as a filterable agent, attempts were made to identify properties of viruses that separated them from other microorganisms. The single defining feature of all viruses is that they are obligate intracellular molecular parasites.

A second inviolate property of viruses is that they do not reproduce by binary fission – a method of asexual reproduction where pre-existing cells split into two identical daughter cells. For viruses, the process of reproduction is akin to an assembly line in which different parts come together to create new viral particles.

In general, viruses contain only one type of nucleic acid (either DNA or RNA) that carries the information necessary for viral replication. Nevertheless, it is clear now that some viruses contain other nucleic acid molecules; for example, in retroviruses, cellular transfer RNAs are essential for the action of the enzyme reverse transcriptase.

The chemical composition of viruses varies between different virus families. For the simplest of viruses, the virion is composed of viral structural proteins and nucleic acid, but the situation becomes more complex with the enveloped viruses. The latter types of viruses are maturing by budding through different cellular membranes that are modified by the insertion of viral proteins.

Several properties should be considered most important in constructing a scheme for classification of all the viruses: the nature of the nucleic acid present in the virion, the symmetry of the protein shell, dimensions of the virus particle and presence or absence of a lipid membrane.

The International Committee on Viral Taxonomy (ICTV), which was given the task of developing a universal taxonomic scheme for all the viruses, has put an emphasis on the viral genome (which is a blueprint for producing new viruses) as a basis of all classification decisions.

In formal virus taxonomy, families, subfamilies and genera are always written in italics, with the first letters of the names capitalized. Instead of formal names (e.g. Parvoviridae), common names are often used for viruses, even in medical literature (e.g. parvoviruses).

Study of Viruses

Even from the earliest times, it was clear that the filterable agents could not be cultivated on artificial media; this particular characteristic has withstood the test of time. Virus isolation in cell culture is still considered the gold standard against which other assays must be compared.

Still, the most obvious method of virus detection and identification is a direct visualization of the agent. The morphology of most viruses is sufficiently characteristic to identify the image as a virus and to assign an unknown virus to the adequate family. Furthermore, certain non-cultivable viruses can be detectable by electron microscopy.

The culture of animal cells typically involves the use of culture medium containing salts, glucose, vitamins, amino acids, antimicrobial drugs, buffers and (usually) blood serum which provides a source of necessary cellular growth factors. For certain cell-lines, defined serum-free media have been developed, which contain specific growth factors.

Serological tests are used to show the presence or absence of antibody to a specific virus. The presence of antibody indicates exposure to the agent, which may be due to a current clinical condition or to an earlier unrelated infection. Some examples are hemagglutination, complement fixation tests, radioimmunoassays, immunofluorescence, enzyme-linked immunosorbent assay (ELISA), radioimmune precipitation and Western blot assays.

Molecular techniques (such as polymerase chain reaction or PCR) rapidly gained acceptance and today have a myriad of applications. They are widely used for diagnostic clinical virology, as well as for research purposes. Such nucleic acid-centered technology also offered substantial advances in detection of viruses involving nucleic acid hybridization techniques.

Mycology

Mycology is the branch of biology that deals with the study of fungi. It includes the research of their genetic and biochemical properties, and their use in medicine and food along with their hazards.

Fungi are found in almost all types of environments and can be highly beneficial. They help to decompose organisms, they are great sources for antibiotics, and they serve as food. However, some fungi can be harmful and poisonous. Therefore, mycology is necessary to fully understand the properties and suitable uses of mushrooms.

The use of mushrooms date back to prehistoric times, however, the first written mention of fungi was during 480-406 BC. Mycology was originally a branch of botany, but this was changed because in terms of evolution fungi are more closely related to animals than to plants. The term mycology and mycologist were introduced into the English vocabulary in 1836. The first person to conduct and publish a proper study on fungi was Raymond Sabouraud in 1910, so he is regarded as the father of medical mycology.

Mycologists have focused on mushrooms that possibly have anti-cancer, hypoglycemic, and immune-system improving activities. Medical mycology is the most rapidly expanding branch of mycology. Mycology also encompasses mycotoxicology, mycoremediation (the process of using fungi to destroy environmental contaminants and pests), and phytopathology (the study of plant diseases).

There are different branches of mycology that specialize in various features of fungi; for example, the study of pathogenic fungi is known as medical mycology. A biologist who specializes in the study of mushrooms and fungi is known as a mycologist.

Parasitology

Parasitology is the study of species that depend on others for survival that cause harm to the species on which it depends. They can be plant or animal. The former is a type of plant that depends on another plant for its life cycle. Those in the animal kingdom are broadly broken down into three groups:

- Protozoa, the animal-like single-cell organisms that can be ingested or transmitted by insect bites. Malaria is an example of this. Mosquitoes carry the protozoa which then infects the host when the mosquito feeds, making its way into the bloodstream.
- Parasitic worms called helminths such as the tapeworm which lives inside the human gut, typically ingested from eating undercooked meat from an infected animal. Some parasitic worms in tropical countries can burrow into the skin of the foot.
- Arthropods such as insects and arachnids that carry disease or through their biological functions, cause them. The most common example in the west today are ticks and flea - human and animal. This group also includes the sexually-transmitted pubic lice.

Parasitic plants have roots that have evolved to penetrate the roots or stems of another plant and leech from its nutrients. Example of parasitic plants includes mistletoe, the Australian Christmas Tree, and Yellow Rattle.

As most parasites are considerably smaller than the species on which they depend, parasitology is a subdiscipline of microbiology which is, in turn, is a subdiscipline of biology. Today, it is a multidisciplinary approach with aspects of epidemiology, genetics and genomics ecology, botany and zoology. Globally, human infection from parasites number in the billions and can be from the near-harmless (taking only nutrients and energy) to killers responsible for the deaths of millions annually. The same applies to plant biodiversity. Some are harmless to plants while others have the ability to destroy an entire season's harvest of agricultural crops. WHO estimates that around 30% of people in the world today are infected with Ascaris lumbricoides, a nematode. Parasitic diseases such as malaria have only increased despite efforts to combat them. Due to the arrival of AIDS, some have seen a resurgence such as cryptosporidiosis.

Parasites

It would be impossible to list all species of parasite here as there are so many. However, some are more prevalent, dangerous or famous than others. Each has been the subject of important studies and are demonstrably ecologically important.

Fleas

Perhaps the most famous of all, and the most visible are fleas although they are not the largest parasites. Fleas affect any species of animal with hair, fur, or feathers. This includes humans, cats and dogs and other pets, wild animals such as mice and rats. Livestock is also prone to flea infestation without treatment. Fleas have eight legs and are therefore arachnids, closer genetically to spiders than any insect. With strong hind legs, they can leap relatively huge distances to spread from host to host. They bite and feed on the blood of an infected host and lay eggs in fur which then hatch to spread elsewhere. Dog and cat fleas can bite humans, but they cannot colonize a human body. Human fleas are relatively rare in the developed world today. They are, however, responsible for one of history's deadliest diseases - The Bubonic Plague which was carried by rat fleas infected with bacteria called Yersinia Pestis. Far more common are secondary infection from scratching and itching and the body not healing properly.

Guinea Worm

The Guinea Worm is a parasitical worm that lays larva in water sources. When consumed (usually drinking from substandard water sources) the worm infects the person or animal. The larva then uses the host body to gestate to the adult form. This is not common in western countries and has been common in poorer African countries although it is declining through improved sanitation. This causes a condition called Guinea Worm Disease or GWD and is one of the world's most deadly neglected diseases. It's also the first to be listed as a parasitical disease for eradication - this will mean the elimination of the disease but not the parasite itself. In 2016, the CDC estimated that it is common in just three countries: South Sudan, Ethiopia, and Chad. There were just 25 human cases of the condition in 2016 with only several thousand dog infestations.

Horsehair Worm

Although they are included in the nematomorph family, they are not true nematodes. They are longer and thinner and look like strands of straw or clumps of horsehair, especially when found in large masses as they tend to be in the spring. They get their name from a traditional belief that they fell from a horse and somehow came to life, but we know now that this is not true. They are parasites but are of no danger to humans, pets, or livestock. Instead, their lifecycle centers on insects. For grasshoppers, these are deadly. Some parasites simply live off the body of another creature or plant. But there are also parasites that use chemicals and pheromones to make the host lifeform to do something against its natural instincts. In the case of the horsehair worm, that is suicide. An infected insect will (when the parasite is ready to become an adult form) throw itself into a body of water all under the direction of chemical influence.

Mistletoe

One of the two most famous plants associated with Christmas and one under which lovers traditionally kiss is parasitical. There are several dozen varieties of mistletoe and every one of them is parasitical, typically living off trees. Mistletoe does not have roots. Instead of the traditional method of leeching nutrients from the soil, it uses an organ or appendage called the "haustorium" which burrows into trees to hijack nutrients. Unlike many other parasitical plants, mistletoe is flowering and has both male and female varieties. Some also have leaves (the best-known kind certainly does) for photosynthesis so that it can produce its own food. A tree with a mistletoe infestation is not problematic in itself but owners are encouraged to manage the growth effectively or the tree will be unable to sustain itself. While most people are interested in mistletoe for its Christmas association, the plant is intriguing for botany because it is largely a semi-parasite, having qualities of traditional plants too, but lacks several key parts to produce enough energy.

Parasitic and Parasitoid Wasps

This is not a single species but a group of some 80,000 species worldwide and probably one of the most diverse groupings of any animal, let alone insect, globally. Most of the 80,000 strong species parasitize just a single species each, usually other insects, which include cockroaches, worms, ants and termites, sometimes spiders. Most are parasitoid rather than parasitic which means they rely on other species for one particular stage of their life. Parasitoid wasps usually use other insects for reproduction but will not feed on them. Like the horseshoe worm mentioned above, these parasitoid wasps use chemicals and pheromones to override the natural instincts of the host, controlling it to engage in activity beneficial to the invading parasite. The Costa Rican parasitoid wasp Hymenoepimecis argyraphaga attaches a larva to the spider's abdomen. The chemical signatures then override the spider's web formation. Instead of designing a web to capture its prey, the spider will alter the pattern to create regular, recurring lines. This is virtually useless for catching flies but makes it heavy enough to hold the developing cocoon inside which will be more of the wasps. The Iseropus wasps (of which there are 9 species) is another similar example. They use caterpillars during their cocoon stage to inject their own larva through the wall of the cocoon.

Rafflesia: The Corpse Flowers

Several plants go by the name "corpse flower" which includes 28 subspecies of Rafflesia. All are found in southeast Asia including Malaysia, Indonesia and Thailand amongst others. Rafflesia has no leaves, roots or even a stem, although it does have a flower, and lives solely on native vines. Like mistletoe mentioned above, it has a haustorium which burrows into the flesh of the host plant from which the Rafflesia extracts nutrients that the host plant has absorbed. It does not become visible until it has flowered so a plant could be infected for a long time before any visible signs appear. It's called the "corpse flower" because it has evolved to emit a smell of rotting flesh and some subspecies even look like putrefying flesh. This serves two purposes: the first is a defense mechanism to prevent herbivorous animals from eating it, the second is a pollination tactic to use carrion flies to reproduce. Another plant which is also called a Corpse Flower (the world's largest flower) - the Titan Arum - is not parasitical.

Tapeworm

Living in the intestines of humans, these are quite common globally in the developing and developed world and usually caused by ingesting undercooked meat. They are called tapeworm because they look similar to the DIY tool, the tape measure. They cannot live without a host and sometimes, surgery is required to remove them although in most cases it will require simple oral medication.

It is one of the reasons why there are such stringent food hygiene and safety laws. Thankfully, they are not as common in the developed world thanks to food safety. Symptoms of infection include appetite loss, weight loss, fatigue, diarrhea, and abdominal pains. Infected people may also see signs of eggs in their stools, or larvae and segments of tapeworm.

Tongue-Eating Louse

Not dangerous to humans, the tongue-eating louse is a fish parasite that swims in through a fish's gills and attaches itself to the tongue. The gender of the parasite can be determined by where it attaches. The female attaches directly to the tongue while the male attaches to the gill's arches. Then it severs the arteries and blood vessels linking the tongue with the rest of the body so that the tongue becomes necrotic and detaches, leaving the parasite to become the fish's new tongue by attaching to the remaining stub. In most cases, it will not damage the fish beyond the removal of the tongue but a fish with more than one parasitical infection can suffer from malnutrition. They are widespread across the tropics and subtropics with some isolated cases in temperature regions. The parasite is not harmful to humans but if still alive when the person or animal eats an infected fish, they can bite.

Varroa Mite

Largely believed to play a part in colony collapse disorder in livestock bees, the varroa mite parasitizes certain species of honeybee where it reproduces. It attaches to the body of a drone and sucks the fat from inside, weakening it. It often leaves a colony weakened over the fall and winter when activity is low, not enough food has been stored. Secondary, the varroa mite brings a cocktail of diseases to a honeybee colony, the most damaging of which is the deformed wing virus which - as the description suggests - stunts wings, making it difficult and near impossible for bees to fly to collect pollen. They enter a bee colony and lay in wait until the larval cell is capped over. Then, behind the protection of the wax, they lay eggs on the developing larva. Bees with a parasitical varroa mite are far more prone to infection and disease.

Zombie Ant Fungus

Most fungi are not parasitical, feeding off dead organic material. Some go against this such as the zombie ant fungus, or to give it the scientific name Ophiocordyceps unilateralis. It is widespread in tropical rainforests where ants are prevalent and a vital part of the ecosystem. Spores infect an ant while it is foraging in a forest canopy. Here, it breaks through an ant's exoskeleton, throughout the body and to the nervous system and the brain. Then it releases chemicals to infect the brain so that the ant will transport the fungus for the next stage of its life. The ant will walk to the warm, moist forest floor and climb up a stalk where it will clamp to a leaf and wait until it dies - usually of starvation. When the parasite eventually kills the host, it gestates to the germination stage and a stalk forces its way out of the ant's dead body ready for the process to begin again. What is remarkable about this fungi is that can identify the ant species and adapt. An experiment showed that the chemicals released differ depending on the species of the host.

Sub-Disciplines of Parasitology

As parasites are so prevalent, it was inevitable that it would break down into subdisciplines to examine different aspects of parasites and their life cycles.

Immuno-parasitology

Immunoparasitology is the study of the immunology and the interactions of the immune system between a host and an invading parasite. It's largely used in the development of vaccines. Researchers look at how the parasite gets around the immune system by use of genetics, specific proteins or chemical reactions of parasites. It's been particularly successful in the last decade with many vaccines successfully developed in veterinary sciences and human health. However, investment is required for continued development in some areas. As discussed above, it is estimated that parasites represent some 50% of all biological life. Not all of these affect humans, our pets, or livestock, but improvements are expected in these areas. There is still much that researchers do not understand and there remains a challenge in examining genetic aspects of the relationship.

Medical Parasitology

This is the branch of parasitology concerned with parasites that affect humans and the medical issues that they cause. Some parasites do not cause disease; they simply leech nutrients from the infected person which cause malnutrition and weight loss. However, some carry virulent and deadly diseases. For example, the ectoparasite called "tick" can carry Lyme Disease - the most prevalent parasite-spread disease in the northern hemisphere. Mosquitoes are not parasites, but they can spread parasitical diseases such as malaria. Medical parasitology is also concerned with the development of medical treatments to hinder or neutralize the disease as well as the study of disease genetics, disease diagnosis, prevention, control, and treatment.

Paleoparasitology

This is the intersection between paleontology and parasitology. Simply, it's the study of parasites in the past, their relationships with host species, and disease vectors, epidemiology, pathology and health, paleoecology and paleoclimatology. As paleontology applies to animal species in the past, paleoparasitology is concerned with parasites of extinct animals or extinct parasites and their interactions.

Archaeoparasitology

There is some debate about whether this is a true subdiscipline of parasitology or whether it should be considered in with paleoparasitology. The "archaeo" gives it an archaeological context, meaning that this area of study concerns the interactions between humans of the past and human ancestor species, and their relationships and ecology with parasites.

Parasitic Conservation Biology

It may seem inconceivable to be concerned about the conservation status and continued existence of parasites, but they have a vital role to play in local and planetary ecology. However, they represent an enormous area of biology and their existence is vital to ecological balance on local and global systems. Those that are responsible for diseases have uses in population control to ensure species do not become imbalanced, leading to ecological damage. Due to the diseases they can cause, some governments have opted for a systemic program of eradication of the parasite itself instead of releasing funding for treatment, prevention, and cure of the conditions for which they are sometimes responsible in people, livestock, and food crops. Conservation biologists attempt to challenge the view of parasites of being (at best) useless and (at worst) a menace that would lead to better health and ecology without them. Conservationists are presently pushing for funding and awareness for greater research and appreciation of this broad area of biological research.

Parasitic Molecular Biology

Molecular biology is a cross disciplinary area of chemistry and biology, with a particular focus on biochemistry, genetics, and biology at a micro level. It examines the interactions between a cell's systems such as RNA and DNA, and the processes of synthesizing proteins with a particular focus on how these interactions take place and their regulation. Everything that applies to regular molecular biology applies here but to parasites and their interactions in their own bodies and in the host-parasite relationship. Many of these interactions have protein and DNA markers, particularly in how the parasite fools the genetic code of the host and the potential problems for the immune system. This is sometimes also known as molecular parasitology.

Parasite Ecology

Ecology is the relationship between living things in an ecosystem. We sometimes perceive it as conservationism; that plays a small part of the ecology, but only one part in a wide web of areas of research. Essentially, ecology is the study of the network of life. In parasite ecology, researchers examine the range of attributes that parasites bring to a biome and the consequences (positive and negative) or their existence. Like all other biological life, parasites have adapted and continue to adapt to changes to their environment including environmental change, immune resistance and other defensive strategies of hosts, their abilities to jump species (where possible) and the relationship between carriers who are not affected by the parasite and the hosts who are affected.

Their complex lifecycle within an environment is also of interest, as is the effects they have on each species within a biome. Finally, researchers examine abundance, prevalence, and the conditions that would lead to change.

Structural Parasitology

To understand parasites, it's necessary to understand many aspects of their anatomy. Parasite molecular biology is one of those areas. Another is structural parasitology which is the study of parasite proteins through the application of technology such as NMR and x-ray crystallography. The aim is to understand 3D modeling of proteins and the molecules in a parasitical relationship - applying to the parasite, its host, and their interactions. One of the major aims here is to examine functional pathways in organisms, comparing them to their hosts. Ultimately, it is hoped that discovering and understanding these links will lead to better vaccines for prevention, treatment of disease, and removal of the symptoms and diseases that they sometimes cause. In particular, it is hoped that the control and treatment of malaria will most benefit from this research area.

Quantitative Parasitology

Quantitative parasite ecology, the term "qualitative" refers to the use of applied mathematics and data modelling, usually involving statistics and distribution, in the examination of something. That is no different for parasitology. Each parasite species will tend to have an aggregated distribution among hosts. What this means is that most potential hosts carry few or no parasites while a relatively small number carry a large number of individual parasites. The bell curve is heavily weighted at the lower end. This goes against the typical balance of ecology and biology in general which is usually slightly skewed to the right (at the higher end). Quantitative parasitology, therefore, utilizes advanced biostatistical methods and applications to understand such distributions.

Parasite Taxonomy and Phylogenetics

- Taxonomy: The science of classifying biological species.
- Phylogeny: the examination of a species evolutionary history and lineage.

In simple terms, parasite taxonomy and phylogenetics are areas of study that concern the evolution of parasites in terms of their relationships to other species, common ancestry and their classification in the phylogenetic tree. This is a complex area prone to complications and confusion as few species (and this applies to botany and zoology broadly, not just parasitology) fit neatly in a single box. Relationships between species are complex because of shared genetic traits and independent traits evolving in separate species. Where parasitology is concerned, it is not just concerned with evolution and relationships between related species, but also in the evolution of species to their hosts - why they affect some and not others.

Veterinary Parasitology

Veterinary sciences concern the interactions between animals and humans that are an every-day part of human life. This includes pets and livestock, for example. Many of these species have the potential to be infected with a parasite and experience disease and disease symptoms when infected. Veterinary parasitology is the study of parasites in animals - mostly species used in a domestic context, but also in some areas of wildlife. As well as the interactions, researchers study the genesis and the development in animal hosts, of their parasites, and the taxonomy, morphology, systematics, and life cycle. This is largely an area of medical science overlapping with treatment and vaccine research to ensure minimal impact of sickness on human health and economics and livelihood. Its primary function is to improve and protect health.

Challenges and Developments of Parasitology

Climate Change Spreading Disease

It's been widely reported that a changing climate is leading to increased instances of some parasite-related diseases and some spreading out of their traditional tropical and subtropical zones. Diseases are already spreading thanks to the warming climate. Malaria cannot survive at temperatures below 20C and as climate change continues to increase warming some researchers expect malaria to spread to North Africa and eventually into southern Europe, similarly with dengue in the Pacific and Caribbean. Some expect it to reach the southern US states if temperatures continue to rise at the rate that they are presently rising. Evidence for this is limited at present, but temperature rises have yet to reach the required averages in the subtropical areas of the Americas, Africa, and southern Europe.

Climate Change Impact on Vulnerable Species

There are hundreds of species on the vulnerable and at-risk register. As parasites depend on these species, their continued existence - and the balance to the ecology that they provide - depend heavily on the existence of their hosts. Despite that many parasites such as those that cause dengue and malaria can expect to benefit from a changing climate, there are many more that are presently at risk and that is not good for their survival. Parasites are likely to become extinct before their host species. With limited numbers of hosts come limited chances for infection and therefore, parasite reproduction. One example of this is the marsupial woylie, also known as the brush-tailed bettong. Just one hundred years ago, it was endemic to Western Australia but with only 5,600 of their numbers, is on the brink of extinction. The situation for its most common parasite is even direr.

Prejudice Towards Parasitism

While researchers understand the importance of parasites to ecology and the food chain, that understanding has not transmitted well outside of the biological research arena. The very word "parasite" has a negative connotation; it is used against people and targeted groups. This continues the prejudice against these remarkable and unique plants and animals, fueling a belief that they not only don't serve a purpose - but also that they provide no benefit, taking and not giving. This has led to low funding and low interest in parasite research outside of the academic community for academia's sake. There is funding for medical research, but not simply to understand lifecycle and genetics. This has led to conservation issues of some critically endangered or vulnerable species.

Emerging Parasites

Because of parasite prevalence, diseases that they can cause is a global health concern. It is complex and resource-intensive to cope with the diseases that we know about. Climate change, resistance to treatment, remission or disappearance of other conditions and a number of other factors is leading to emerging or re-emerging diseases. trypanosomiasis, schistosomiasis, and leishmaniasis have all been in remission for generations but trypanosomiasis (resulting from the infection of a protozoan that spreads sleeping sickness to humans) is now in epidemic proportions in Africa in the last few years. Leishmaniasis, a New World disease spread by a protozoan called Leishmania spread by sand flies) is reaching out into new areas that had not previously experienced them. Malaria is no emerging but it is spreading and now kills more people than all other diseases except tuberculosis.

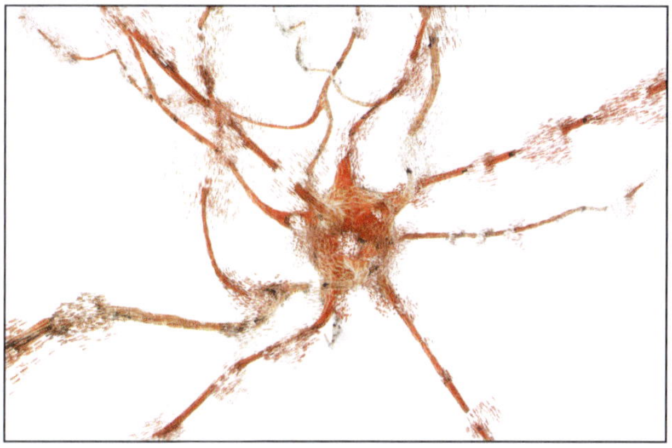

One Health Approach

Traditionally, health research and treatment has existed within each of its individual bubbles unless there was a clear and defined relationship. However, that is not and has never been the reality of any biological research. As ecology becomes more integral and the importance of relationships between biological systems becomes clearer, more overlaps and cross-disciplinary approaches are required and created. Today, there is a move towards what is called a "one health" approach. Essentially, this is treating health as a global network. The pursuit of human and veterinary health should not lead to poor ecological health. This is one area where parasitology has been at loggerheads with health. Today, with a greater understanding of the importance of parasites to the environment, there is a change of attitude towards co-existence and eradication of disease, but not of the species.

Parasites as Vital for Sustainability

Parasites are perhaps not the first thing that come to mind when considering sustainability of ecology, ecosystems, and conservation issues. We know they are now vital for ecological health, species health (population control as a vector for ensuring populations do not grow out of control) by picking off the weakest in a population, much as predators do. There has been a fundamental shift in the last couple of decades to begin to see parasites through a similar lens with which we view predators. That means, researchers perceive them as vital for sustainability.

Use of GIS in Parasite Management

Geographical Information Systems has been a boon to many areas of research. Utilizing qualitative big data analytics which it then outputs into a map form for digestible analysis by a variety of stakeholders, GIS has also made some great strides in and for parasitology. It provides functionality that conventional cartography could rarely achieve. In the last decade alone, major studies using GIS have been successfully conducted on ticks as vectors for Lyme Disease and Encephalitis, Malarial and Rift Valley Disease-carrying mosquitoes, midges, and insects that carry liver fluke. As certain diseases re-emerge or spread due to climate change, it is expected to become a fundamental tool in the planning of relief services and management of such conditions. This has also included remote sensing, and spatial data analysis, just as it has continued to be for general health issues, disaster management, and public health.

Protozoology

The science of the study of the protozoa is called Protozoology. Protozoans are the first group of organized but simplest animals in the animal kingdom. Modern Zoologists thought that the higher animals have been evolved from the single celled protozoans through the evolutionary processes.

Hence protozoans are the first group of animals that appear on this earth surface. So they are regarded as the most primitive animals in the animal kingdom. They are unicellular without any tissue grade of organization but multiply asexually by binary or multiple fission or by budding, in some cases sexual reproduction by conjugation.

Morphology of Protozoology

They are very minute and microscopic animals. The structure of a protozoan "cell" consists of cytoplasm body and nucleus. The cytoplasm is differentiated into two parts— the outer hyaline portion, ectoplasm; function is locomotive, protective and sensory and the inner granular portion, the endoplasm, function is nutritive and reproductive.

Generally the protozoans possess definite locomotory organs, such as flagella, cilia or pseudopodia. In some of them cysts are formed during extreme condition. In protozoa free living, parasitic or symbiotic nature of animals are noticed. Contractile vacuole is present in many protozoans. Nutrition in protozoans generally takes place by holozoic process, but in some cases saprozoic or holophytic, even parasitic mode of nutrition also occur.

Life Cycle of Protozoology

Protozoan can multiply by asexual method of reproduction for a long time. They may switch on to sexual method or undergoes cystic change to its host. In parasitic protozoa both asexual and sexual reproductions occur in a different host. This process is known as alternation of generation.

Amoebae, intestinal flagellates, Balantidium coli, etc. are monogenetic i.e. require one host to complete their life cycles. The parasites may adopt in a different host as well and transfer themselves passively in cystic stage. Again some parasitic protozoa like Leishmania, Trypanosoma and Plasmodia, etc. are digenetic i.e. require two hosts to complete their life cycles.

In Plasmodia infection (Malaria), man is the secondary or intermediate host but mosquito is the definitive or primary host. Tsetse fly is the intermediate host and man is the definitive host in Trypanosomiasis. In Leishmaniasis sand-fly is the intermediate host and man is the definitive host.

References

- Bacteriology, cell-biology, biology-and-genetics: encyclopedia.com, Retrieved 29 June, 2019
- What-is-Virology, health: news-medical.net, Retrieved 22 January, 2019
- Mycology: maximumyield.com, Retrieved 16 April, 2019
- Parasitology: environmentalscience.org, Retrieved 23 July, 2019
- Protozoology-definition-morphology-and-life-cycle, protozoan-cell, parasitology: biologydiscussion.com, Retrieved 28 May, 2019

4

Microbial Physiology and Ecology

The study of the structure of cells along with the growth and metabolism of microorganisms within a living system is known microbial physiology. Microbial ecology is a domain that deals with the relationship of one microorganism with one another as well as with their environment. All the diverse principles of microbial physiology and ecology have been carefully analyzed in this chapter.

Microbial Cell

Microbial cell is a pathogenic bacterium, which is a living thing that is too small to be seen with the naked eye. We can see them by using a microscope. The term microbial cells are very general. This term used to describe many different types of life forms, with dramatically different sizes and characteristics: bacteria, archaea, fungi, and protists.

The Prokaryotic Cell

The members of the prokaryotic world make up a vast heterogeneous group of very small unicellular organisms. This group includes eubacteria, or true bacteria, and archaebacteria. Bacteria are one of the most important groups of the microbial world.

The chief distinguishing characteristics of prokaryotic cells are:

- Their genetic material (DNA) is not enclosed within a membrane.
- They lack other membrane-bounded organelles.
- Their DNA is not associated with histone proteins (special chromosomal proteins found in eukaryotes).
- Their cell walls most of the time contains the complex polysaccharide peptidoglycan.
- They usually divide by binary fission. During this process, the DNA is copied and the cell splits into two cells.

Size, Shape and Arrangement of Bacterial Cells

Size

Prokaryotes are among the smallest of all organisms (0.5 to 2.0 μm). Because of their small size, bacteria have a large surface-to-volume ratio. The smallest member of the genus is about 0.3μm in diameter. Even smaller cells have been reported like the nanobacteria or ultramicrobacteria appear to range from around 0.2μm to not less than 0.05μm. E. coli, a bacillusof about average size is

1.1 to 1.5 µm wide by 2.0 to 6.0 µm long. Spirochaetes occasionally reach 500 µm in length and the cyanobacterium Oscillatoria is about 7 µm in diameter. The bacterium, Epulosiscium fishelsoni, can be seen with the naked eye (600 µm long by 80 µm in diameter). Thus a few bacteria are much larger than the average eukaryotic cell (typical plant and animal cells are around 10 to 50 µm in diameter).

Shape and Arrangement

Typically bacteria have three basic shapes – spherical, rod like and spiral.

Spherical Bacteria

Coccus (pluralcocci) in pairs– diplococcic; in chains– Streptococci (Streptococcus); in cube like groups of eight– Sarcinae (Sarcina); in grape like structures– Staphylococci (Staphylococcus).

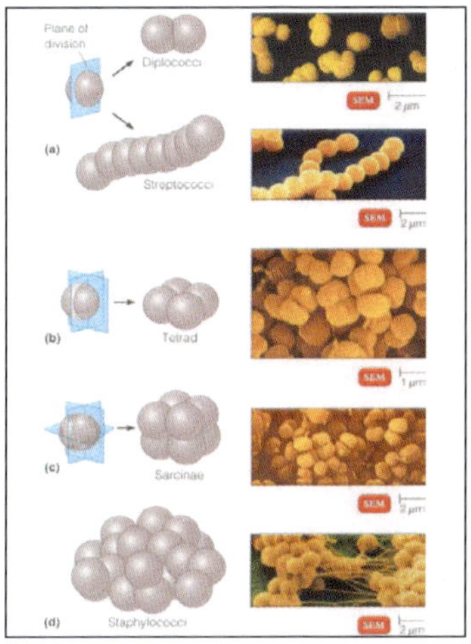

Arrangement of spherical cells

Rod like Bacteria

Bacillus (pluralbacilli) in pairs –diplococci; in chains – streptococci. Still others look like cocci and are called coccobacill.

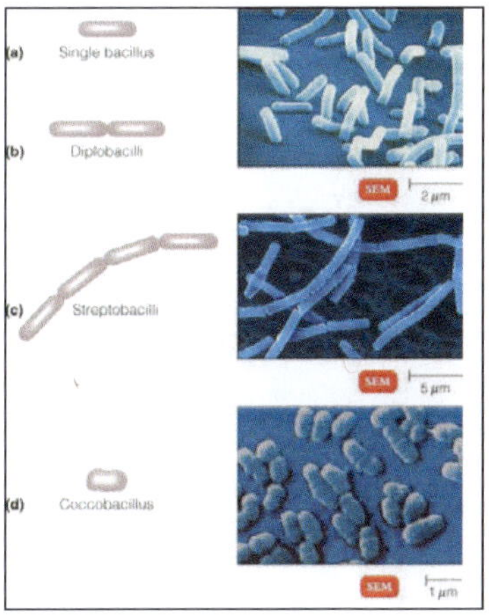

Arrangement of rod like cells

Spiral Bacteria

Comma shaped – Vibrio (Vibrio); Helical, long and curved – Spirilla (Rhodospirillium); Helical and flexible – Sprirochetes (Treponema and Borrelia) movement by axial Filaments.

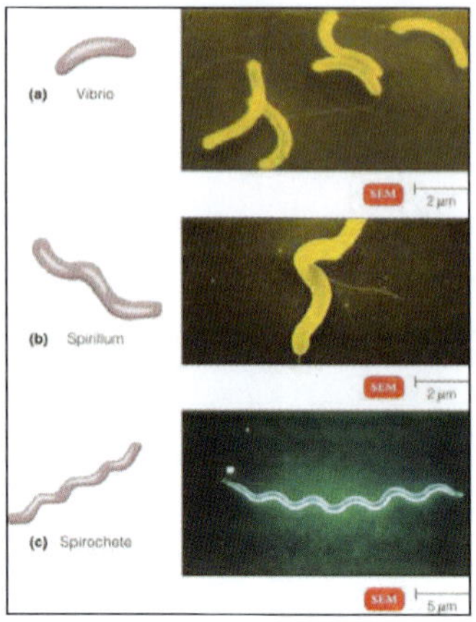

Sprial bacteria

In addition there are: Star shaped cells (genus Stella); Square flat cells (halophilic archaebacteria); Triangular cells (Haloarcula); Branching mycelium – like filaments streptomyces.

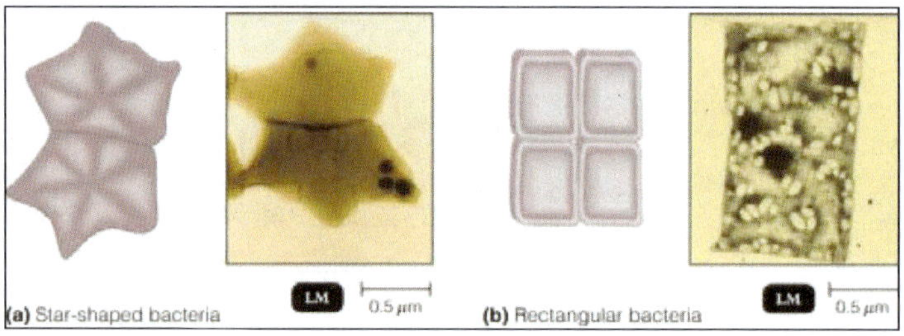

Star shaped and rectangular shaped bacteria

Genetically most bacteria are monomorphic (single shape), but some have many shapes and are called pleomorphic (Rhizobium and Corynebacterium).

Structure

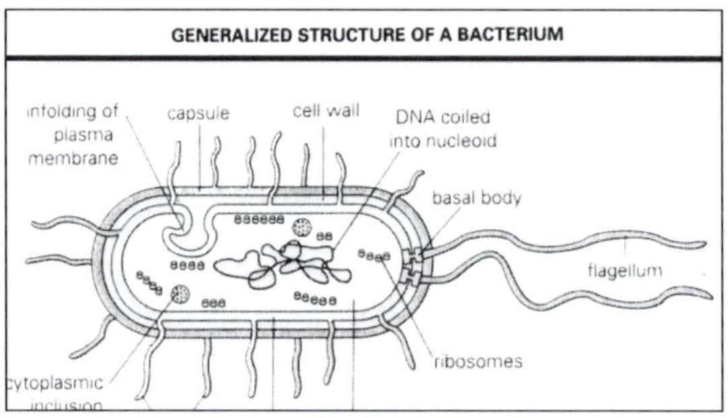

Generalized structure of a bacterium

A variety of structures are found in prokaryotic cells. The cells are always bounded by a chemically complex cell wall. In between the cell wall and the periplasmic space lies the plasma membrane. Internal membrane bound organelles are absent in prokaryotes and hence the internalmorphologically is very simple.The genetic material is localized ina discrete region called nucleoid and not separated from the surrounding cytoplasm by membrane. Ribososmes and inclusions are found scattered in the cytoplasmic matrix. Bacteria use flagella for locomotion, and in addition, many cells are surrounded by a capsule or slime layer external to the cell wall.

The Bacterial Cell Wall

Is a complex, semi-rigid structure that surrounds the underlying plasma membrane and protects the cell and the internal parts from adverse surrounding environment. Except for Mycoplasmas and some Achaea most bacteria have strong walls that give them shape and protect them from osmotic lysis; wall shape and strength is primarily due to peptidoglycan. The cell walls of pathogenic bacteria have components that contribute to their pathogenicity and protect a cell from toxic substances and are the site of action of several antibiotics. Major functions – prevent cells from rupturing, maintains shape and serves as a point of anchorage for flagella. Christian Gram developed the Gram stain in 1884, and it became evident that bacteria could be divided into two major groups based on their response to staining. Gram positive bacteria stained purple and the gram negative ones colored pink or red by the technique. The gram positive cell wall consists of a single thick layer of 20 to 80 nm of peptidoglycan or murein layer lying outside the cell membrane. In gram negative cells, the peptidoglycan layer is thin of about 2 to 7 nm surrounded by a 7 to 8 nm thick outer membrane. The gram positive cells are stronger than gram negative ones because of the peptidoglycan layer.

A space is seen between the plasma membrane and the outer membrane in electron micrographs of gram negative bacteria and a similar one but smaller gap may be observed between the plasma membrane and wall in gram positive bacteria. This space is called periplasmic space.The substance that occupies the space is the periplasm. The periplasmic space in gram negative bacteria ranges from 1nm to as great as 71 nm. It contains many proteins and participates in nutrient acquisition, hydrolytic enzymes attacking nucleic acids and phosphorylated molecules. The periplasmic space contains enzymes involved in peptidoglycan synthesis and modification of toxic compounds that could harm the cell. Gram positive bacteria may not have a visible periplasmic space but appear to have as many periplasmic proteins and several enzymes that are secreted are often called exozymes.

The Archaea differs from other prokaryotes. Thewalls lack peptidoglycan and are composed of proteins, glycoprotein or polysaccharides.

Structure of Peptidoglycan

The cell wall of bacteria is composed of a macromolecular network of polymer called peptidoglycan (murein). The polymer contains two sugar derivatives, Monosaccharides- N-acetylglucosamine (NAG) and N-acetylmuramicacid (NAM). In addition, several different amino acids, three of which – D-Glutamic acid, D-alanine, and mesodiaminopimelic acid – not found in proteins are present. The pretense of D-amino acids protects against attack by most peptidases. The backbone of this polymer is composed of alternating N-acetylglucosamine and N-acetylmuramic acid residues. A peptide chain of four alternating D- and L- amino acids is connected to the carboxyl group of N-acetylmuramic acid. Many bacteria substitute another diaminoacid, usually L-Lysine, in the third position for meso-diaminopimelic acid. Chains of linked peptidoglycan subunits are joined by cross-links between the peptides. Often the carboxyl group of the terminal D-alanine is connected directly to the amino group of diaminopimelic acid, but a peptide interbridge may be used instead. Most gramnegative cell wall peptidoglycan lacks the peptide interbridge.

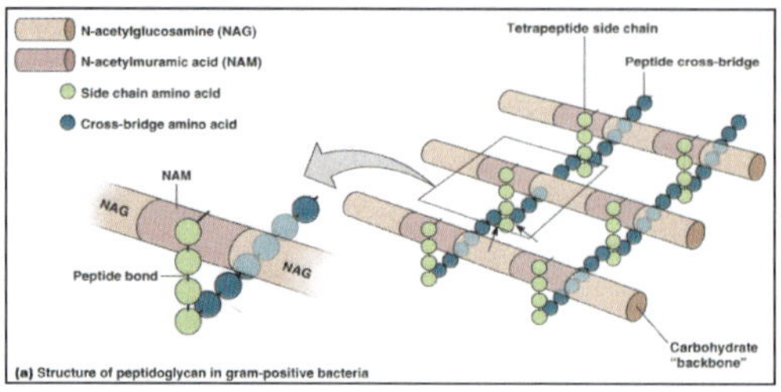

Peptidoglycan structure in gram-positive cell

Gram Positive Cell Walls

The cell wall consists of many layers of peptidoglycan, forming a thick, rigid structure, which contains a peptide interbridge. Gram positive cell walls also contain teichoic acids, polymers of glycerol and ribitol joined by phosphate groups.To the glycerol and ribotol groups, the amino acids such as D-alanine or sugars like glucose are attached. There are two classes to teichoic acids; lipoteichoic acid (if they are attached to the lipid of the plasma membrane) and wall teichoic acid (which extend to the surface of the peptidoglycan and are negatively charged). Teichoic acids are not found in gram-negative bacteria.

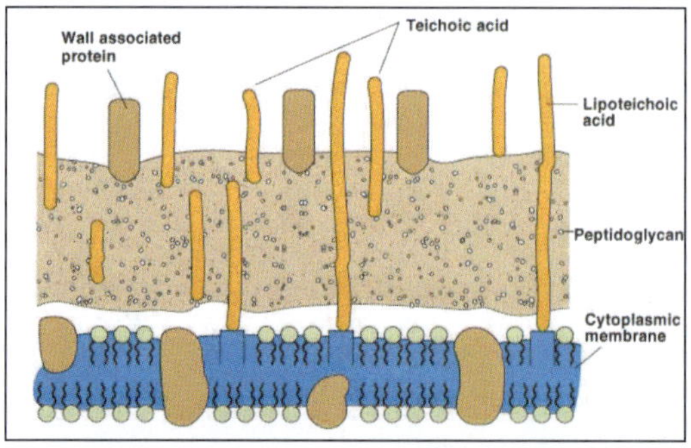

Cell wall of Gram positive bacteria

Some bacteria like Staphylococci and most other gram-positive bacteria have a layer of proteins on the surface of their cell wall peptidoglycan. Some are noncovalently attached by binding to the peptidoglycan, teichoic acids and other receptors and these proteins are involved in the interactions of the cell with its environment. Ex. S-layer proteins. Other surface proteins are covalently attached to the peptidoglycan and in gram positive pathogens these have roles such as aiding in adhesion to host tissues, preventing opsonization and blocking phagocytosis.

Gram Negative Cell Walls

The cell wall contains only a thin layer of peptidoglycan. In E. coli, it is about 2- 3nmthick and contains only one or two layers of sheets of peptidoglycan. The peptidoglycan is bounded to lipoproteins in the outer membrane and is embedded in a soft material, the periplasmic gel. Gram negative cell walls do not contain teichoic acids. They are more susceptible to mechanical breakage because of small amount of peptidoglycan.Outer membrane - found primarily in gram-negative cell consists of lipoproteins, liposaccharides and phospholipids and lies outside the thin peptidoglycan layer. The most abundant membrane protein is Braun's lipoprotein, a small lipoprotein covalently joined to the underlying peptidoglycan and embedded in the outer membrane by its hydrophobic end. The outer membrane e and plasma membrane appear to be in direct contact at many locations in the cell wall. The adhesion sites may be regions of direct contact or possibly true membrane fusions. It is proposed that substances can move into the cell through these adhesions sites rather than travelling through the periplasm.

The most unusual constituents of the outermembrane are its lipopolysaccharides. (LPSs). These contain both lipid and carbohydrate and consist of three parts: 1. Lipid A, 2. the core polysaccharide, and 3. the O side chain. The lipid A region contains two glucosamine sugar derivatives, each with three fatty acids and phosphate or pyrophosphate attached. This region is buried in the outer membrane and the remaining portion projects from the surface. The second region or the core polysaccharide is joined to lipid A and the third region or the O side chain which is a polysaccharide chain extends outward from the core. The O side chain is readily recognized by host antibodies, gram negative bacteria may thwart host defenses by readily changing the nature of their O side chains to avoid detection. The core polysaccharide contains charged sugars and phosphate, LPS contributes to the negative charge on the bacterial surface. Lipid A is a major constituent of the outer membrane, and the LPS helps stabilize the membrane structure. Lipid A is often toxic and as a result LPS can act as an endotoxin and cause some symptoms that arise in gram-negative bacterial infections. The function of the outer membrane is to serve as a protective barrier. It prevents or slows the entry of bile salts, antibiotics, and other toxic substances that might kill or injure the bacterium. Provides a barrier to certain antibiotics (like penicillin), lysozyme detergents, heavy metals, bile salts and digestive. Porins are proteins that permit small molecules to pass through the outer membrane (about 800 MW). Specific channel proteins – allow specific substances like vitamin B12, iron, nucleotides and maltose.

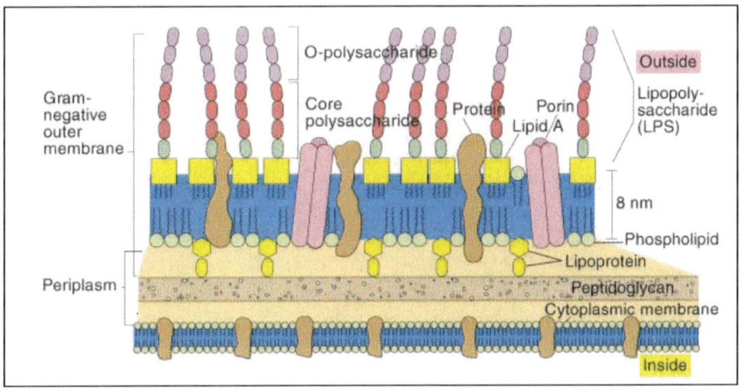

Cell wall of gram negative bacteria

Atypical Cell Walls

- Mycoplasma – is a bacterial genus that naturally lacks cell walls. Their plasma membranes have lipids called sterols, which protect them from osmoticlysis.
- Archaeabacteria – have pseudomurein (N-acetylalosaminuronic acid) but no peptidoglycan.
- L forms - are mutant bacteria with defective cell walls.

Damage to the Cell Wall

The presence of cell wall is essential to protect bacteria against destruction by osmotic pressure. The bacterial cytoplasm is much more concentrated with solutes than in most microbial habitats which are hypotonic. During osmosis, water moves across selectively permeable membranes such as the plasma membrane from dilute solutions (higher water concentration) to more concentrated solutions (lower water concentration). Usually water generally enters the bacterial cells and the osmotic pressure may reach 20 atmospheres. Plasma membrane cannot resist such high pressures and the cell will swell and be physically disrupted and destroyed, a process called lysis. In hypertonic habitats, the water flows outward, and the cytoplasm shrivels up and pulls away from the cell wall. This phenomenon is called plamolysis and is useful in food preservation because many microorganisms cannot grow in dried foods and jellies as they cannot avoid plasmolysis. The importance of the cell wall in protecting bacteria can be demonstrated by treatment with lysozyme (naturally occurs in eukaryotic cells and is a constituent of tears, mucus and saliva), which attacks the peptidoglycan by hydrolyzing the bond that connects NAM and NAG units. Penicillin inhibits peptidoglycan synthesis. Gram positive cell walls are destroyed and the remaining cellular contents are referred to as protoplast.Gram negative cells are not completely destroyed and the remaining cellular contents are referred to as spheroplast. Protoplast and spheroplast are subject to osmotic lysis. Antibiotics such as penicillin destroy bacteria by interfering with the formation of the peptide cross bridges of peptidoglycan and ultimately cell wall synthesis.

Structures Internal to the Cell Wall

They include plasma membrane, organelles in the cytoplasm like nuclear area, ribosomes, inclusion bodies and endospores.

Plasma (Cytoplasmic) Membrane

Membranes are absolute requirement of all living organisms. It is the chief point of contact with the cell's environment and thus is responsible for much of its relationship with the outside world. Plasma membrane – encloses the cytoplasm and consists of phospholipids and proteins (fluid mosaic model). Most membrane-associated lipids are structurally asymmetric with polar and non-polar ends. The polar ends interact with water and are hydrophilic and the nonpolar hydrophobic ends are insoluble in water. The lipid composition of bacterial membranes varies with environmental temperature in such a way that the membrane remains fluid during growth. Bacterial membranes usually differ from eukaryotic membranes in lacking sterols such as cholesterol and they contain pentacyclic sterol-like molecules called hopanoids and these are said to stabilize the bacterial membranes. Cell membranes are very thin structures about 5 to 10 nm thick and can be seen only with electron microscope. Plasma membranes have a complex internal structure; the small globular particles seen in these membranes are thought to be membrane proteins that lie within the membrane lipid bilayer.

The most widely accepted current model for membrane structure is the fluid mosaic model of S. Jonathan Singer and Garth Nicholson. Two types of membrane proteins are seen, Peripheral proteins - which are loosely connected to the membrane and can be easily removed and are soluble in aqueous solutions and make up about 20 to 30% of total membrane protein. About 70 to 80% of membrane proteins are integral proteins. These cannot be easily extracted from membranes and are insoluble in aqueous solutions when freed of lipids. Integral proteins, like membrane lipids are amphipathic; their hydrophobic regions are buried in the lipid while the hydrophilic portions project from the membrane surface. The plasma membrane retains the cytoplasm, particularly in cells without cell walls, and separates it from the surroundings. Plasma membranes serve as a selectively permeable barrier; it allows particular ions and molecules to pass, either into or out of the cell, while preventing the movement of others. Transport systems can be used for such tasks as nutrient uptake, waste excretion, and protein secretion. The plasma membrane also is the location of a variety of crucial metabolic processes; respiration, photosynthesis, the synthesis of lipids and cell wall constituents, and probably chromosome segregation.

The bacterial plasma membrane can be destroyed by alcohols and polymixins which cause leakage of intracellular contents and subsequent cell death of the organism.

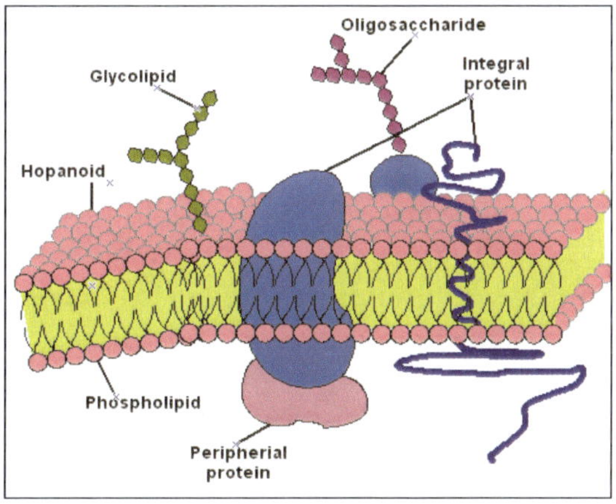

Plasma membrane structure

Internal Membrane Systems

Prokaryotes do not contain complex membrane systems as present in eukaryotes like chloroplast and mitochondria. They contain membranous structures like the one observed most is mesosome. Mesosomes – irregular infoldings or invaginations of the plasma membrane in the shape of vesicles, tubules, or lamellae. They can be seen in both gram positive and gram-negative bacteria. These are often found next to the septa or cross-walls in dividing bacteria and sometimes seems attached to the bacterial chromosome. Thus they seem to be involved in cell wall formation during division or play a role in chromosome replication and distribution to daughter cells.

Some bacteria have internal membrane systems quite different from the mesosomes. The infoldings of the plasma membrane can become extensive and complex in photosynthetic bacteria such as the cyanobacteria and purple bacteria or in bacteria with very high respiratory activity like the nitrifying bacteria. They may be aggregates of spherical vesicles, flattened vesicles, or tubular membranes. Their function may be to provide a larger membrane surface for greater metabolic activity.

Cytoplasm

Is the fluid component inside the plasma membrane. Cytoplasm is about 80% water and contains primarily proteins (enzymes), carbohydrates, lipids, inorganic ions and many low MW compounds. Major structures in the cytoplasm are DNA, ribosomes and inclusions.

Nuclear Area

The striking difference between prokaryotic and eukaryotic systems is the way in which their genetic material is packaged. Prokaryotes lack a membrane-delimited nucleus. Contains a single long circular molecule of double-stranded DNA bacterial chromosomes do not include histones and are not surrounded by a nuclear envelope and located in an irregularly shaped region called nucleiod. The nuclear area can be spherical, elongated, or dumb-bell shaped. It is attached to the plasma membrane and proteins of plasma membrane are believed to be responsible for replication of the DNA. It has been discovered recently that Vibrio cholera has more than one chromosome. Electron microscope studies have shown the nucleiod in contact with either the mesosome or the plasma membrane and hence evidence that the membranes may be involved in the separation of DNA into daughter cells during division. Chemical analysis reveals that they nucleoids are composed to about 60% DNA, 30% RNA and 10% protein. E. coli, which is about 2 to 6um long, the closed DNA circle measures approximately 1400 um. Hence, it is evident that the DNA is efficiently packaged to fit within the nucleoid and the DA is looped and coiled extensively.

Many bacteria possess extra chromosomal double stranded, circular DNA molecules called plasmids in addition to their chromosome.They replicate independently and are associated with plasma membrane proteins. Plasmids usually contain from five to 100 genes. Plasmids may carry genes for such activities as antibiotic resistance, tolerance to toxic metals, production of toxins, and synthesis of enzymes. Plasmid DNA is used for gene manipulation in biotechnology. Because plasmids move between different bacteria, drug resistance can spread throughout a population.

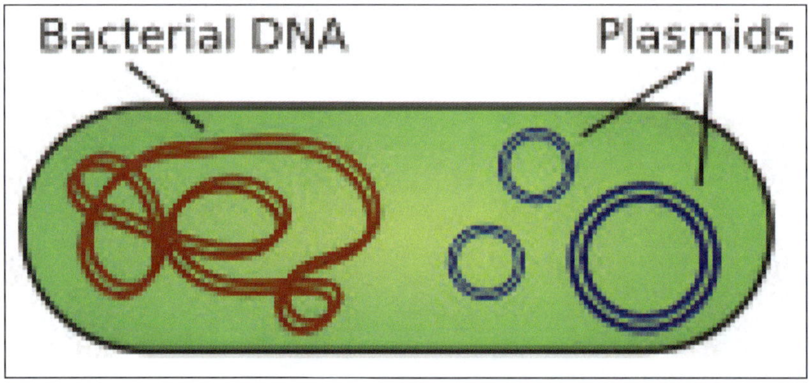

Bacterial DNA and plasmids

Ribosomes

The cytoplasmic matrix is also packaged with ribosomes, they also may be loosely attached to the plasma membrane. They look like small, featureless particles at low magnification electron microscope. They are made up of both protein and ribonucleic acid (RNA). Ribosome's function as sites of protein synthesis; matrix ribosomes synthesize proteins destined to remain within the cell, whereas the plasma membrane ribosomes make proteins for transport to the outside.The shape of each protein is determined by its amino acids sequence and the special proteins called molecular chaperones or chaperones aid the polypeptide in folding to its proper shape. Prokaryotic ribosomes are smaller than eukaryotic ribosomes. Ribosomes are composed of two subunits, each subunit being composed of protein and a type of RNA called ribosomal RNA (rRNA).They are comm. Only 70S: 30S subunit (1 molecule of rRNA) and 50S subunit (2 molecules of rRNA) and have dimensions of about 14 to 15 nm, a molecular weight of approximately 2.7 million.The S in 70S stands for Svedberg value or sedimentation coefficient. It is the sedimentation velocity in a centrifuge; the faster a particle travels when centrifuged, the greater is its Svedberg value. The Sedimentation coefficient is a function of a particle's molecular weight, volume and shape. Several antibiotics, such as streptomycin, neomycin and tetracyclines, exert their antimicrobial effects by inhibiting protein synthesis on ribosomes.

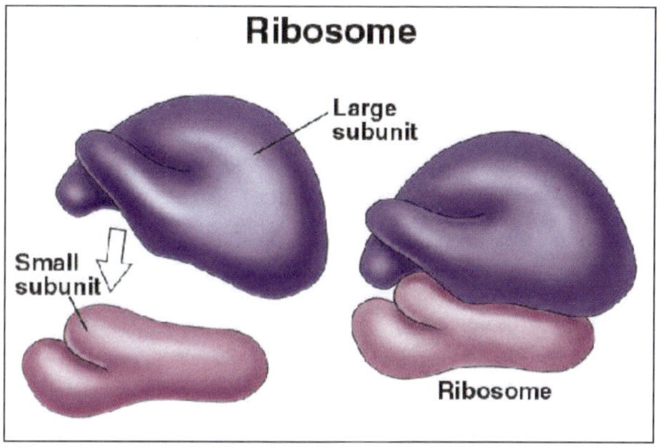

Ribosmes in bacteria

Inclusions

Inclusion bodies can be divided into two types:

- Inclusion bodies not bounded by a membrane and lie free in the cytoplasm. Ex. Polyphosphate granules, cyanophycingranules and some glycogen granules.
- Inclusion bodies enclosed by a membrane about 2-4nm thick. Ex.PolyBhydroxybutyrate granules, some glycogen and sulfur granules, carboxysomes and gas vacuoles.

Organic inclusion Bodies

Glycogen

Polymer of glucose units composed of long chains formed by alpha (1-4) glycosidic bonds and branching chains connected to themby alpha (1-6) glycosidic bonds. Ex. glycogen and starch, and their presence can be demonstrated when iodine is applied to the cells (glycogen granules appear reddish brown and starch granules appear blue).

Poly B-hydroxybutyrate

Contains beta-hydroxybutyrate molecules joined by ester bonds between the carboxyl and hydroxyl groups of adjacent molecules. Appear in various species of Mycobacterium, Bacillus, Azotobacter, Spirillum and other genera. Lipid inclusions are revealed by use of fat-soluble dyes, such as Sudan dyes.

Glycogen and PHB are carbon storage reservoirs providing material for energy and biosynthesis.

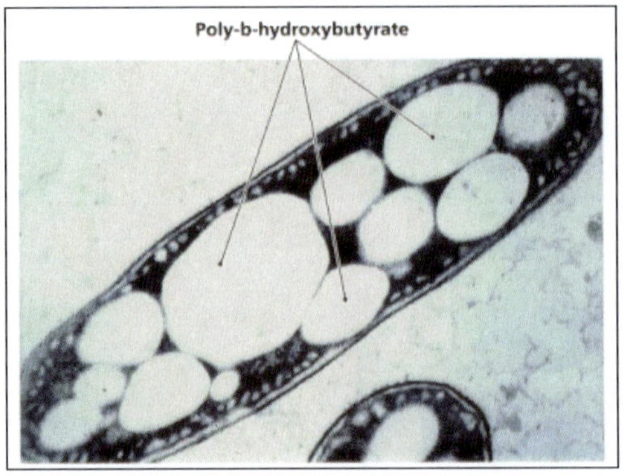

Poly B- hydroxybutyrateinclusions

Cyanophycin Granules

Cyanobacteria are composed of large amino acids containing approximately equal amounts of amino acids arginine and aspartic acid. These are used to store extra nitrogen for the bacteria.

Carboxysomes

These are polyhedral and hexagonal inclusions that contain the enzyme ribulose 1,5-diphosphate carboxylase. Bacteria that use carbon dioxide as their sole source of carbon require this enzyme for carbon dioxide fixation during photosynthesis (Ex. nitrifying bacteria, cyanobacteria, and Thiobacilli).

Gas Vacuoles

These are hollow cavities found in many aquatic prokaryotes, including cyanobacteria, anoxygenic photosynthetic bacteria and halobacteria. Each vacuole consists of rows of several individual gas vesicles, which are hollow cylinders covered by protein. Their function is to maintain buoyancy so that the cells can remain at the depth in the water appropriate for them to receive sufficient amounts of oxygen, light and nutrients. They are impermeable to water and permeable to atmospheric gases.

Inorganic inclusion Bodies

Polyphosphate Granules or Metachromatic Granules

Linear polymer of organo phosphates joined by ester bonds. Reservoirs for phosphate, an important component of cell nucleic acids and also energy reserves. Represents a reserve of inorganic phosphate (polyphosphate) that can be used in the synthesis of ATP. Stain red with certain blue dyes, such as methylene blue, and are collectively known as volutin. Found in algae, fungi and protozoans, as well as bacteria.

These granules are quite large and are characteristic of Corynebacterium diphtheriae, the causative agent of diphtheria, thus they have diagnostic significance.

Sulphur Granules

Sulphur bacteria, which belong to the genus Thiobacillus, derive energy by oxidizing sulfur and sulfur containing compounds. These bacteria may deposit sulfur granules in the cell, where they serve as an energy reserve. Purple photosynthetic bacteria use H2S as electron donor and accumulate resulting sulfurin either the periplasmic space or in special cytoplasmic globules.

Magnetosomes

Not for storage, but these are used by some bacteria to orient in the earth's magnetic field. These inclusion bodies contain iron in the form of magnetite (greigite or pyrite). Ex. Aquaspirillum magnetotacticum. Also present in heads of birds, dolphins, and turtles etc which aid in navigation.

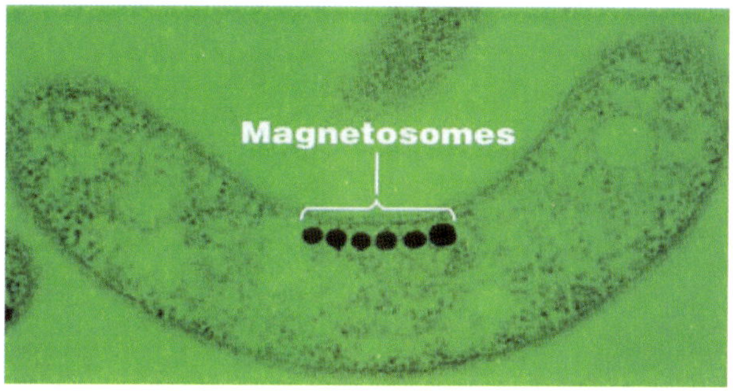

Magnetosomes

Endospores: An Escape Pod for DNA

Endospores are a survival mechanism:they are triggered to form during adverse environmental conditions. They are NOT reproductive structures as only one cell gives rise to one spore and endospores can be identified with special stains and differentiated from the vegetative cell. Endospores are resistant to: heat: withstand boiling for over one hour; desiccation, UV radiation and chemical disinfectants. The resistance of these spores has serious consequence and some very pathogenic bacteria have the ability of produce such spores.

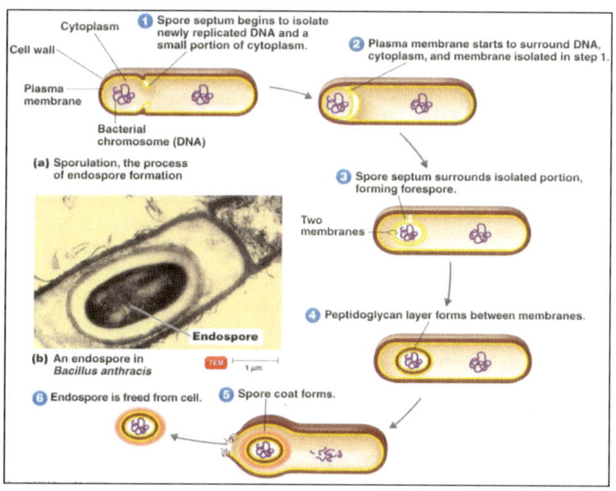

Endospore formation

Structures External to the Cell Wall

This includes glycocalyx, fimbriae, pili, flagella, axial filaments.

Glycocalyx (Capsules, Slime Layers and S-layers)

It is a viscous (sticky), gelatinous polymer composed of polysaccharide, polypeptide or both. If the substance is organized and is firmly attached to the cell wall, the glycocalyx is described as a capsule (negative staining). If the substance is unorganized and only loosely attached to the cell wall, the glycocalyx is described as a slime layer. Capsules protect pathogenic bacteria from phagocytosis (process by which certain white blood cells engulf and destroy microbes) and contribute to virulence. Unencapsulated Streptomyces pneumoniae and Bacillus anthracis does not cause disease because the cells are readily phagocytosized. This allows the bacteria to attach to various surfaces, such as rocks in fast-moving streams, plant roots, human tooth and tissues and even other bacteria. Capsules also contain water which prevents them from desiccation. Other examples are Streptococcus mutans (dental caries), Klebsiella pneumoniae (respiratory tract). These can protect a cell against dehydration. Capsules and slime layers usually are made up of polysaccharides, but they may be constructed of othermaterial, like Bacillus anthracis has a capsule of poly D-glutamic acid. Capsules are clearly visible in the light microscope by using stains or special capsule stains.

A regularly structured layer called S-layer is usually seen in many gram positive and gram negative bacteria. It consists of proteins or glycoproteins and resembles a pattern something similar to floor tiles. The S-layer adheres directly to the outer membrane in case of gram negative bacteria and with the peptidoglycan surface in gram positive bacteria. These protect the bacteria against ion and pH fluctuations, osmotic stress, enzymes, or the predacious bacterium Bdellovibrio. The S layer also hclps maintain the shape and envelope rigidity of at least bacterial cells and also promotes cell adhesion to surfaces. Sometimes, the layer also seems to protect some pathogens against complement attack and phagocytosis, thus contributing to their virulence.

Fimbriae and Pili

Many gram negative bacteria have hairlike appendages that are shorter, straighter and thinner than flagella and are used for attachment rather than for motility. They are usually called fimbriae. These structures contain a protein called pilin. Fimbriae - occur at the poles of the bacterial cell, or they can be evenly distributed over the entire surface of the cell. Fimbriae of Neisseria gonorrhoeae the causative agent of gonorrhea help the microbe to colonize mucous membranes to cause the disease. At least some types of fimbriae attach bacteria to solid surfaces such as rocks in streams and host tissues.

Pilior sex pili or pilus- usually longer than fimbriae and number only one to ten per cell. Pili function to join bacterial cells prior to the transfer to DNA from one cell to another (sometimes called sex pili). They are genetically determined by sex factors or conjugative plasmids and are required for bacterial mating. Some bacterial viruses attach specifically to receptors on sex pili at the start of their reproductive cycle.

Flagella

Motile bacteria move by use of flagella, threadlike locomotor appendages extending outward from the plasma membrane and cell wall. They are slender, rigid structures, about 20 nm across and up to 15 or 20 μm long. Bacterial species often differ distinctively in their patterns of flagella distribution.

- Monotrichous- single polar flagellum located at one end Amphitrochous.
- With two flagella, one at each end Lophotrichous.
- With two or more flagella at one or both ends Peritrichous.
- Flagella all over the surface Atrichous.
- Bacteria without flagella (Cocci rarely have flagella).

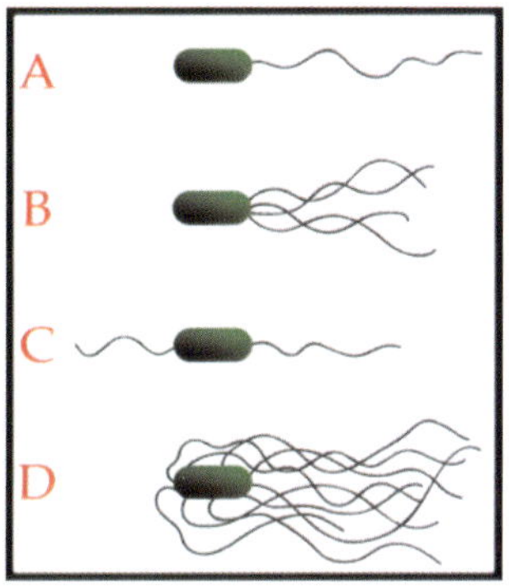

Flagellar arrangement. A. Monotrichous B. Lophotrichous C. Amphitrichous D. Peritrichous

Structure

Transmission electron microscopic studies have shown that the bacterial flagellum is composed of three parts: 1) Filament– outermost region and contain the globular protein flagellin. 2) Hook– the filament is attached to hook, which consists of a different protein. 3) Basal body- which anchors the flagellum to the cell wall and plasma membrane. It consists of a small central rod inserted into it are a series of rings. The filament is a hollow, rigid cylinder constructed of a single protein called flagellin (MW from 30,000 to 60,000). Some bacteria have sheaths surrounding their flagella. For example Bdellovibrio has a membranous structure surrounding the filament. Vibrio cholerae has a lipopolysaccharide sheath.

The hook and basal body are quite different from the filament. Slightly wider than the filament, the hook is made of different protein subunits. The basal body is the most complex structure of the flagellum. In E.coli and Gram negative bacteria, the body has four rings connected to central rod. The outer L and P rings associate with the lipopolysaccharide and peptidoglycan layers. The inner M ring connects the plasma membrane.Gram positive have only twp basal body rings, an inner ring connected to the plasma membrane and an outer one probably attached to the peptidoglycan.

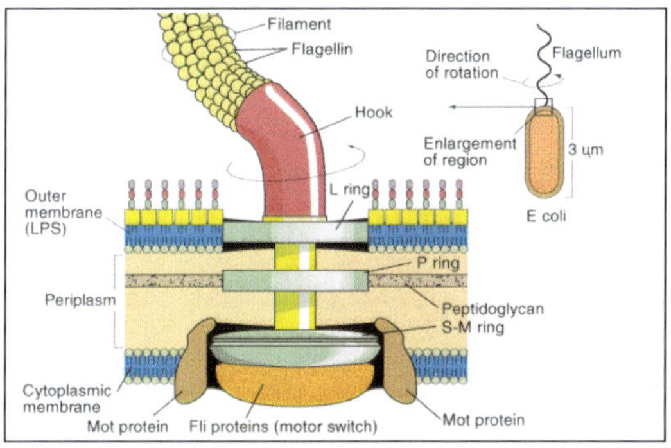

Structure of bacterial flagella (Gram negative)

The synthesis of flagella is a complex process involving atleast 20 to 30 genes. Flagellin subunits are transported through the filament's hollow internal core. When they reach the tip, the subunits spontaneously aggregate under the direction of a special filament cap so that the filament grows at its tip rather than at the base. Filament synthesis is an excellent example of self-assembly.

Flagellar Movement

The mechanism of flagellar movement in prokaryotes is different from eukaryotic flagella. The bacterium moves when the helix rotates as the filament is in the shape of rigid helix. The flagella act just like propellers on a boat. The direction of flagellar rotation determines the nature of bacterial movement. The movement in monotrichous bacteria stop and tumble randomly by reversing the flagellar rotation. The polar flagella, rotate counter clockwise during normal forward movement, whereas the cell itself rotates slowly clockwise. Peritrichous bacteria also operate in a similar way. To move forward, the flagella rotate counter clockwise. As they do so, they bend at their hooks to for a rotating bundle that propels them forward. Clockwise rotation of the flagella disrupts the bundle and the cell tumbles.

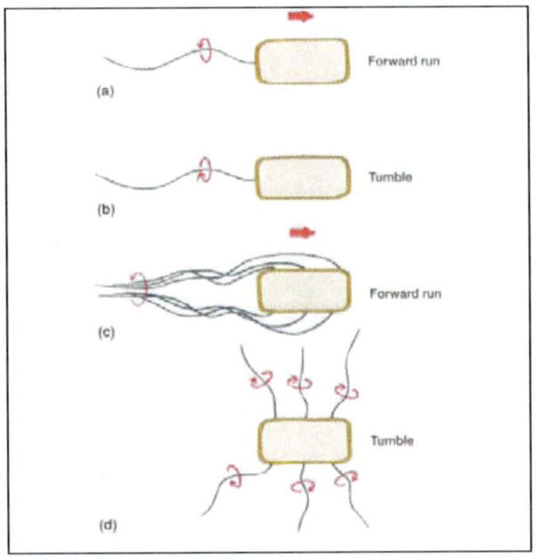

Bacterial flagellar movement

Motility enables the bacterium to move toward a favorable environment or away from a particular stimulus called taxis. Chemotaxis (include chemicals) and phototaxis(include light). Bacteria do not always swim aimlessly but are attracted by such nutrients as sugars and amino acids, and are repelled by many harmful substances and bacterial waste products. Movement toward chemical attractants and away from repellents is known as chemotaxis. The mechanism ofchemotaxis in E.coli has been studied most. Forward swimming is due to counterclockwise rotation of the flagellum, whereas tumbling results from clockwise rotation. The bacteria must be able to avoid toxic substances and collect in nutrient-rich regions and at the proper oxygen levels. E.coli has four different chemoreceptors that recognize serine, aspartate and maltose, ribose and galactose and dipeptides respectively. These chemoreceptors often are called methyl-accepting chemotaxis proteins (MCPs).

Some bacteria can move bymechanisms other than flagellar rotation. Spirochetes are a group of bacteria that have unique structure and motility (Treponemapallidum, the causative agent of syphilis and Borrelia burgdorgeri, the causative agent of Lyme disease).Spirochetes travel through viscous substances such as mucus or mud by flexing and spinning movement caused by special axial filaments - bundles of fibrils that arise at the ends of the cell beneath the outer sheath and spiral around the cell.The rotation of the filaments produces an opposing movement of the outer sheath that propels the spirochetes by causing them to move like corkscrews.

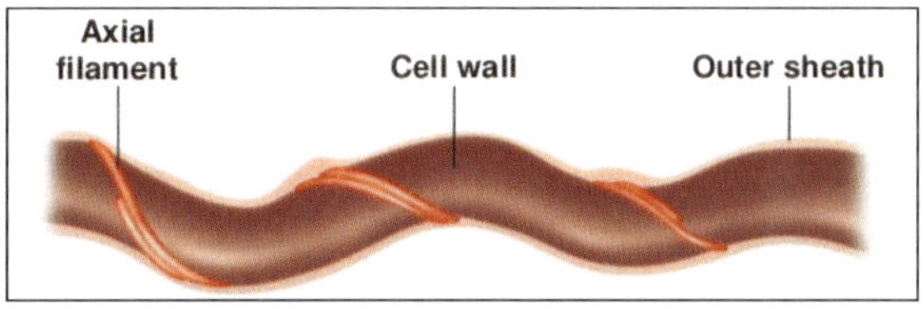

Axial filaments seen in spricohetes

A very different type of motility, gliding motility, is employed by many bacteria; cyanobacteria, myxobacteria and cytophagas and some mycoplasmas.

Morphology of Microbial Cell

Size and Shape of Microbial Cell

The size, shape and arrangement of microbial cells vary with species to which they belong. Bacteria are of about 0.1 to 60 × 6 µm in size. However, there is variation in dimension of bacilli (5 × 0.4-0.7 µm), pseudomonads (0.4-0.7 µm diameter, 2-3 µm length) and micrococci (about 0.5 µm diameter).

It is the rigid cell wall that determines the shape of bacterial cell. Generally, the bacterial cells are spherical (coccus, plural cocci which mean berries), elongated rods (bacillus, plural bacilli), helical rods (Spirillum, plural spirilli), pear-shaped (Pasteurid), lobed spheres (Sulfolobus), rods with squared ends (Bacillus anthracis), rods with helically sculptured surface (Seliberid) and of changing shape (pleomorphic), etc.

The unicellular cyanobacterial cells are usually spherical (Chroococcus, Scenedesmus, Anacystis), some are elongated and multicellular.

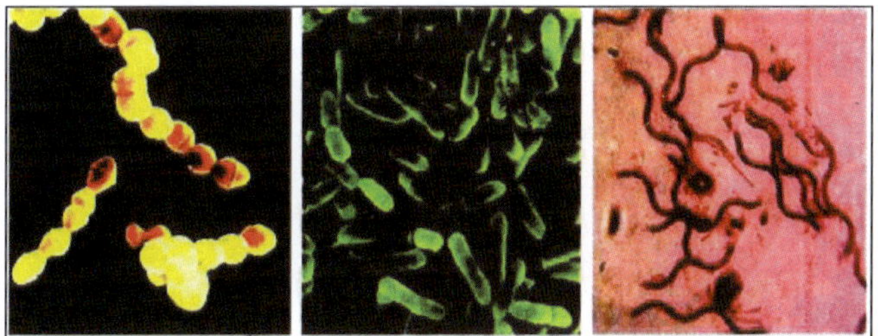

(a) Cells of *Pseudomonas* (cylindrical) (b) *Streptococcus* (spherical) (c) Spirilla (twisted shape)

Arrangement of Microbial Cell

The arrangement of cells is more complex in cocci than bacilli. The arrangement of cells depends upon adherence of cells together after the cell division.

Different forms of arrangement are given below:

Coccus Forms

There are several groups of cocci based on the number and arrangement of cells.

- Diplococcus: Cells divide in one plane and get attached permanently in pairs.
- Streptococcus: Cells divide in one plane and remain attached to form a linear chain of cells.
- Tetracocci: Cells divide in two planes and form groups of four cells.
- Straphylococci: Cells divide in three planes in an irregular pattern producing bunches of cocci.
- Sarcinae: Cells divide in three planes in regular pattern producing bunches of cocci.

Forms of Bacillus

There are a few groups of bacilli unlike cocci as the former divide across their short aries.

- Mono-bacillus: The single elongated cells freely present in nature are mono-bacillus.
- Diplobacillus: After division the cells remain adhered and appear in paired form.
- Streptobacillus: After division the cells remain attached in chains appearing like straws.
- Coccobadllus: The oval cells looking like cocci are called coccobacilli.

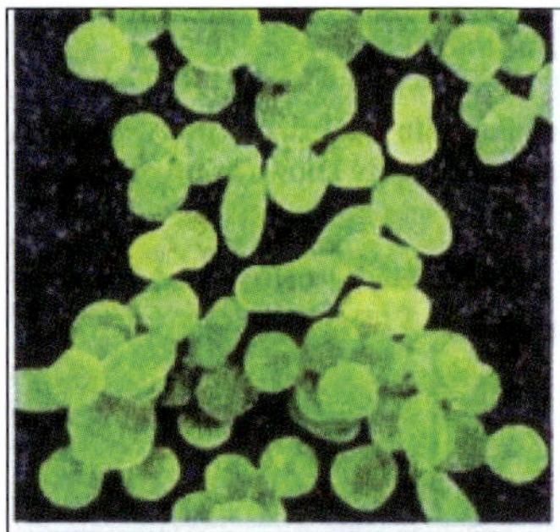

A Coccobacllus prokaryote.

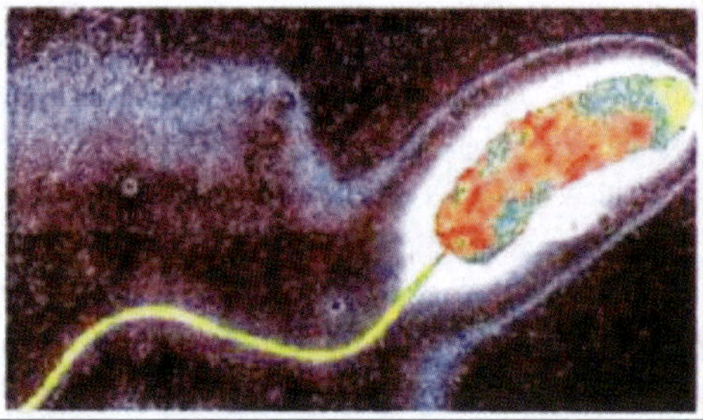

Vibrio cholerae.

There is two meaning of bacillus, one is the form and the second is the genus. For example the bacterium Bacillus anthracis causes anthrax disease.

Forms of Spirilli

- Vibrioid: Bacterial cells having less than one complete twist form vibrioid shape e.g. Vibrio cholerae.
- Helical: Cells that have more than one twist form a distinct helical shape e.g. Spirillum (with flagella).

Other Forms

- Pleomorphic: Of changing forms e.g. Rhizobium, Mycoplasma, etc.
- Trichomes: Cells divide in one plane forming a chain which has much larger area of contact between the adjacent cells e.g. Baggiatoa, Saprospria.

- Palisade: The cells are arranged laterally (side by side) to form a match sticks like structure and at angles to one another e.g. Corynebacterimn diphtheriae.

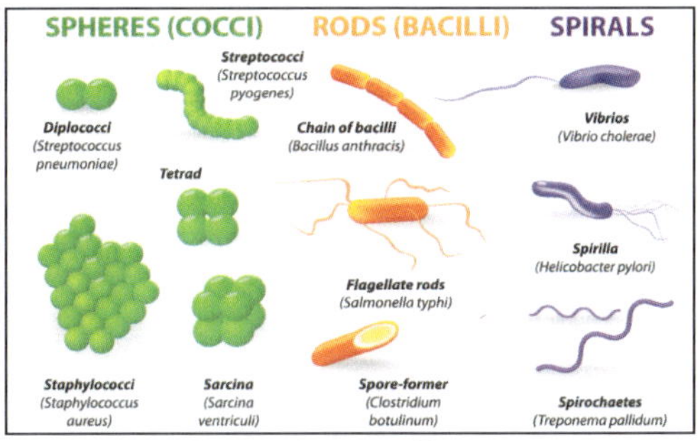

Different forms of bacteria.

Hyphae

Some microorganisms form the multicellular, thin-walled, profusely branched filaments called hyphae. The interwoven hyphae are collectively known as mycelium e.g. Streptomyces, Aspergillus, and Penicillium. The cyanobacterial cells are comparatively larger than the bacterial cells.

In addition, the cell size of eukaryotes such as algae, fungi, protozoa, etc. is several times greater than the cells of prokaryotes. The unicellular cyanobacteria are usually spherical or elongated. The fungi are either unicellular (yeast, Candida) or multicellular hyphal (Fusarium, Aspergillus) forms.

Virus Structure

Viruses are not plants, animals, or bacteria, but they are the quintessential parasites of the living kingdoms. Although they may seem like living organisms because of their prodigious reproductive abilities, viruses are not living organisms in the strict sense of the word.

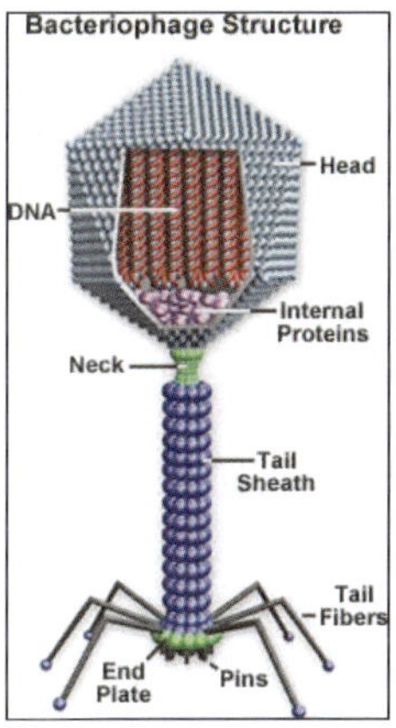

Without a host cell, viruses cannot carry out their life-sustaining functions or reproduce. They cannot synthesize proteins, because they lack ribosomes and must use the ribosomes of their host cells to translate viral messenger RNA into viral proteins. Viruses cannot generate or store energy in the form of adenosine triphosphate (ATP), but have to derive their energy, and all other metabolic functions, from the host cell. They also parasitize the cell for basic building materials, such as amino acids, nucleotides, and lipids (fats). Although viruses have been speculated as being a form of protolife, their inability to survive without living organisms makes it highly unlikely that they preceded cellular life during the Earth's early evolution. Some scientists speculate that viruses started as rogue segments of genetic code that adapted to a parasitic existence.

All viruses contain nucleic acid, either DNA or RNA (but not both), and a protein coat, which encases the nucleic acid. Some viruses are also enclosed by an envelope of fat and protein molecules. In its infective form, outside the cell, a virus particle is called a virion. Each virion contains at least one unique protein synthesized by specific genes in its nucleic acid. Viroids (meaning "viruslike") are disease-causing organisms that contain only nucleic acid and have no structural proteins. Other viruslike particles called prions are composed primarily of a protein tightly integrated with a small nucleic acid molecule.

Viruses are generally classified by the organisms they infect, animals, plants, or bacteria. Since viruses cannot penetrate plant cell walls, virtually all plant viruses are transmitted by insects or other organisms that feed on plants. Certain bacterial viruses, such as the T4 bacteriophage, have evolved an elaborate process of infection. The virus has a "tail" which it attaches to the bacterium surface by means of proteinaceous "pins." The tail contracts and the tail plug penetrates the cell wall and underlying membrane, injecting the viral nucleic acids into the cell. Viruses are further classified into families and genera based on three structural considerations: 1) the type and size of their nucleic acid, 2) the size and shape of the capsid, and 3) whether they have a lipid envelope surrounding the nucleocapsid (the capsid enclosed nucleic acid).

There are predominantly two kinds of shapes found amongst viruses: rods, or filaments, and spheres. The rod shape is due to the linear array of the nucleic acid and the protein subunits making up the capsid. The sphere shape is actually a 20-sided polygon (icosahedron).

The nature of viruses wasn't understood until the twentieth century, but their effects had been observed for centuries. British physician Edward Jenner even discovered the principle of inoculation in the late eighteenth century, after he observed that people who contracted the mild cowpox disease were generally immune to the deadlier smallpox disease. By the late nineteenth century, scientists knew that some agent was causing a disease of tobacco plants, but would not grow on an artificial medium (like bacteria) and was too small to be seen through a light microscope. Advances in live cell culture and microscopy in the twentieth century eventually allowed scientists to identify viruses. Advances in genetics dramatically improved the identification process.

- Capsid - The capsid is the protein shell that encloses the nucleic acid; with its enclosed nucleic acid, it is called the nucleocapsid. This shell is composed of protein organized in subunits known as capsomers. They are closely associated with the nucleic acid and reflect its configuration, either a rod-shaped helix or a polygon-shaped sphere. The capsid has three functions: 1) it protects the nucleic acid from digestion by enzymes, 2) contains special sites on its surface that allow the virion to attach to a host cell, and 3) provides proteins that enable the virion to penetrate the host cell membrane and, in some cases, to inject the infectious nucleic acid into the cell's cytoplasm. Under the right conditions, viral RNA in a liquid suspension of protein molecules will self-assemble a capsid to become a functional and infectious virus.

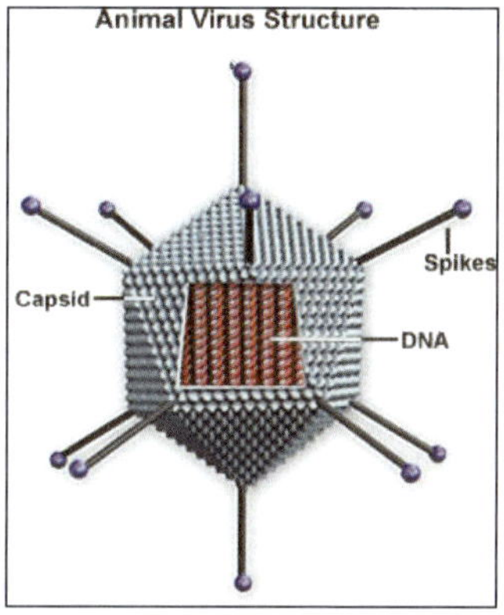

- Envelope - Many types of virus have a glycoprotein envelope surrounding the nucleocapsid. The envelope is composed of two lipid layers interspersed with protein molecules (lipoprotein bilayer) and may contain material from the membrane of a host cell as well as that of viral origin. The virus obtains the lipid molecules from the cell membrane during the viral budding process. However, the virus replaces the proteins in the cell membrane with its own proteins, creating a hybrid structure of cell-derived lipids and virus-derived proteins. Many viruses also develop spikes made of glycoprotein on their envelopes that help them to attach to specific cell surfaces.

- Nucleic Acid - Just as in cells, the nucleic acid of each virus encodes the genetic information for the synthesis of all proteins. While the double-stranded DNA is responsible for this in prokaryotic and eukaryotic cells, only a few groups of viruses use DNA. Most viruses maintain all their genetic information with the single-stranded RNA. There are two types of RNA-based viruses. In most, the genomic RNA is termed a plus strand because it acts as messenger RNA for direct synthesis (translation) of viral protein. A few, however, have negative strands of RNA. In these cases, the virion has an enzyme, called RNA-dependent RNA polymerase (transcriptase), which must first catalyze the production of complementary messenger RNA from the virion genomic RNA before viral protein synthesis can occur.

1. The Influenza (Flu) Virus - Next to the common cold, influenza or "the flu" is perhaps the most familiar respiratory infection in the world. In the United States alone, approximately 25 to 50 million people contract influenza each year. The symptoms of the flu are similar to those of the common cold, but tend to be more severe. Fever, headache, fatigue, muscle weakness and pain, sore throat, dry cough, and a runny or stuffy nose are common and may develop rapidly. Gastrointestinal symptoms associated with influenza are sometimes experienced by children, but for most adults, illnesses that manifest in diarrhea, nausea, and vomiting are not caused by the influenza virus though they are often inaccurately referred to as the "stomach flu." A number of complications,

such as the onset of bronchitis and pneumonia, can also occur in association with influenza and are especially common among the elderly, young children, and anyone with a suppressed immune system.

2. The Human Immunodeficiency Virus (HIV) - The virus responsible for HIV was first isolated in 1983 by Robert Gallo of the United States and French scientist Luc Montagnier. Since that time, a tremendous amount of research focusing upon the causative agent of AIDS has been carried out and much has been learned about the structure of the virus and its typical course of action. HIV is one of a group of atypical viruses called retroviruses that maintain their genetic information in the form of ribonucleic acid (RNA). Through the use of an enzyme known as reverse transcriptase, HIV and other retroviruses are capable of producing deoxyribonucleic acid (DNA) from RNA, whereas most cells carry out the opposite process, transcribing the genetic material of DNA into RNA. The activity of the enzyme enables the genetic information of HIV to become integrated permanently into the genome (chromosomes) of a host cell.

Microbial Cell Culture

A microbiological culture, or microbial culture, is a method of multiplying microbial organisms by letting them reproduce in predetermined culture media under controlled laboratory conditions Microbial cultures are foundational and basic diagnostic methods used extensively as a research tool in molecular biology. Microbial cultures are used to determine the type of organism, its abundance in the sample being tested, or both. It is one of the primary diagnostic methods of microbiology and used as a tool to determine the cause of infectious disease by letting the agent multiply in a predetermined medium. It is often essential to isolate a pure culture of microorganisms. A pure (or axenic) culture is a population of cells or multi cellular organisms growing in the absence of other species or types. A pure culture may originate from a single cell or single organism, in which case the cells are genetic clones of one another. For the purpose of gelling the microbial culture, the medium of agarose gel (agar) is used. Agar is a gelatinous substance derived from sea weed. A cheap substitute for agar is guar gum, which can be used for the isolation and maintenance. Culture media contains nutrients and physical growth parameters necessary for microbial growth. All microorganisms cannot grow in a single culture medium and in fact many can't grow in any known culture medium.

Basic Constituents of Media:

- Inorganic salts
- Carbohydrates
- Amino Acids
- Vitamins
- Fatty acids and lipids
- Proteins and peptides

- Serum
- Trace Elements

Each type of constituent performs a specific function as outlined below:

- Inorganic Salts: The inclusion of inorganic salts in media performs several functions. Primarily they help to retain the osmotic balance of the cells and help regulate membrane potential by provision of sodium, potassium and calcium ions. All of these are required in the cell matrix for cell attachment and as enzyme cofactors.
- Carbohydrates: The main source of energy is derived from carbohydrates generally in the form of sugars. The major sugars used are glucose and galactose, however, some media contain maltose or fructose. The concentration of sugar varies from basal media containing 1g/L to 4.5g/L in some more complex media. Media containing the higher concentration of sugars are able to support the growth of a wider range of cell types. Pyruvate is included in the formulation of some media, as an alternative energy source.
- Amino Acids: Amino acids are the building blocks of proteins. 'Essential' amino acids must be added to culture media as cells are not able to synthesize these themselves. The concentration of amino acids in the culture medium will determine the maximum cell density that can be achieved - once depleted the cells will no longer be able to proliferate.

In relation to cell culture, glutamine, an essential amino acid, is particularly significant. In liquid media or stock solutions glutamine degrades relatively rapidly. Optimal cell performance usually requires supplementation of the media with glutamine prior to use. Some media formulations include L-alanyl glutamine which is a more stable form of glutamine, and do not require supplementation. Adding supplements of non-essential amino acids to media both stimulates growth and prolongs the viability of the cells in culture.

- Vitamins: Serum is an important source of vitamins in cell culture. However, many media are also enriched with vitamins making them consistently more suitable for a wider range of cell lines. Vitamins are precursors for numerous co-factors. Many vitamins, especially B group vitamins, are necessary for cell growth and proliferation and for some lines the presence of B12 is essential. Some media also have increased levels of vitamins A and E. The vitamins commonly used in media include riboflavin, thiamine and biotin.
- Proteins and Peptides: These are particularly important in serum free media. The most common proteins and peptides include albumin, transferrin, fibronectin and fetuin and are used to replace those normally present through the addition of serum to the medium.
- Fatty Acids and Lipids: Like proteins and peptides these are important in serum free media since they are normally present in serum e.g. cholesterol and steroids essential for specialised cells.
- Trace Elements: These include trace elements such as zinc, copper, selenium and tricarboxylic acid intermediates. Selenium is a detoxifier and helps remove oxygen free radicals.
- Organisms that cannot grow in artificial culture medium are known as obligate parasites

Mycobacterium leprae, rickettias, Chlamydias, and Treponema pallidum are obligate parasites. Bacterial culture media can be distinguished on the basis of composition, consistency and purpose.

Classification of culture media used in microbiology laboratory on the basis of consistency:

- Solid medium: Solid medium contains agar at a concentration of 1.5-2.0% or some other, mostly inert solidifying agent. Solid medium has physical structure and allows bacteria to grow in physically informative or useful ways (e.g. as colonies or in streaks). Solid medium is useful for isolating bacteria or for determining the colony characteristics of the isolate.

- Solid Media: They are prepared with agar at concentrations of 0.5% or less. They have soft-custard like consistency and are useful for the cultivation of microaerophilic bacteria or for determination of bacterial motility.

- Liquid (Broth) medium: These media contain specific amounts of nutrients but don't have trace of gelling agents such as gelatin or agar. Broth medium serves various purposes such as propagation of large number of organisms, fermentation studies, and various other tests. e.g. sugar fermentation tests, MR-VR both.

Classification of culture media based on the basis of composition:

- Synthetic or chemically defined medium: A chemically defined medium is one prepared from purified ingredients and therefore whose exact composition is known. Synthetic medium may be simple or complex depending up on the supplement incorporated in it.

- Non-synthetic or chemically undefined medium: Nonsynthetic medium contains at least one component that is neither purified nor completely characterized nor even completely consistent from batch to batch. Often these are partially digested proteins from various organism sources. Nutrient broth, for example, is derived from cultures of yeasts.A simple non-synthetic medium is capable of meeting the nutrient requirements of organisms requiring relatively few growth factors whereas complex non-synthetic medium support the growth of more fastidious microorganisms.

Classification of bacterial culture media based on the basis of purpose/functional use/application:

Many special purpose media are needed to facilitate recognition, enumeration, and isolation of certain types of bacteria. To meet these needs, numerous media are available.

- General purpose media/Basic media: Basal media are basically simple media that supports most non-fastidious bacteria. Peptone water, nutrient broth and nutrient agar are considered as basal medium. These media are generally used for the primary isolation of microorganisms.

- Enriched medium (Added growth factors): Addition of extra nutrients in the form of blood, serum, egg yolk etc, to basal medium makes them enriched media. Enriched media are used to grow nutritionally exacting (fastidious) bacteria. Blood agar, chocolate agar, Loeffler's serum slope etc are few of the enriched media. Blood agar is prepared by adding 5-10% (by volume) blood to a blood agar base. Chocolate agar is also known as heated blood agar or lysed blood agar.

- Selective and enrichment media are designed to inhibit unwanted commensal or contaminating bacteria and help to recover pathogen from a mixture of bacteria. While selective media are agar based, enrichment media are liquid in consistency. Both these media serve the same purpose. Any agar media can be made selective by addition of certain inhibitory agents that don't affect the pathogen of interest. Various approaches to make a medium selective include addition of antibiotics, dyes, chemicals, alteration of pH or a combination of these.

Selective Medium

Principle: Differential growth suppression Selective medium is designed to suppress the growth of some microorganisms while allowing the growth of others. Selective medium is agar based (solid) medium so that individual colonies may be isolated.

Examples of Selective Media Include:

- Thayer Martin Agar used to recover N.gonorrhoeae contains antibiotics; vancomycin, colistin and nystatin.
- Mannitol Salt Agar and Salt Milk Agar used to recover S.aureus contains 10% NaCl. c) Potassium tellurite medium used to recover.
- Diphtheriae contains 0.04% potassium tellurite.
- Mac Conkey's Agar used for Enterobacteriaceae members contains bile salt that inhibits most gram positive bacteria.
- Pseudosel Agar (Cetrimide Agar) used to recover P. aeruginosacontainscetrimide (antiseptic agent).
- Crystal Violet Blood Agar used to recover S. pyogenes contains 0.0002% crystal violet.
- Lowenstein Jensen Medium used to recover M.tuberculosis is made selective by incorporating malachite green.
- Wilson and Blair's Agar for recovering S. typhi is rendered selective by the addition of dye brilliant green.
- Selective media such as TCBS Agar used for isolating V. cholerae from fecal specimens have elevated pH (8.5- 8.6), which inhibits most other bacteria.

Enrichment Culture Medium

Enrichment medium is used to increase the relative concentration of certain microorganisms in the culture prior to plating on solid selective medium. Unlike selective media, enrichment culture is typically used as broth medium. Enrichment media are liquid media that also serves to inhibit commensals in the clinical specimen. Selenite F broth, tetrathionate broth and alkaline peptone water (APW) are used to recover pathogens from fecal specimens.

Differential/Indicator Medium: Differential Appearance

Certain media are designed in such a way that different bacteria can be recognized on the basis of their colony colour. Various approaches include incorporation of dyes, metabolic substrates etc, so that those bacteria that utilize them appear as differently coloured colonies. Such media are called differential media or indicator media. Differential media allow the growth of more than one microorganism of interest but with morphologically distinguishable colonies.

Examples of differential media include:

a. Mannitol salts agar (mannitol fermentation = yellow).

b. Blood agar (various kinds of hemolysis i.e. α, β and γ hemolysis).

c. Mac Conkey agar (lactose fermenters, pink colonies whereas non- lactose fermenter produces pale or colorless colonies).

d. TCBS (Vibrio cholerae produces yellow colonies due to fermentation of sucrose).

e. Anaerobic media: Anaerobic bacteria need special media for growth because they need low oxygen content, reduced oxidation –reduction potential and extra nutrients.

Media for anaerobes may have to be supplemented with nutrients like hemin and vitamin K. Such media may also have to be reduced by physical or chemical means. Boiling the medium serves to expel any dissolved oxygen. Addition of 1% glucose, 0.1% thioglycollate, 0.1% ascorbic acid, 0.05% cysteine or red hot iron filings can render a medium reduced. Before use the medium must be boiled in water bath to expel any dissolved oxygen and then sealed with sterile liquid paraffin.

Robertson Cooked Meat (RCM) medium that is commonly used to grow Clostridium spps contains a 2.5 cm column of bullock heart meat and 15 ml of nutrient broth. Thioglycollate broth contains sodium thioglycollate, glucose, cystine, yeast extract and casein hydrolysate. Methylene blue or resazurin is an oxidation-reduction potential indicator that is incorporated in the medium. Under reduced condition, methylene blue is colorless.

Assay Media

These media are used for the assay of vitamins, amino acids and antibiotics. E.g. antibiotic assay media are used for determining antibiotic potency by the microbiological assay technique.

Preparation and Storage of Culture Media

Care must be taken to adjust the pH of the medium before autoclaving. Various pH indicators that are in use include phenol red, neutral red, bromothymol blue, bromocresol purple etc. Dehydrated media are commercially available and must be reconstituted as per manufacturers' recommendation. Most culture media are sterililized by autoclaving. Certain media that contain heat labile components like glucose, antibiotics, urea, serum, blood are not autoclaved. These components are filtered and may be added separately after the medium is autoclaved. Certain highly selective media such as Wilson and Blair's medium and, trypticase soya broth (TSB) agar need not be sterilized. It is imperative that a representation from each lot be tested for performance and contamination before use. Once prepared, media may be held at 4-5 °C in the refrigerator for 1-2 weeks. Certain liquid media in screw capped bottles or tubes or cotton plugged can be held at room temperature for weeks.

1. Sterilization: The media and culture vessel have to be sterilized to prevent the growth of unwanted microorganisms and thus contamination. If laboratoryscale experiments are carried out in 100 to 1000 ml flasks, or in lesser volumes such as 50 ml or 10 ml, the media along with the culture flasks or vials can be steam-sterilized with an autoclave. Depending on the quantity of the materials autoclaved, the sterilization can be carried out alternativelyin a pressure cooker of convenient size. Steam-sterilization with an autoclave or pressure cooker is carried out at 120 °C for 15 to 20 minutes under 15 psi pressure. When microbes are cultivated in a fermentor for large-scale operation, it is convenient to sterilize the fermentoras a whole with or without media. Media may be sterilized separately or in situ, in the fermentor itself. Steam is used for the sterilization of the media and fermentor, by passing the steam through the sterilization jacket or the coil around the fermentor. When the fermentor is sterilized without media in it, steam can be sparged into the vessel through all openings, allowing it to exit very slowly. Sparging is a process by which sterile air or steam is allowed to pass through the medium in the vessel with the help of a sparging device placed at the bottom of the fermentor. The steam pressure is held at 15 psi for 20 to 30 minutes while circulating or holding the steam within the vessel or in the jacket.

2. Environment for microbial growth: The nutrient composition of the medium, the ionic concentration of salts, pH, and temperature influence the growth of microorganisms in the culture and its metabolic state. Most of the bacteria grow at neutral pH, whereas yeast and fungi prefer acidic pH. Similarly, different organisms prefer different optimum temperatures for active growth and multiplication. The optimum temperature has to be maintained in the culture with the help of an incubator in the case of small-scale cultures and circulating water of the appropriate temperature through the jacket of the fermentor.

3. Aeration and mixing: Mixing of the broth is essential for the uniform distribution of the nutrients and the microbial population in the culture. Aeration is needed for the easy gas exchange between the medium and the environment. Aerated medium will be rich in oxygen. Aeration and mixing can be easily achieved by shaking the medium on a shaker in the case of small scale cultures (shake flasks cultures). In large-scale cultivation in bioreactors the transfer of oxygen to organisms is very difficult because it requires proper mixing. In fermentors, the proper mixing of cells, media components, and oxygen is achieved by stirring the medium with the help of a mechanical stirrer with baffles attached to it. Baffles help in maintaining turbulence. Microbial-free air passed through the media ensures proper aeration, and this forced aeration also helps in the mixing of media, cells, and oxygen.

4. Fermentors: These are bioreactors used for the cultivation of microbial cells on large scale under controlled conditions for industrial purposes. This closed metallic or glass vessel has the adequate arrangement for aeration, mixing of media by agitation, temperature control, pH control, anti-foaming, control of overflow, sterilization of media and vessel, cooling, and sampling (removal of sample, while the fermentor is on). Agitation of the media in the bioreactor may be through stirring or aeration or both. This equipment is convenient for operation

continuously for a number of days. The essential parts of a laboratory fermentor are given in figure below:

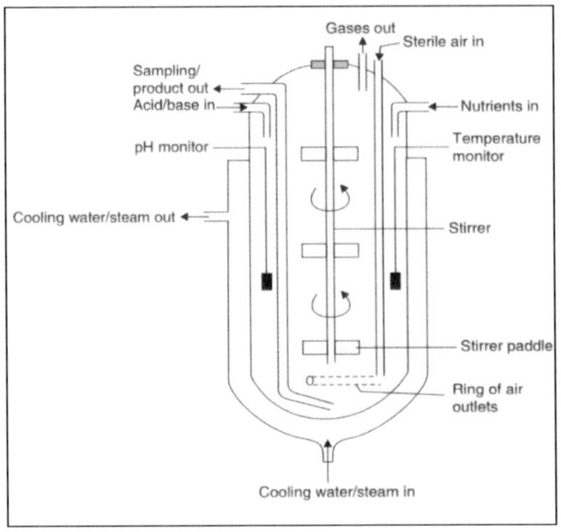

As indicated in the figure, the bioreactors are provided with controls for monitoring and adjusting the many physical and chemical parameters such as temperature, pH, nutrient composition, foaming, etc. Maximum cell growth and product formation can be achieved by controlling these parameters that assist cell growth and metabolism leading to high output of the product. A stirred tank bio-reactors the most commonly used bioreactor for microbial cultivation, in which the microbial medium is stirred with an impeller. A high density of metabolically active cells in the medium can result in sudden depletion of dissolved oxygen creating an anaerobic condition in the medium. This can result in serious consequences in the quality of the product or even in the type of products formed in the fermentation reaction. Similarly, the cell growth and product formation can alter pH of the medium, which can also create problems in the further growth and metabolism of the cell cultures. R apid growth also results in the depletion of essential nutrients that directly link to the growth and metabolism that causes the production of the product. All these changes are monitored by the accessories of the fermentor or bioreactor and are accordingly indicated automatically. For example, whenever there is a change in pH from the optimum value, automatically a sufficient amount of acid or alkali is added to the media to keep the optimum pH constant. Similarly, if there is foaming in the media the sensor will detect the foam formation and accordingly, the antifoam agent is delivered into the medium to prevent the foaming. In addition to the industrial type of bioreactors or fermentors, there are fermentors of small volumes suitable for operating in the laboratories, known as laboratory fermentors. These laboratory fermentors are for 10 to 100 liters of volume and are used for optimizing culture conditions and nutritional parameters for better growth of cells and production of metabolites for conducting research studies in the laboratory. Types of Microbial Cultures The culturing of the microbial system can be achieved in different ways. The type of culture method sometimes depends on the type of the microbial system or on the type of the product that we expect. For example, one can get two entirely different products from the same organism by changing the nutritional and other parameters or even culturing vessels.

1. Batch culture: This is a small-scale laboratory experiment in which a microbial culture is growing in a small volume flask. It consists of a limited volume of broth culture in a flask

inoculated with the bacterial or microbial inoculum and follows a normal growth phase. It is a closed-culture system because the medium contains a limited amount of nutrients and will be consumed by the growing microorganisms for their growth and multiplication with the excretion of certain metabolites as products. In batch cultures, the nutrients are not renewed and the exponential growth of cells is limited to a few generations. The growth phase of the culture consists of an initial lag phase, a log phase or the exponential growth phase, and a stationary phase. During the log phase the consumption of the nutrients will be the maximum resulting in the maximum biomass output with the excretion of the product. At the stationary phase the rate of growth decreases and becomes zero. This is because at the stationary phase the cells are exposed to a changed environment where there is only a small amount of nutrients and more cells along with the accumulation of metabolites, which may have a negative effect on the growth of the cells.

2. Fed-batch culture: The batch culture can be made into a semi-continuous culture or fed-batch culture by feeding it with fresh media sequentially at the end of the log phase or in the beginning of the stationary phase without removing cells. Because of this the volume of the culture will go on increasing as fresh media is added. This method is specially suited for cultures in which a high concentration of substrate is inhibitory to cell multiplication and biomass formation. In such situations the substrate can be fed at low concentrations to achieve cell growth. This method can easily produce a high cell density in the culture medium, which may not be possible in a batch fermentor or shake flask culture. This is especially important when the product formation is intracellular to achieve maximum product output per biomass.

3. Continuous culture: Bacterial cultures can be maintained in a state of exponential growth over long periods of time using a system of continuous culture, designed to relieve the conditions that stop exponential growth in batch cultures. Continuous culture, in a device called a chemo-stat, can be used to maintain a bacterial population at a constant density, a situation that is, in many ways, more similar to bacterial growth in natural environments. This is a very convenient method to get continuous cell growth and product formation over a long period of time. In continuous culture, the nutrient medium including the raw material is supplied at a rate that is equal to the volume of media with cells and product displaced or removed from the culture. The volume removed and the volume added is the same. In effect there is no change in the net volume as well as the chemical environment of the culture. In a chemo-stat, the growth chamber is connected to a reservoir of sterile medium.

Once the growth is initiated, fresh medium is continuously supplied from the reservoir, the volume of fluid in the growth chamber is maintained at a constant level by some sort of overflow drain. Fresh medium is allowed to enter into the growth chamber at a rate that limits the growth of the bacteria. The bacterial cells grow (cells are formed) at the same rate at which bacterial cells (and spent medium) are removed by the overflow. The rate of addition of the fresh medium determines the rate of growth because the fresh medium always contains a limiting amount of an essential nutrient. Thus, the chemostat relieves the insufficiency of nutrients, the accumulation of toxic substances, and the accumulation of excess cells in the culture, which are the parameters that initiate the stationary phase of the growth cycle. The bacterial culture can be grown and maintained at relatively constant conditions, depending on the flow rate of the nutrients. Density is constant in a turbidostat culture, which is also a continuous culture. Since the culture is fed with the fresh medium at specific rate, a steady state of growth and metabolism is achieved. At a steady state, the cell multiplication and substrate consumption for growth and product formation occur at a fixed rate. The growth rate is maintained constantly. The formation of new biomass is balanced with the removal of cells from the outlet. Continuous culture is very suitable for the production of cell biomass and products, if it is excreted into the medium. It is widely used for the production of single-cell protein from liquid effluents as a byproduct of the waste treatment. The organic waste present in the effluent is converted into microbial biomass, which is known as single-cell proteins.

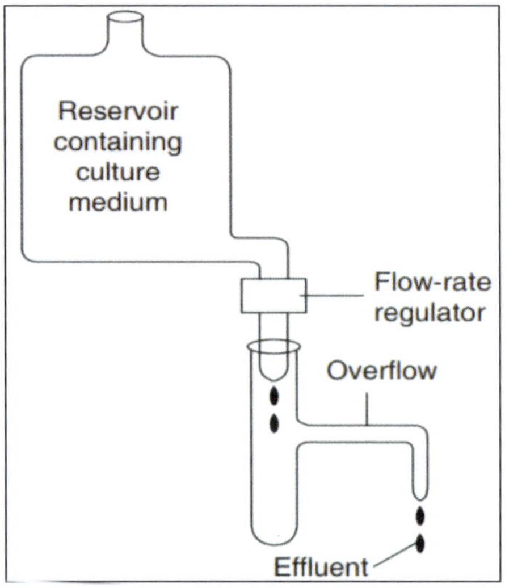

Schematic diagram of a chemostat, a device used for the continuous culture of microbes. The chemostat relieves the environmental conditions that restrict growth by continuously supplying nutrients to cells andremoving waste substances and spent cells from the culture medium.

4. Synchronous cultures: Synchronous cultures are composed of the population of cells that arethe same stage of their life cycle. All the cells in the culture will divide at the same time, will grow for a generation time, and all will divide again at the same time. Thus, the entire population is kept uniform with respect to growth and division. It is not possible to analyze a single bacterial cell to obtain the information about growth behavior, Organization, differentiation and macromolecular synthesis. Synchronous culture provides the entire cell crop in the same stage of growth. Measurement made on such cultures are equivalent to the measurement made individual cells. Synchronous cultures of bacteria can be obtained by a number of techniques. Two fundamentally different experiment approaches have employed. In the first approach, a Synchronous population of the cells can be sorted out according to age or size by physical separation of cells. In methods of the second type, a culture is induced by manipulating the physical environment or chemical composition of the medium to obtain asynchronously dividing the population. The techniques based on selection are preferable to those based on induction since induction is likely to introduce distortions in physiologic state of the cells.

Microbial Growth

Prokaryotes, such as bacteria, propagate by binary fission. For unicellular organisms, cell division is the only method used to produce new individuals. In both prokaryotic and eukaryotic cells, the outcome of cell reproduction is a pair of daughter cells that are genetically identical to the parent cell. In unicellular organisms, daughter cells are individuals.

Due to the relative simplicity of the prokaryotes, the cell division process, or binary fission, is a less complicated and much more rapid process than cell division in eukaryotes. The single, circular DNA chromosome of bacteria is not enclosed in a nucleus, but instead occupies a specific location, the nucleoid, within the cell. Although the DNA of the nucleoid is associated with proteins that aid in packaging the molecule into a compact size, there are no histone proteins and thus, no nucleosomes in prokaryotes. The packing proteins of bacteria are, however, related to the cohesin and condensin proteins involved in the chromosome compaction of eukaryotes.

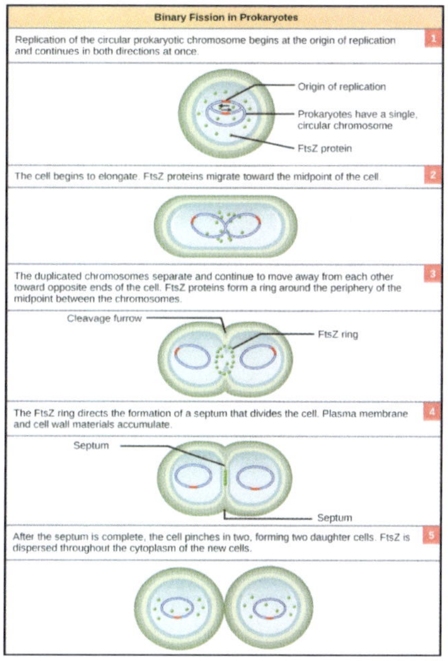

Binary Fission: These images show the steps of binary fission in prokaryotes.

The bacterial chromosome is attached to the plasma membrane at about the midpoint of the cell. The starting point of replication, the origin, is close to the binding site of the chromosome at the plasma membrane. Replication of the DNA is bidirectional, moving away from the origin on both strands of the loop simultaneously. As the new double strands are formed, each origin point moves away from the cell wall attachment toward the opposite ends of the cell. As the cell elongates, the growing membrane aids in the transport of the chromosomes. After the chromosomes have cleared the midpoint of the elongated cell, cytoplasmic separation begins. The formation of a ring composed of repeating units of a protein, FtsZ, directs the partition between the nucleoids. Formation of the FtsZ ring triggers the accumulation of other proteins that work together to recruit new membrane and cell wall materials to the site. A septum is formed between the nucleoids, extending gradually from the periphery toward the center of the cell. When the new cell walls are in place, the daughter cells separate.

Mitotic Spindle Apparatus

The precise timing and formation of the mitotic spindle is critical to the success of eukaryotic cell division. Prokaryotic cells, on the other hand, do not undergo karyokinesis and, therefore, have no need for a mitotic spindle. However, the FtsZ protein that plays such a vital role in prokaryotic cytokinesis is structurally and functionally very similar to tubulin, the building block of the microtubules that make up the mitotic spindle fibers that are necessary for eukaryotes. FtsZ proteins can form filaments, rings, and other three-dimensional structures that resemble the way tubulin forms microtubules, centrioles, and various cytoskeletal components. In addition, both FtsZ and tubulin employ the same energy source, GTP (guanosine triphosphate), to rapidly assemble and disassemble complex structures.

FtsZ and tubulin are homologous structures derived from common evolutionary origins. In this example, FtsZ is the ancestor protein to tubulin (a modern protein). While both proteins are found in extant organisms, tubulin function has evolved and diversified tremendously since evolving from its FtsZ prokaryotic origin. A survey of mitotic assembly components found in present-day unicellular eukaryotes reveals crucial intermediary steps to the complex membrane-enclosed genomes of multicellular eukaryotes.

Fts Proteins and Cell Division

FtsZ is a protein encoded by the ftsZ gene that assembles into a ring at the future site of the septum of bacterial cell division.

FtsZ is a protein encoded by the ftsZ gene that assembles into a ring at the future site of the septum of bacterial cell division. This is a prokaryotic homologue to the eukaryotic protein tubulin. FtsZ has been named after "Filamenting temperature-sensitive mutant Z". The hypothesis was that cell division mutants of E. coli would grow as filaments due to the inability of the daughter cells to separate from one another.

FtsZ was the first protein of the prokaryotic cytoskeleton to be identified. During cell division, FtsZ is the first protein to move to the division site, and is essential for recruiting other proteins that produce a new cell wall between the dividing cells. FtsZ's role in cell division is analogous to that of actin in eukaryotic cell division, but unlike the actin-myosin ring in eukaryotes, FtsZ has no known motor protein associated with it. The origin of the cytokinetic force thus remains unclear, but it is believed that the localized synthesis of new cell wall produces at least part of this force. It is interesting to note that L-form bacteria that lack a cell wall do not require FtsZ for division, which implies that bacteria may have retained components of an ancestral mode of cell division.

Much is known about the dynamic polymerization activities of tubulin and microtubules, but little is known about these activities in FtsZ. While it is known that single-stranded tubulin protofilaments form into 13 stranded microtubules, the multistranded structure of the FtsZ-containing Z-ring is not known. It is only speculated that the structure consists of overlapping protofilaments. Recently, proteins similar to tubulin and FtsZ have been discovered in large plasmids found in Bacillus species. They are believed to function as components of segrosomes, which are multiprotein complexes that partition chromosomes/plasmids in bacteria. The plasmid homologs of tubulin/FtsZ seem to have conserved the ability to polymerize into filaments.

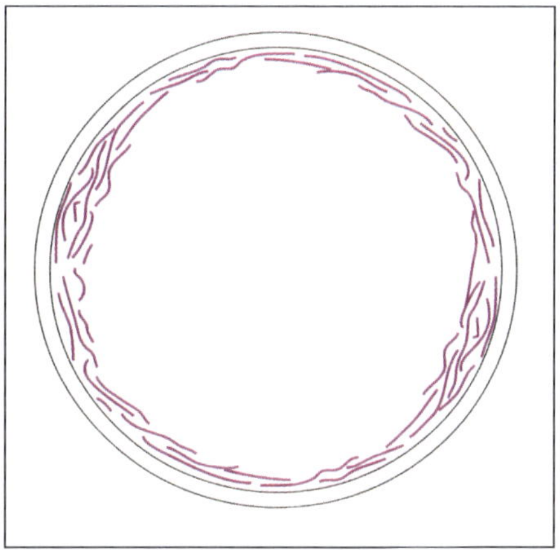

FtsZ Filaments: The Z-ring forms from smaller subunits of FtsZ filaments. These filaments may pull on each other and tighten to divide the cell.

FtsZ has the ability to bind to GTP, and also exhibits a GTPase domain that allows it to hydrolyze GTP to GDP and a phosphate group. In vivo, FtsZ forms filaments with a repeating arrangement of subunits, all arranged head-to-tail. These filaments form a ring around the longitudinal midpoint, or septum, of the cell. This ring is called the Z-ring. The GTP hydrolyzing activity of the protein is not essential to the formation of filaments or division. Mutants lacking the GTPase domain form twisted and disordered septa. These cells with irregular septa can still divide, although abnormally. It is unclear as to whether FtsZ actually provides the physical force that results in division or serves as a marker for other proteins to execute division.

The Z-ring forms from smaller subunits of FtsZ filaments. These filaments may pull on each other and tighten to divide the cell. If FtsZ does provide force that divides the cell, it may do so through the relative movement of subunits. In this model, FtsZ scission force comes from the relative lateral movement of subunits. Lines of FtsZ would line up together parallel and pull on each other creating a "cord" of many strings that tightens itself. In other models, FtsZ does not provide the contractile force but provides the cell a spatial scaffold for other proteins to execute the division of the cell. This is akin to the creating of a temporary structure by construction workers to access hard-to-reach places of a building. The temporary structure allows unfettered access and ensures that the workers can reach all places. If the temporary structure is not correctly built, the workers will not be able to reach certain places, and the building will be deficient.

This "scaffold theory" is supported by information that shows that the formation of the ring and localization to the membrane requires the concerted action of a number of accessory proteins. ZipA or the actin homologue FtsA permit initial FtsZ localization to the membrane. Following localization to the membrane, division proteins of the Fts family are recruited for ring assembly. Many of these proteins, such as FtsW, FtsK, and FtsQ are involved in stabilization of the Z ring and may also be active participants in the scission event. The formation of the Z-ring closely coincides with cellular processes associated with replication.

MreB and Determinants of Cell Morphology

MreB is a protein found in bacteria homologous to actin.

MreB is a protein found in bacteria that has been identified as a homologue of actin, as indicated by similarities in tertiary structure and conservation of active site peptide sequence. The conservation of protein structure suggests the common ancestry of the cytoskeletal elements formed by actin and MreB, found in prokaryotes. Indeed, recent studies have found that MreB proteins polymerize to form filaments that are similar to actin microfilaments.MreB controls the width of rod-shaped bacteria, such as Escherichia coli. A mutant E. coli that creates defective MreB proteins will be spherical instead of rod-like. Also, bacteria that are naturally spherical do not have the gene encoding MreB. Prokaryotes carrying the mreB gene can also be helical in shape. MreB has long been thought to form a helical filament underneath the cytoplasmic membrane. However, this model has been brought into question by three recent publications showing that filaments cannot be seen by electron cryotomography and that GFP-MreB can be seen as patches moving around the cell circumference. It has also been shown to interact with several proteins that are proven to be involved in length growth (for instance PBP2). Therefore, MreB probably directs the synthesis and insertion of new peptidoglycan building units into the existing peptidoglycan layer to allow length growth of the bacteria.

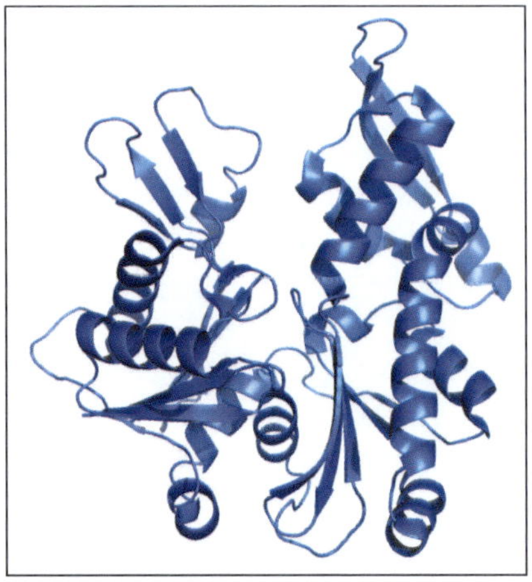

Atomic structure of MreB, a prokaryotic structural protein: Procaryotic MreB in cartoon representation. The fold of the protein is similar to its eukaryotic counterpart.

MreB is a cytoskeleton element that assembles into filamentous structures within the bacterial cytoplasm. MreB and its homologs have been shown to interact and co-localize with cytoplasmic protein(MurB-G), membrane-imbedded proteins (MreD, MraY and RodA), as well as other molecules with large periplasmic domain in organism. Recent research shows that peptidoglycan precursors are inserted into cell wall following helical pattern which is dependent on MreB, and it's reported that MreB also promote the GT activity of PBPs. This ability of MreB is because of RodZ, an inner membrane protein containing an 80-residue, N-terminal cytoplasmic region, and a 200-amino acid periplasmic C-terminal tail. RodZ co-localizes with MreB helices in a manner that is strictly dependent on its cytoplasmic region. MreB- RodZ complexes act as a major stabilizing factor in bacterial cell wall and ensure the insertion of new peptidoglycan in a spiral like fashion into the cell wall.

Peptidoglycan Synthesis and Cell Division

Peptidoglycan, also known as murein, is a polymer and consists of sugars and amino acids which form the cell walls of bacteria.

Peptidoglycan, also known as murein, is a polymer consisting of sugars and amino acids that forms a mesh-like layer outside the plasma membrane of bacteria (but not Archaea), forming the cell wall. The sugar component consists of alternating residues of β-(1,4) linked N-acetylglucosamine and N-acetylmuramic acid. Attached to the N-acetylmuramic acid is a peptide chain of three to five amino acids. The peptide chain can be cross-linked to the peptide chain of another strand forming the 3D mesh-like layer. Some Archaea have a similar layer of pseudopeptidoglycan or pseudo-murein, where the sugar residues are β-(1,3) linked N-acetylglucosamine and N-acetyltalosamin-uronic acid. That is why the cell wall of Archaea is insensitive to lysozymes, which are present in human sweat and tears as part of innate immunity.

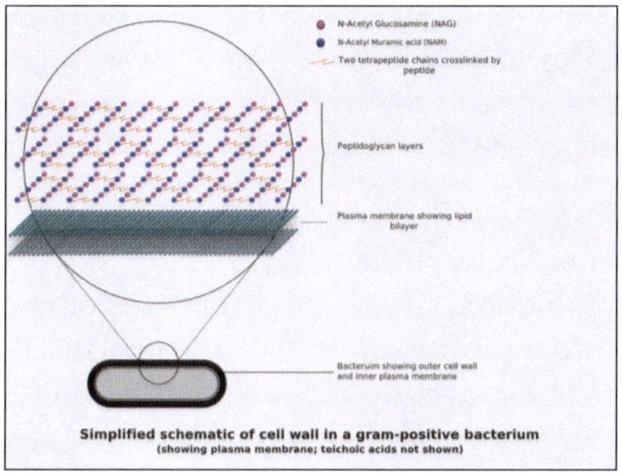

Simplified sc hematic of a cell wall in a Gram-positive bacteria: Cross-linking between amino acids in different linear amino sugar chains occurs with the help of the enzyme transpeptidase and results in a 3-dimensional structure that is strong and rigid.

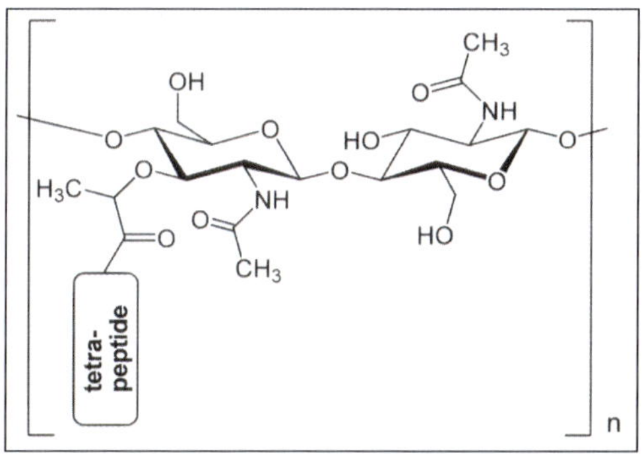

Peptidoglycan structure: The peptidoglycan layer in the bacterial cell wall is a crystal lattice structure formed from linear chains of two alternating amino sugars, namely N-acetylglucosamine (GlcNAc or NAG) and N-acetylmuramic acid (MurNAc or NAM).

Peptidoglycan serves a structural role in the bacterial cell wall giving it strength, as well as counteracting the osmotic pressure of the cytoplasm. A common misconception is that peptidoglycan gives the cell its shape. However, it is actually the MreB protein that facilitates cell shape. Peptidoglycan is also involved in binary fission during bacterial cell reproduction.

The peptidoglycan layer is substantially thicker in Gram-positive bacteria (20 to 80 nanometers) than in Gram-negative bacteria (7 to 8 nanometers), with the attachment of the S-layer. Peptidoglycan forms around 90% of the dry weight of Gram-positive bacteria but only 10% of Gram-negative strains. Thus, presence of high levels of peptidoglycan is the primary determinant of the characterisation of bacteria as gram-positive. In Gram-positive strains, it is important in attachment roles and stereotyping purposes. For both Gram-positive and Gram-negative bacteria, particles of approximately 2 nm can pass through the peptidoglycan. Gram-positive and Gram-negative bacteria are sensitive to different types of antiobiotics.

Generation Time

Bacterial growth occurs by the division of one bacterium into two daughter cells in a process called binary fission.

Bacterial growth is the division of one bacterium into two daughter cells in a process called binary fission. Providing no mutational event occurs the resulting daughter cells are genetically identical to the original cell. Therefore, "local doubling" of the bacterial population occurs. Both daughter cells from the division do not necessarily survive. The doubling time is the generation time of the bacteria. If the number surviving exceeds unity on average, the bacterial population undergoes exponential growth.

The measurement of an exponential bacterial growth curve in batch culture was traditionally a part of the training of all microbiologists. The basic means requires bacterial enumeration (cell counting) by direct and individual (microscopic, flow cytometry), direct and bulk (biomass), indirect and individual (colony counting), or indirect and bulk (most probable number, turbidity, nutrient uptake) methods. In autecological studies, bacterial growth in batch culture can be modeled with four different phases: lag phase, exponential or log phase, stationary phase, and death phase.

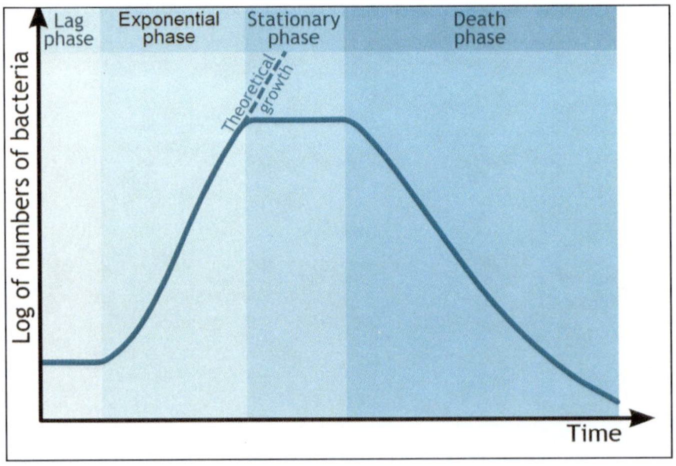

Bacterial Growth Curve: This chart shows the logarithmic growth of bacteria. Note the Y-axis scale is logarithmic meaning that the number represents doubling. The phases of growth are labelled on top.

During lag phase, bacteria adapt themselves to growth conditions. It is the period where the individual bacteria are maturing and not yet able to divide. During this phase of the bacterial growth cycle, synthesis of RNA, enzymes, and other molecules occurs. The exponential phase (sometimes called the log phase or the logarithmic phase) is a period characterized by cell doubling. The number of new bacteria appearing per unit time is proportional to the present population. If growth is not limited, doubling will continue at a constant rate so both the number of cells and the rate of population increase doubles with each consecutive time period. For this type of exponential growth, plotting the natural logarithm of cell number against time produces a straight line. The slope of this line is the specific growth rate of the organism, which is a measure of the number of divisions per cell per unit time.

The actual rate of this growth (i.e. the slope of the line in the figure) depends upon the growth conditions, which affect the frequency of cell division events and the probability of both daughter cells surviving. However, exponential growth cannot continue indefinitely because the medium is soon depleted of nutrients and enriched with wastes. Finally, the stationary phase is due to a growth-limiting factor, such as depletion of a nutrient and/or the formation of inhibitory products such as organic acids. Death of cells as a function of time is rather unpredictable and very difficult to explain. At death phase, bacteria run out of nutrients and die. This basic batch culture growth model draws out and emphasizes aspects of bacterial growth which may differ from the growth of macrofauna. It emphasizes clonality, asexual binary division, the short development time relative to replication itself, the seemingly low death rate, the need to move from a dormant state to a reproductive state or to condition the media, and finally, the tendency of lab adapted strains to exhaust their nutrients. In reality, even in batch culture, the four phases are not well defined.

Microbial Metabolism

Metabolism is the dynamic balance between difference types of reaction(catabolism+ anabolism) that provide the cell with energy or building materials & those to utilize them. It consists of two general activities:

- Energy Production
- Energy Utilization

Metabolic pathways are treated as a sequence of enzymes functioning as a unit, with each enzyme using as its substrate a product of the preceding enzyme-catalyzed reaction. This picture of metabolic pathways is incomplete because we will usually ignore the regulation of pathway operation for the sake of space and simplicity. However, one should keep in mind that both regulation of the activity of individual pathways and coordination of the action of separate sequences are essential to the existence of life. Cells become disorganized and die without adequate control of metabolism, and regulation is just as important to life as is the efficient use of energy.

The Role of ATP in Metabolism: Many reactions in the cell are endergonic and will not proceed far toward completion without outside assistance. One of ATP's major roles is to drive such endergonic reactions more to completion. ATP is a high-energy molecule. That is, it breaks down or hydrolyzes almost completely to the products ADP and Pi with a Go′ of 7.3 kcal/mole. ATP + H_2O ADP + Pi.

In reference to ATP the term high-energy molecule does not mean that there is a great deal of energy stored in a particular bond of ATP. It simply indicates that the removal of the terminal phosphate goes to completion with a large negative standard free energy change, or the reaction is strongly exergonic. In other words, ATP has a high phosphate group transfer potential; it readily transfers its phosphate to water. The phosphate group transfer potential is defined as the negative of ᴸGo′ for the hydrolytic removal of phosphate. A molecule with a higher group transfer potential will donate phosphate to one with a lower potential.

Thus ATP is ideally suited for its role as an energy currency. It is formed in energy-trapping and -generating processes such as photosynthesis, fermentation, and aerobic respiration. In the cell's economy, exergonic ATP breakdown is coupled with various endergonic reactions to promote their completion In other words ATP links energy-generating reactions, which liberate free energy, with energy-using reactions, which require free energy input to proceed toward completion. Facilitation of chemical work is the focus of the preceding example, but the same principles apply when ATP is coupled with endergonic processes involving transport work and mechanical work.

Structure and Classification of Enzymes

Enzymes may be defined as protein catalysts that have great specificity for the reaction catalyzed and the molecules acted on. A catalyst is a substance that increases the rate of a chemical reaction without being permanently altered itself. Thus enzymes speed up cellular reactions. The reacting molecules are called substrates, and the substances formed are the products.

Many enzymes are indeed pure proteins. However, many enzymes consist of a protein, the apoenzyme, and also a non protein component, a cofactor, required for catalytic activity.

The complete enzyme consisting of the apoenzyme and its cofactor is called the holoenzyme. If the cofactor is firmly attached to the apoenzyme it is a prosthetic group. Often the cofactor is loosely attached to the apoenzyme. It can even dissociate from the enzyme protein after products have been formed and carry one of these products to another enzyme. Such a loosely bound cofactor is called a coenzyme. For example, NAD^+ is a coenzyme that carries electrons within the cell. Many vitamins that humans require serve as coenzymes or as their precursors. Niacin is incorporated into NAD^+ and riboflavin into FAD. Metal ions may also be bound to apoenzymes and act as cofactors.

The Mechanism of Enzyme Reactions

It is important to keep in mind that enzymes increase the rates of reactions but do not alter their equilibrium constants. If a reaction is endergonic, the presence of an enzyme will not shift its equilibrium so that more products can be formed. Enzymes simply speed up the rate at which a reaction proceeds toward its final equilibrium. How do enzymes catalyze reactions? Although a complete answer would be long and complex, some understanding of the mechanism can be gained by considering the course of a normal exergonic chemical reaction. A + B, C + D When molecules A and B approach each other to react, they form a transition-state complex, which resembles both the substrates and the products . The activation energy is required to bring the reacting molecules together in the correct way to reach the transition state. The transition-state complex can then decompose to yield the products C and D. The difference in free energy level between reactants and products is Go′. Thus the equilibrium in our example will lie toward the products because Go′ is negative (i.e the products are at a lower energy level than the substrates).

Clearly A and B will not be converted to C and D if they are not supplied with an amount of energy equivalent to the activation energy. Enzymes accelerate reactions by lowering the activation energy; therefore more substrate molecules will have sufficient energy to come together and form products. Even though the equilibrium constant (or └Go′) is unchanged, equilibrium will be reached more rapidly in the presence of an enzyme because of this decrease in the activation energy.

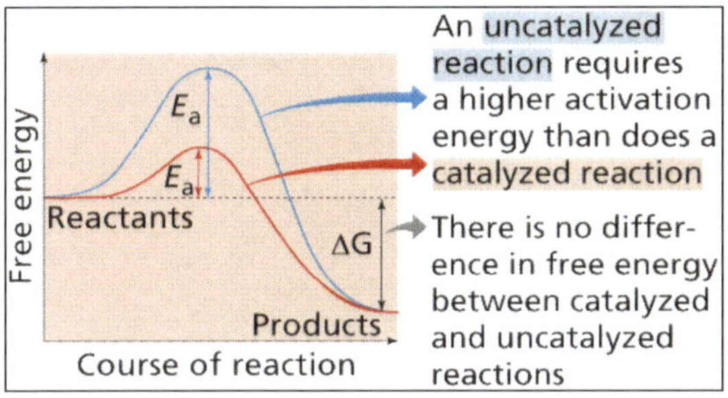

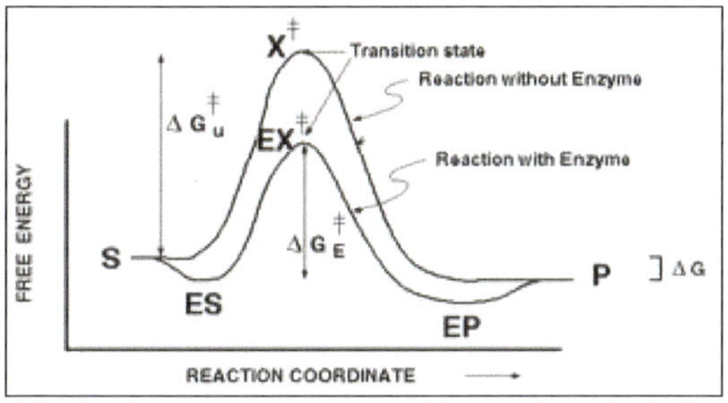

Researchers have expended much effort in discovering how enzymes lower the activation energy of reactions, and the process is becoming clearer. Enzymes bring substrates together at a special place on their surface called the active site or catalytic site to form an enzyme-substrate complex. The enzyme can interact with a substrate in two general ways. It may be rigid and shaped to precisely fit the substrate so that the correct substrate binds specifically and is positioned properly for reaction. This mechanism is referred to as the lock-and-key model. An enzyme also may change shape when it binds the substrate so that the active site surrounds and precisely fits the substrate.

This has been called the induced fit model and is used by hexokinase and many other enzymes. The formation of an enzyme-substrate complex can lower the activation energy in many ways. For example, by bringing the substrates together at the active site, the enzyme is, in effect, concentrating them and speeding up the reaction. An enzyme does not simply concentrate its substrates, however. It also binds them so that they are correctly oriented with respect to each other in order to form a transition-state complex. Such an orientation lowers the amount of energy that the substrates require to reach the transition state. These and other catalytic site activities speed up a reaction hundreds of thousands of times, even though microorganisms grow between 20 °C and approximately 113 °C. These temperatures are not high enough to favor most organic reactions in the absence of enzyme catalysis, yet cells cannot survive at the high temperatures used by an organic chemist in routine organic syntheses. Enzymes make life possible by accelerating specific reactions at low temperatures.

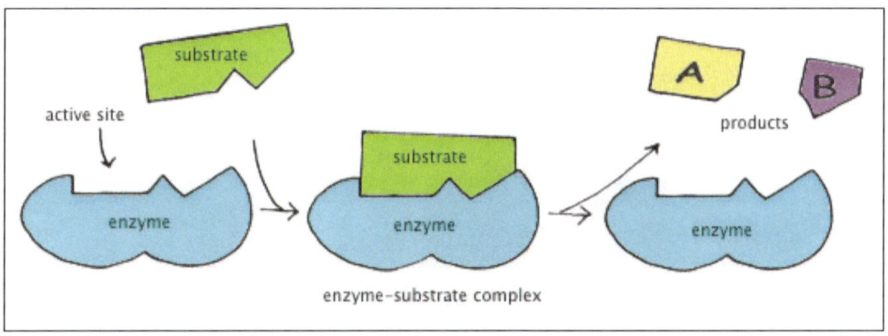

An example of Enzyme-Substrate Complex.

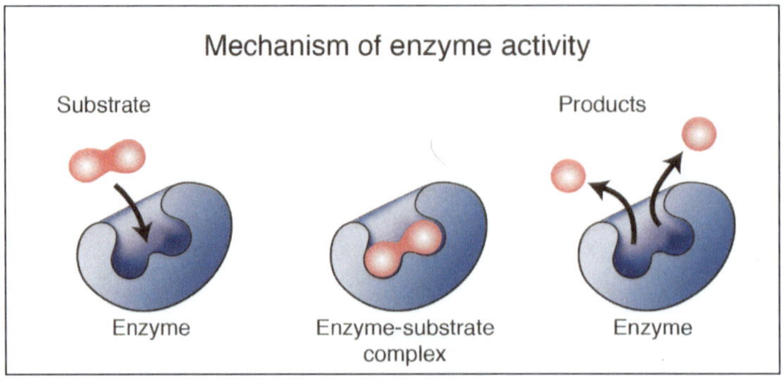

How Do Enzymes Work.

Metabolism without Oxygen

Anaerobic Cellular Respiration

The production of energy requires oxygen. The electron transport chain, where the majority of ATP is formed, requires a large input of oxygen. However, many organisms have developed strategies to carry out metabolism without oxygen, or can switch from aerobic to anaerobic cell respiration when oxygen is scarce.

During cellular respiration, some living systems use an organic molecule as the final electron acceptor. Processes that use an organic molecule to regenerate NAD^+ from NADH are collectively referred to as fermentation. In contrast, some living systems use an inorganic molecule as a final electron acceptor. Both methods are called anaerobic cellular respiration, where organisms convert energy for their use in the absence of oxygen.

Certain prokaryotes, including some species of bacteria and archaea, use anaerobic respiration. For example, the group of archaea called methanogens reduces carbon dioxide to methane to oxidize NADH. These microorganisms are found in soil and in the digestive tracts of ruminants, such as cows and sheep. Similarly, sulfate-reducing bacteria and archaea, most of which are anaerobic, reduce sulfate to hydrogen sulfide to regenerate NAD^+ from NADH.

Eukaryotes can also undergo anaerobic respiration. Some examples include alcohol fermentation in yeast and lactic acid fermentation in mammals.

Lactic Acid Fermentation

Anaerobic bacteria: The green color seen in these coastal waters is from an eruption of hydrogen sulfide-producing bacteria. These anaerobic, sulfate-reducing bacteria release hydrogen sulfide gas as they decompose algae in the water.

The fermentation method used by animals and certain bacteria (like those in yogurt) is called lactic acid fermentation. This type of fermentation is used routinely in mammalian red blood cells and in skeletal muscle that has an insufficient oxygen supply to allow aerobic respiration to continue (that is, in muscles used to the point of fatigue). The excess amount of lactate in those muscles is what causes the burning sensation in your legs while running. This pain is a signal to rest the overworked muscles so they can recover. In these muscles, lactic acid accumulation must be removed by the blood circulation and the lactate brought to the liver for further metabolism. The chemical reactions of lactic acid fermentation are the following:

Pyruvic acid + NADH $\leftrightarrow$ lactic acid + NAD^+

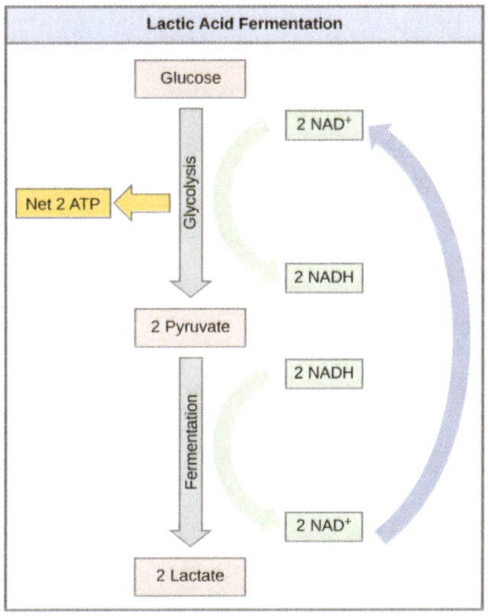

Lactic acid fermentation: Lactic acid fermentation is common in muscle cells that have run out of oxygen.

The enzyme used in this reaction is lactate dehydrogenase (LDH). The reaction can proceed in either direction, but the reaction from left to right is inhibited by acidic conditions. Such lactic acid accumulation was once believed to cause muscle stiffness, fatigue, and soreness, although more recent research disputes this hypothesis. Once the lactic acid has been removed from the muscle and circulated to the liver, it can be reconverted into pyruvic acid and further catabolized for energy.

Alcohol Fermentation

Another familiar fermentation process is alcohol fermentation, which produces ethanol, an alcohol. The use of alcohol fermentation can be traced back in history for thousands of years. The chemical reactions of alcoholic fermentation are the following (CO_2 does not participate in the second reaction):

Pyruvic acid $\rightarrow$ CO_2 + acetaldehyde + NADH $\rightarrow$ ethanol + NAD^+

The first reaction is catalyzed by pyruvate decarboxylase, a cytoplasmic enzyme, with a coenzyme of thiamine pyrophosphate (TPP, derived from vitamin B_1 and also called thiamine). A carboxyl group is removed from pyruvic acid, releasing carbon dioxide as a gas. The loss of carbon dioxide reduces the size of the molecule by one carbon, making acetaldehyde. The second reaction is catalyzed by alcohol dehydrogenase to oxidize NADH to NAD^+ and reduce acetaldehyde to ethanol.

Alcohol Fermentation: Fermentation of grape juice into wine produces CO2 as a byproduct. Fermentation tanks have valves so that the pressure inside the tanks created by the carbon dioxide produced can be released.

The fermentation of pyruvic acid by yeast produces the ethanol found in alcoholic beverages. Ethanol tolerance of yeast is variable, ranging from about 5 percent to 21 percent, depending on the yeast strain and environmental conditions.

Other Types of Fermentation

Various methods of fermentation are used by assorted organisms to ensure an adequate supply of NAD+ for the sixth step in glycolysis. Without these pathways, that step would not occur and no ATP would be harvested from the breakdown of glucose.Other fermentation methods also occur in bacteria. Many prokaryotes are facultatively anaerobic. This means that they can switch between aerobic respiration and fermentation, depending on the availability of oxygen. Certain prokaryotes, like Clostridia, are obligate anaerobes. Obligate anaerobes live and grow in the absence of molecular oxygen. Oxygen is a poison to these microorganisms, killing them on exposure.

It should be noted that all forms of fermentation, except lactic acid fermentation, produce gas. The production of particular types of gas is used as an indicator of the fermentation of specific carbohydrates, which plays a role in the laboratory identification of the bacteria.

Cellular Respiration

Microorganisms such as cyanobacteria can trap the energy in sunlight through the process of photosynthesis and store it in the chemical bonds of carbohydrate molecules. The principal carbohydrate formed in photosynthesis is glucose. Other types of microorganisms such as nonphotosynthetic bacteria, fungi, and protozoa are unable to perform this process. Therefore, these organisms must rely upon preformed carbohydrates in the environment to obtain the energy necessary for their metabolic processes.

Cellular respiration is the process by which microorganisms obtain the energy available in carbohydrates. They take the carbohydrates into their cytoplasm, and through a complex series of metabolic processes, they break down the carbohydrate and release the energy. The energy is generally not needed immediately, so it is used to combine ADP with phosphate ions to form ATP molecules. During the process of cellular respiration, carbon dioxide is given off as a waste product. This carbon dioxide can be used by photosynthesizing cells to form new carbohydrates. Also in the process of cellular respiration, oxygen gas is required to serve as an acceptor of electrons. This oxygen gas is identical to the oxygen gas given off in photosynthesis.

The overall mechanism of cellular respiration involves four subdivisions: glycolysis, in which glucose molecules are broken down to form pyruvic acid molecules; the Krebs cycle, in which pyruvic acid is further broken down and the energy in its molecule is used to form high-energy compounds such as NADH; the electron transport system, in which electrons are transported along a series of coenzymes and cytochromes and the energy in the electrons is released; and chemiosmosis, in which the energy given off by electrons is used to pump protons across a membrane and provide the energy for ATP synthesis.

Glycolysis: The process of glycolysis is a multistep metabolic pathway that occurs in the cytoplasm of microbial cells and the cells of other organisms. At least six enzymes operate in the metabolic pathway.

In the first and third steps of the pathway, ATP is used to energize the molecules. Thus, two molecules of ATP must be expended in the process. Further along in the process, the six-carbon glucose molecule is converted into intermediary compounds and then is split into two three-carbon compounds. The latter undergo additional conversions and eventually form pyruvic acid at the conclusion of the process.

During the latter stages of glycolysis, four ATP molecules are synthesized using the energy given off during the chemical reactions. Thus, four ATP molecules are synthesized and two ATP molecules are inserted into the process for a net gain of two ATP molecules in glycolysis.

Also during glycolysis, another of the reactions yields enough energy to convert NAD toNADH. The reduced coenzyme (NADH) will later be used in the electron transport system, and its energy will be released. During glycolysis, two NADH molecules are produced.

As glycolysis does not use oxygen, the process is considered to be anaerobic. For certain anaerobic organisms, such as certain bacteria and fermentation yeasts, glycolysis is the sole source of energy. It is a somewhat inefficient process because much of the cellular energy remains in the two molecules of pyruvic acid.

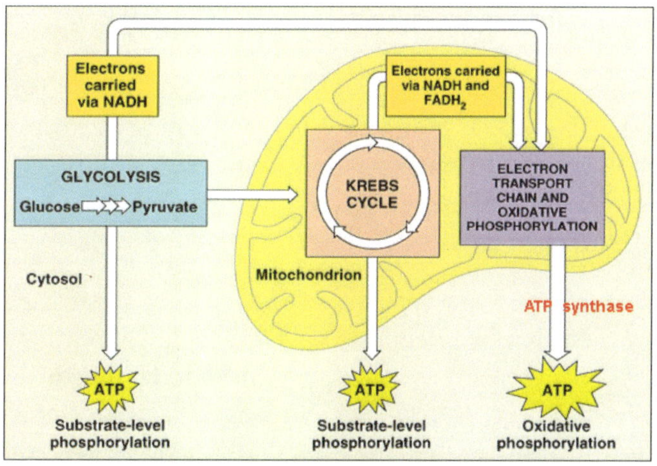

An overview of the processes of cellular respiration showing the major pathways and the places where ATP is synthesized.

The Krebs cycle: Following glycolysis, the mechanism of cellular respiration then involves another multistep process called the Krebs cycle, also called the citric acid cycle and the tricarboxylic acid cycle. The Krebs cycle uses the two molecules of pyruvic acid formed in glycolysis and yields high-energy molecules of NADH and FADH and some ATP and carbon dioxide.

The Krebs cycle occurs at the cell membrane of bacterial cells and in the mitochondriaof eukaryotic cells. Each of these sausage-shaped organelles of eukaryotic microorganisms possesses inner and outer membranes, and therefore an inner and outer compartment. The inner membrane is folded over itself many times; the folds are called cristae. Along the cristae are the important enzymes necessary for the proton pump and for ATP production.

Prior to entering the Krebs cycle, the pyruvic acid molecules are processed. Each three-carbon molecule of pyruvic acid undergoes conversion to a substance called acetyl-coenzyme A, or acetyl-CoA. In the process, the pyruvic acid molecule is broken down by an enzyme, one carbon atom is released in the form of carbon dioxide, and the remaining two carbon atoms are combined with a coenzyme called coenzyme A. This combination forms acetyl-CoA. In the process, electrons and a hydrogen ion are transferred to NAD to form high-energy NADH.

Acetyl-CoA now enters the Krebs cycle by combining with a four-carbon acid called oxaloacetic acid. The combination forms the six-carbon acid called citric acid. Citric acid undergoes a series of enzyme-catalyzed conversions. The conversions, which involve up to 10 chemical reactions, are all brought about by enzymes. In many of the steps, high-energy electrons are released to NAD. The NAD molecule also acquires a hydrogen ion and becomes NADH. In one of the steps, FAD serves as the electron acceptor, and it acquires two hydrogen ions to become $FADH_2$. Also, in one of the reactions, enough energy is released to synthesize a molecule of ATP. Since there are two pyruvic acid molecules entering the system, two ATP molecules are formed.

Also during the Krebs cycle, the two carbon atoms of acetyl-CoA are released and each forms a carbon dioxide molecule. Thus, for each acetyl-CoA entering the cycle, two carbon dioxide molecules are formed. Since two acetyl-CoA molecules enter the cycle, and each has two carbon atoms, four carbon dioxide molecules will form. Add these four molecules to the two carbon dioxide molecules formed in the conversion of pyruvic acid to acetyl-CoA, and the total is six carbon dioxide molecules. These six CO_2 molecules are given off as waste gas in the Krebs cycle. They represent the six carbons of glucose that originally entered the process of glycolysis.

At the end of the Krebs cycle, the final product formed is oxalo-acetic acid, identical to the oxalo-acetic acid which begins the cycle. The molecule is now ready to accept another acetyl-CoA molecule to begin another turn of the cycle. All told, the Krebs cycle forms (per two molecules of pyruvic acid) two ATP molecules, a large number of NADH molecules, and some $FADH_2$ molecules. The NADH and the $FADH_2$ will be used in the electron transport system.

The electron transport system: The electron transport system occurs at the bacterial cell membrane and in the cristae of the mitochondria in eukaryotic cells. Here, a series of cytochromes (cell pigments) and coenzymes exist. These cytochromes and coenzymes act as carrier molecules and transfer molecules. They accept high-energy electrons and pass the electrons to the next molecule in the system. At key proton-pumping sites, the energy of the electrons is used to transport protons across the cell membrane or into the outer compartment of the mitochondrion.

Each NADH molecule is highly energetic. It accounts for the transfer of six protons across the membrane. Each $FADH_2$ molecule accounts for the transfer of four protons. Electrons pass from NAD to FAD, to other cytochromes and coenzymes, and eventually they lose much of their energy. The final electron acceptor is an oxygen atom. The electron-oxygen combination then takes on two protons to form a molecule of water (H_2O). As a final electron receptor, oxygen is responsible for removing electrons from the system. If oxygen were not available, electrons could not be passed among the coenzymes, the energy in electrons could not be released, the proton pump could not be established, and ATP could not be produced.

Chemiosmosis: The actual production of ATP in cellular respiration takes place during chemiosmosis. As previously noted, chemiosmosis involves the pumping of protons through special channels in the membranes of mitochondria from the inner to the outer compartment. In bacteria, the pumping occurs at the cell membrane. The pumping establishes a proton gradient. Once the gradient is established, protons pass down the gradient through molecular particles. In these particles, the energy of the protons is used to generate ATP, using ADP and phosphate ions as the starting points.

The energy production in cellular respiration during chemiosmosis is substantial. Most biochemists agree that in prokaryotic microorganisms, a total of 36 molecules of ATP can be produced during cellular respiration. In eukaryotic cells, the number is 34 molecules of ATP. Two molecules of ATP are produced as the net gain of glycolysis, so the grand total is 38 molecules of ATP (36 in eukaryotes). These ATP molecules may then be used in the cell for its needs.

Fermentation: Fermentation is an anaerobic process in which energy can be released from glucose even though oxygen is not available. Fermentation occurs in yeast cells, and a form of fermentation takes place in bacteria.

In yeast cells, glucose can be metabolized through cellular respiration, as in other cells. When oxygen is lacking, however, glucose is still changed to pyruvic acid via glycolysis. The pyruvic acid is first converted to acetaldehyde and then to ethyl alcohol. The net gain of ATP to the yeast cell is two molecules--the two molecules of ATP normally produced in glycolysis.

Yeasts are able to participate in fermentation because they have the necessary enzyme to convert pyruvic acid to ethyl alcohol. This process is essential because it removes electrons and hydrogen ions from NADH during glycolysis. The effect is to free the NAD so that it can participate in future reactions of glycolysis. Yeasts are therefore used in both bread and alcohol production. Alcohol fermentation is the process that yields beer, wine, and other spirits. The carbon dioxide given off supplements the carbon dioxide given off during the Krebs cycle and causes bread to rise.

Microbial Ecology

Microbial ecology is the study of the interactions of microorganisms with their environment, each other, and plant and animal species. It includes the study of symbioses, biogeochemical cycles and the interaction of microbes with anthropogenic effects such as pollution and climate change.

Microbes in Ecosystems

Microbes interact with other microorganisms, higher organisms and the inorganic world (abiotic world):

- Microorganisms can form biofilms called microbial mats and thereby create their own microenvironment and niche.
- The roles of microorganisms in the larger ecosystem include:
    - Primary producers – fix carbon dioxide,
    - Decomposition of organic matter,
    - Nutrient cycling,
    - Disease.
- Movement of microorganisms:
    - Soil can blow around,
    - Rivers move microorganisms,
    - Animals move microorganisms around,
    - Human activities.
- Extremophiles are microbes that live in places where they are the only organisms capable of surviving.

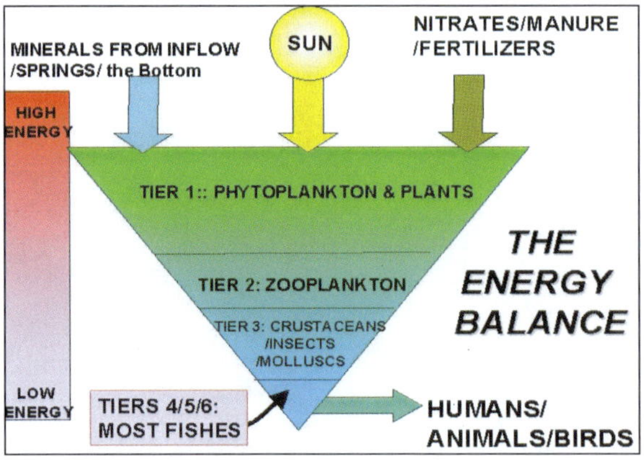

Aquatic and Marine Ecosystems

- There is a wide variety of microorganisms in aquatic systems:
    - Viruses,
    - Bacteria,
    - Diatoms,
    - Algae,

    - Protozoa,
    - Some limited fungi - more in freshwater.
- Cyanobacteria, diatoms and algae serve as major producers and sources of oxygen.
- Protozoa serve as consumers at the lower ends of the food web.
- Bacteria and fungi serve as decomposers.
- Viruses are parasites that can infect other cells.

Soil Ecosystems - There are millions and millions of microbes naturally present in soil.

- The types of microbes in soil:
    - Bacteria,
    - Archaea,
    - Fungi - lots of fungi in the soil,
    - Protozoa,
    - There is a much more diverse microbial communities in soil than fresh and saltwater.
- Soil Formation:
    - Soil contains both abiotic and biotic components.
    - Soil microbes play a major role in nutrient cycling.
    - Soil animals – insects and worms break down larger organic components and increase availability for microbes.
    - The microbes become food for the small animals.

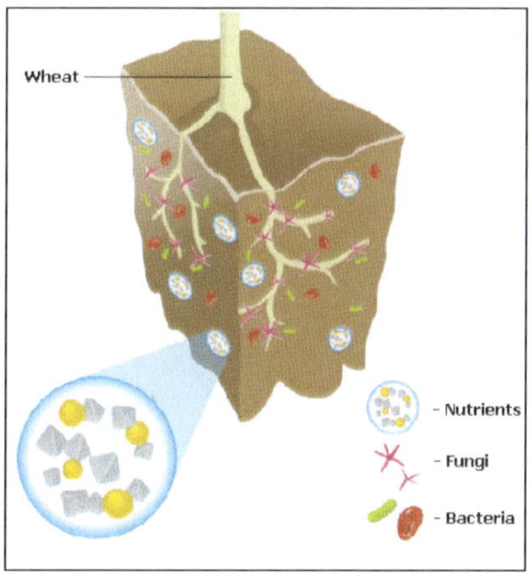

- Roles of microbes in soil include:
    - Breakdown rock particles and convert nutrients such as Phosphorous, Calcium, Manganese, Silica, Zinc, Copper etc. into plant soluble form using acid secretions and hold for later use by the plant.

  - Fungal filaments called hyphae act as extensions to the roots. The surface area of the root system becomes much greater allowing for far better access to moisture and nutrients from the soil. This extended root system acts to aerate the soil.
  - A number of microbes produce natural plant stimulants like enzymes, vitamins and acids which are necessary for growth and vigor. This reduces the need for Nitrogen.
  - Secretions from different microbes help to form soil aggregates which improves soil structure.
- Human Pathogens in Soil- there are a number of human bacterial, fungal, and protozoan pathogens found in soil.

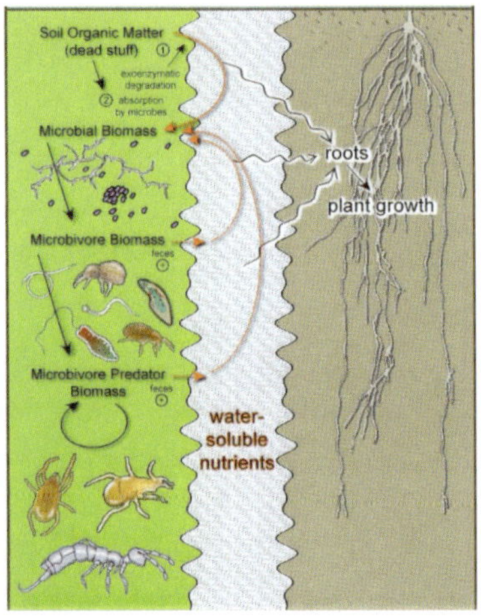

Dry Land Ecosystems

- Cyanobacteria or algae and fungi live in a mutualistic relationship to form lichens.
- Lichens are the major produces in the tundra.
- They are also play a key role in the taiga.

Symbiotic Relations

Relationships between microbes and other organisms:

- Mutualism – both the microbes and the other organism benefit.

- Commensalism - microbes that benefit from the relationship but the other organism is not helped or harmed.
- Parasitism – the microbe benefits and the other organism is harmed. A successful parasite does not harm its host enough to kill it or it loses its source of benefit. Disease is an example of microbe parasitism.
- Predation – the microbe attacks and kills the host. Sometimes there can be beneficial effects of predation:
    - Use to treat patients with antibiotic resistant bacteria.
    - Increased rates of nutrient cycling.
    - Develop a better understanding of pathogens inside the prey – *Legionnella inside* of protozoa.

Microbial Communities within Plants

Samples of mutualism are:

- Methanotrophic organisms and shagnum moss – the methanotrophic bacteria are associated with the moss – the bacteria oxidize methane and release carbon dioxide the moss uses.
- Mycorrhiza- Fungal symbiotic association with plants which helps plants to absorb phosphorus from soil.
- Some examples of commensalism are where the waste products of one organism is utilized by another.
- Nitrosomonas oxidize ammonium to nitrite.
- Nitrobacter oxidize nitrite to nitrate.
- Bacteria alter their environment and allows other bacteria to grow – organisms in sauerkraut.

Microbial Communities within Animals

Samples of mutualism are:

- Aphids and Buchnera aphidicola live in the cytoplasm of the aphid cells and provide 10 amino acids to the aphid – neither will grow apart - coevolution.
- Protozoan – termite relationship – the protozoa breakdown cellulose and release acetate which can be absorbed and used by the termite.
- Many marine invertebrates have endosymbiotic dinoflagellates.
- Endosymbionts in tube worms fix carbon dioxide by the Calvin cycle and provide fixed carbon for the worms.
- Methanotrophic organisms and shagnum moss – the methanotrophic bacteria are associated with the moss – the bacteria oxidize methane and release carbon dioxide the moss uses.

- The rumen of ruminants are filled with microorganisms and they help breakdown the plant material the animals eat.

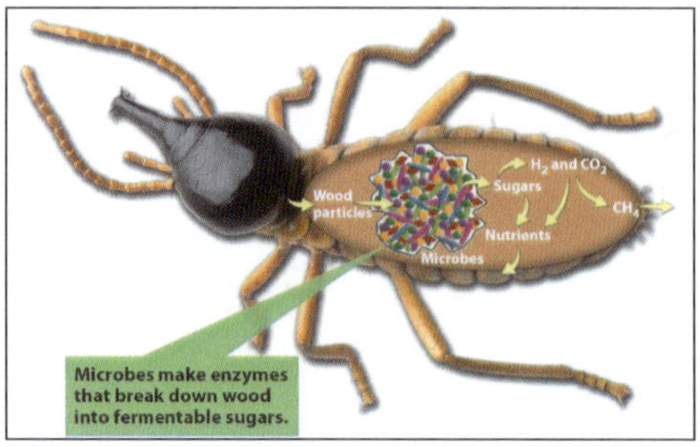

Biogeochemical Cycles - Microbes are key in maintaining ecological balance on Earth.

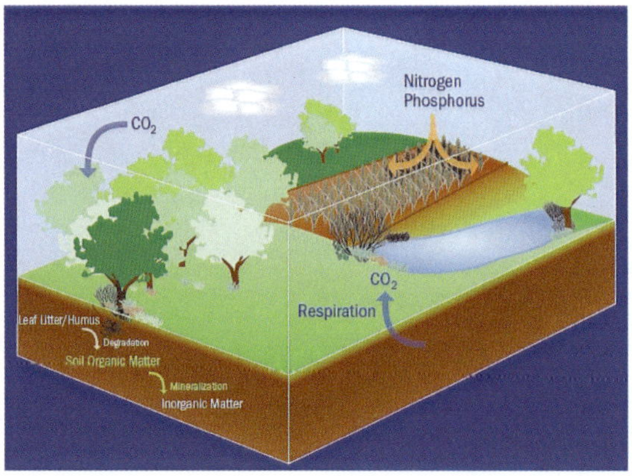

Oxygen Cycle

- Oxygen is a key element for chemical reactions of aerobic respiration.
- Photosynthesis releases oxygen as a product.
- Algae and Cyanobacteria along with plants are the key organisms to carry out photosynthesis.

Phosphorus Cycle

- Microbes use phosphorus in the form of calcium phosphate, magnesium phosphate and iron phosphate.
- Phosphorous from these forms are assimilated as phosphate into DNA, RNA, and other organic compounds.
- When the organisms are used as food by other organisms, it is released into the food chain.
- Decomposers release the phosphates into the soil so it can be absorbed by the plants.
- Phosphate is often used in fertilizer and if excess phosphate from fertilizer or industry is drained into water supplies, it can cause algal blooms.

Carbon Cycle

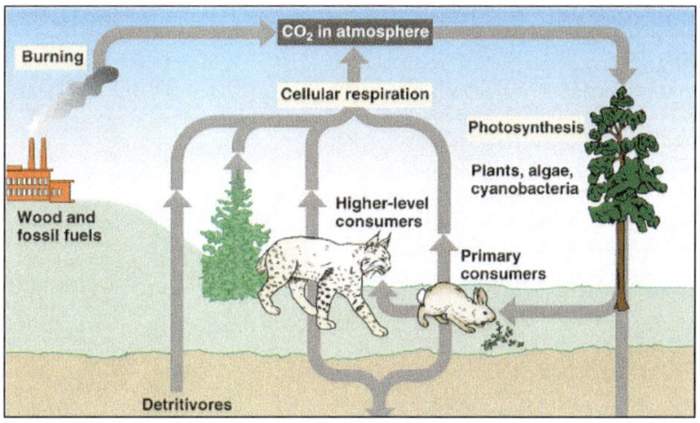

- Cyanobacteria and algae along with plants do the photosynthesis.
- The breakdown of organic matter is a major part of the carbon cycle.
- Bacteria and fungi are the major decomposers.

Nitrogen Cycle is primarily a prokaryotic activity.

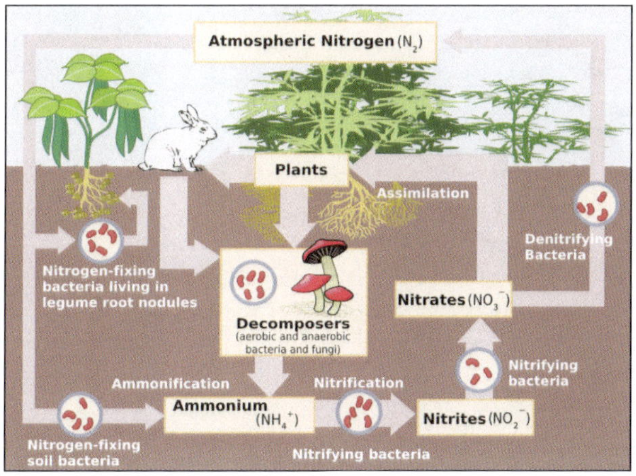

- Bacteria use their metabolic activities during numerous parts of the cycle.
- Bacteria and fungi act as decomposers of plant and animal matter to form ammonia.
- Nitrogen fixation – converting nitrogen gas from atmosphere into ammonia and bacteria fix atmospheric nitrogen in legume root nodules.
- Many bacteria and Cyanobacteria can live apart from legumes and can fix nitrogen in the soil.
- Nitrification – Conversion of ammonia to nitrate – Nitrification occurs in soils, water and marine environments where the nitrifying bacteria live.
- Denitrification – nitrate is converted into gaseous nitrogen. Numerous types of denitrifying bacteria are involved in reducing nitrates to nitrites to nitrous oxide and then to gaseous nitrogen.

Iron Cycle

- The iron cycle revolves around the conversion of ferrous iron to ferric iron and back again.
- Ferric iron reduction is part of the respiratory pathway of some bacteria.
- Some bacteria can produce magnetite.
- Some bacteria can use ferrous iron as an electron donor.

Sulfur Cycle

- Many bacteria play important roles in the sulfur cycle.
- Sulfur compounds are a key part of proteins.
- Sulfur-reducing bacteria during decomposition reduce sulfate compounds to hydrogen sulfide.
- Photosynthetic sulfur bacteria grow anaerobically and oxidize hydrogen sulfide releasing elemental sulfur.
- Species of colorless sulfur bacteria grow in hydrogen sulfide converting the sulfide to sulfate where it is available to plants and then to the rest of the food chain.

Metal toxicity – The toxicity of metals such as mercury or zinc can be enhanced by microbes and end up in the food chain and poison humans.

References

- Morphology-of-microbial-cell-microbiology, microbial-cells, cell: biologydiscussion.com, Retrieved 8 August, 2019
- Virus, cells: magnet.fsu.edu, Retrieved 11 February, 2019
- Microbial-growth, microbiology: lumenlearning.com, Retrieved 23 May, 2019
- Metabolism-without-oxygen: lumenlearning.com, Retrieved 27 August, 2019
- Cellular-respiration, microbiology, biology: cliffsnotes.com, Retrieved 7 March, 2019
- Microbes-and-ecology: soinc.org, Retrieved 21 January, 2019
- What-microbial-ecology: isme-microbes.org, Retrieved 18 April, 2019

5

Microbial Genetics

The subject which combines microbiology with genetic engineering is known as microbial genetics. It deals with the study of microorganisms such as bacteria, virus, fungi and archaea. It has played an important role in the study of evolution, as microbes have a rapid growth rate and a short generation time. This chapter has been carefully written to provide an easy understanding of the varied facets of microbial genetics.

Microbial genetics is a branch of genetics concerned with the transmission of hereditary characters in microorganisms. Within the usual definition, microorganisms include prokaryotes like bacteria, unicellular or mycelial eukaryotes e.g., yeasts and other fungi, and viruses, notably bacterial viruses (bacteriophages). Microbial genetics has played a unique role in developing the fields of molecular and cell biology and also has found applications in medicine, agriculture, and the food and pharmaceutical industries.

Because of their relative simplicity, microbes are ideally suited for combined biochemical and genetic studies, and have been successful in providing information on the genetic code and the regulation of gene activity. The operon model formulated by French biologists François Jacob and Jacques Monod in 1961, is one well known example. Based on studies on the induction of enzymes of lactose catabolism in the bacterium Escherichia coli, the operon has provided the groundwork for studies on gene expression and regulation, even up to the present day. The many applications of microbial genetics in medicine and the pharmaceutical industry emerge from the fact that microbes are both the causes of disease and the producers of antibiotics . Genetic studies have been used to understand variation in pathogenic microbes and also to increase the yield of antibiotics from other microbes.

Hereditary processes in microorganisms are analogous to those in multicellular organisms. In both prokaryotic and eukaryotic microbes, the genetic material is DNA; the only known exceptions to this rule are the RNA viruses. Mutations, heritable changes in the DNA, occur spontaneously and the rate of mutation can be increased by mutagenic agents. In practice, the susceptibility of bacteria to mutagenic agents has been used to identify potentially hazardous chemicals in the environment. For example, the Ames test was developed to evaluate the mutagenicity of a chemical in the following way. Plates containing a medium lacking in, for example, the nutrient histidine are inolculated with a histidine requiring strain of the bacterium Salmonella typhimurium. Thus, only cells that revert back to the wild type can grow on the medium. If plates are exposed to a mutagenic agent, the increase in the number of mutants compared with unexposed plates can be observed and a large number of revertants would indicate a strong mutagenic agent. For such studies, microorganisms offer the advantage that they have short mean generation times, are easily cultured in a small space under controlled conditions and have a relatively uncomplicated structure.

Microorganisms, and particularly bacteria, were generally ignored by the early geneticists because of their small in size and apparent lack of easily identifiable variable traits. Therefore, a method of identifying variation and mutation in microbes was fundamental for progress in microbial genetics.

As many of the mutations manifest themselves as metabolic abnormalities, methods were developed by which microbial mutants could be detected by selecting or testing for altered phenotypes. Positive selection is defined as the detection of mutant cells and the rejection of unmutated cells. An example of this is the selection of penicillin resistant mutants, achieved by growing organisms in media containing penicillin such that only resistant colonies grow. In contrast, negative selection detects cells that cannot perform a certain function and is used to select mutants that require one or more extra growth factors. Replica plating is used for negative selection and involves two identical prints of colony distributions being made on plates with and without the required nutrients. Those microbes that do not grow on the plate lacking the nutrient can then be selected from the identical plate, which does contain the nutrient.

The first attempts to use microbes for genetic studies were made in the USA shortly before World War II, when George W. Beadle and Edward L. Tatum employed the fungus, Neurospora, to investigate the genetics of tryptophan metabolism and nicotinic acid synthesis. This work led to the development of the "one gene one enzyme" hypothesis. Work with bacterial genetics, however, was not really begun until the late 1940s. For a long time, bacteria were thought to lack sexual reproduction, which was believed to be necessary for mixing genes from different individual organisms—a process fundamental for useful genetic studies. However, in 1947, Joshua Lederberg working with Edward Tatum demonstrated the exchange of genetic factors in the bacterium, Escherichia coli. This process of DNA transfer was termed conjugation and requires cell-to-cell contact between two bacteria. It is controlled by genes carried by plasmids, such as the fertility (F) factor, and typically involves the transfer of the plasmid from donor torecipient cell. Other genetic elements, however, including the donor cell chromosome, can sometimes also be mobilized and transferred. Transfer to the host chromosome is rarely complete, but can be used to map the order of genes on a bacterial genome.

Other means by which foreign genes can enter a bacterial cell include transformation, transfection, and transduction. Of the three processes, transformation is probably the most significant. Evidence of transformation in bacteria was first obtained by the British scientist, Fred Griffith in the late 1920s working with Streptococcus pneumoniae and the process was later explained in the 1930s by Oswald Avery and his associates at the Rockefeller Institute in New York. It was discovered that certain bacteria exhibit competence, a state in which cells are able to take up free DNA released by other bacteria. This is the process known as transformation, however, relatively few microorganisms can be naturally transformed. Certain laboratory procedures were later developed that make it possible to introduce DNA into bacteria, for example electroporation, which modifies the bacterial membrane by treatment with an electric field to facilitate DNA uptake. The latter two processes, transfection and transduction, involve the participation of viruses for nucleic acid transfer. Transfection occurs when bacteria are transformed with DNA extracted from a bacterial virus rather than from another bacterium. Transduction involves the transfer of host genes from one bacterium to another by means of viruses. In generalized transduction, defective virus particles randomly incorporate fragments of the cell DNA; virtually any gene of the donor can be transferred, although the efficiency is low. In specialized transduction, the DNA of a temperate virus excises incorrectly and brings adjacent host genes along with it. Only genes close to the integration point of the virus are transduced, and the efficiency may be high.

After the discovery of DNA transfer in bacteria, bacteria became objects of great interest to geneticists because their rate of reproduction and mutation is higher than in larger organisms; i.e., a mutation occurs in a gene about one time in 10,000,000 gene duplications, and one bacterium may produce 10,000,000,000 offspring in 48 hours. Conjugation, transformation, and transduction have been important methods for mapping the genes on the chromosomes of bacteria. These techniques, coupled with restriction enzyme analysis, cloning DNA sequencing, have allowed for the detailed studies of the bacterial chromosome. Although there are few rules governing gene location, the genes encoding enzymes for many biochemical pathways are often found tightly linked in operons in prokaryotes. Large scale sequencing projects revealed the complete DNA sequence of the genomes of several prokaryotes, even before eukaryotic genomes were considered.

Gene Regulation in Bacteria

The genome of E. coli is about 4.6 million base pairs in length, encoding about 4,300 proteins. At any one time, only a fraction of these proteins are being made. How is the expression of these genes regulated.

Promoters and Terminators

In a bacterium such as E. coli, each gene is flanked by a promoter and a terminator. The promoter is simply the site where the enzyme RNA polymerase binds to the DNA and starts making complementary messenger RNA (mRNA). This mRNA carries the instructions required to make a protein and it is later 'read' by the bacterium's ribosomes, where the protein is strung together from amino acids.

The terminator is a similar instruction in the DNA which tells the RNA polymerase to stop transcribing mRNA and to dissociate from the DNA.

It might be tempting to think of the promoter as the 'on' switch and the terminator as the 'off' switch, but this would be wrong. The promoter and terminator just mark the start and end of the DNA sequence that encodes one or more proteins – the promoter and terminator don't tell the cell when to start and stop making those proteins.

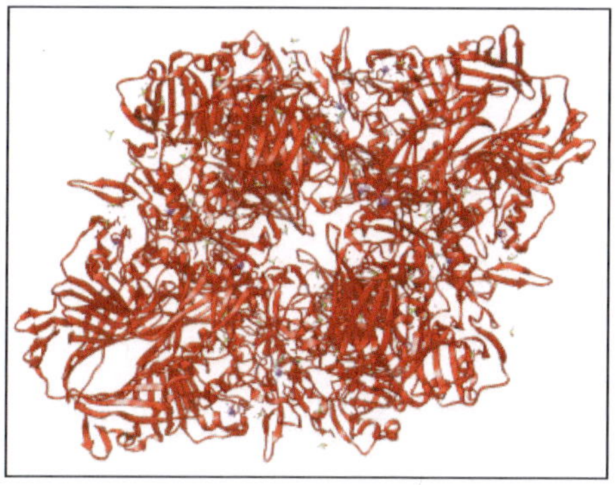

A molecular model of the enzyme β-galactosidase (lactase). The enzyme breaks down lactose to glucose and galactose.

Operons

Sometimes, in bacteria, clusters of two or more related genes are under the control of a single promoter and terminator. A single stretch of mRNA is transcribed from them, allowing the different proteins encoded by the genes to be produced at the same time. For example, there is a group of five genes in E. coli encoding enzymes that are needed for synthesising the amino acid tryptophan.

A grouping like this is called an 'operon'. There is a similar group of three genes encoding enzymes that are needed to break down the sugar arabinose and another group of three genes that are needed to absorb and metabolise lactose. hence the groups of genes above are called the 'tryptophan operon', the 'arabinose operon' and the 'lactose (or Lac) operon'.

In eukaryotes, in contrast, there are no 'operons' — every gene has its own unique promoter and terminator.

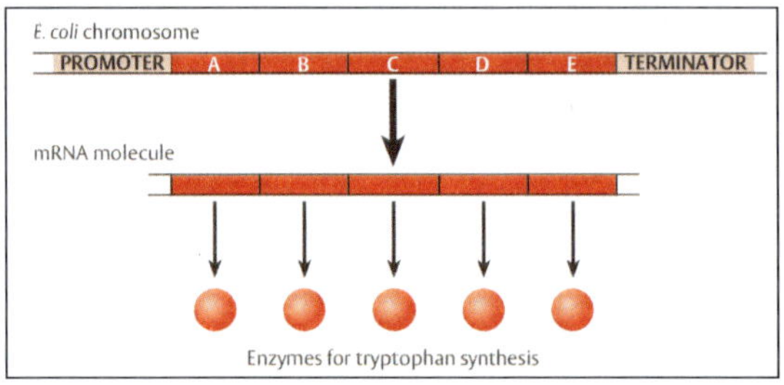

Sometimes a group of genes has just a single promoter and terminator. For example, there are five genes (labelled A to E) that E. coli needs to synthesize the amino acid tryptophan. A group of related genes like this is called an 'operon'.

Constitutive Promoters

Some proteins are made continuously by the cell. These are usually essential 'housekeeping' proteins that the cell needs to stay alive. Their genes are switched on all of the time, and they are said to be expressed 'constitutively'. A constitutive promoter is therefore one which keeps the gene (or genes) downstream of it switched on all the time and is under no higher level of control.

In technical terms, constitutive promoters have two sites in the DNA (at 10 and 35 bases before the start of the protein-encoding sequence) that a subunit of the RNA polymerase always recognises and binds to. The sequence of bases there hardly ever varies: at –10 bases it is TATAAT and at –35 bases it is TTGACA. The strength of a promoter depends upon how closely the DNA matches these two sequences. A strong promoter will have a perfect, or almost perfect, match, and the RNA polymerase will bind to it every time.

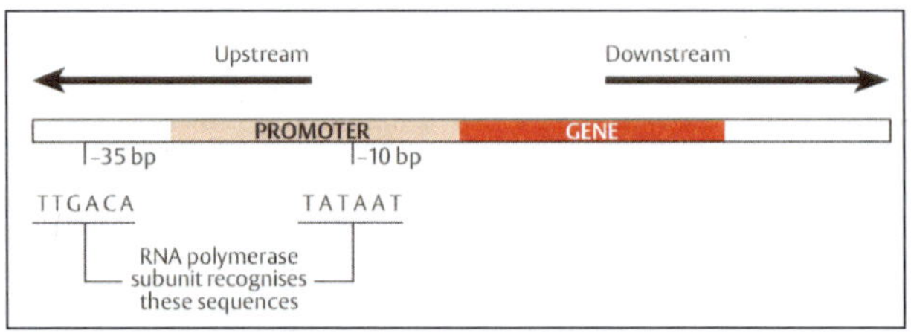

In figure, genes that are switched on all the time are under control of socalled constitutive promoters. RNA polymerase binds to two sites before the gene(s) to be transcribed. If the sequences of DNA in the promoter are 'perfect', then the promoter is strong. If the DNA has undergone mutations, however, the RNA polymerase won't bind so well and the promoter will be weaker.

Positive Regulation Gene Activator Proteins

Some proteins are only required at certain times. For instance, it would be wasteful if a bacterium made enzymes for metabolising maltose if this sugar was not present in its surroundings. Gene activator proteins can switch on protein production in such circumstances. How do these proteins work.

The DNA sequences in promoters before genes that are not continuously active are generally more variable than they are in constitutive promoters, so they are less easily recognised by the RNA polymerase subunit. A gene activator protein helps the RNA polymerase bind to the DNA. The gene activator protein itself first binds to the DNA, then it guides the RNA polymerase into place.

The activator protein, however, is not normally able to bind to DNA. Only in the presence of a signal molecule can this be done. When the signal molecule binds to the activator protein, it forces the protein to change shape and this new shape is able to bind to the DNA [proteins whose activity is altered by binding to another molecule like this are called allosteric].

The signal molecule is often the sugar, for instance, that is broken down by the enzymes that are encoded downstream of the promoter. The presence of the sugar therefore acts as a switch to trigger the production of the proteins that metabolise it.

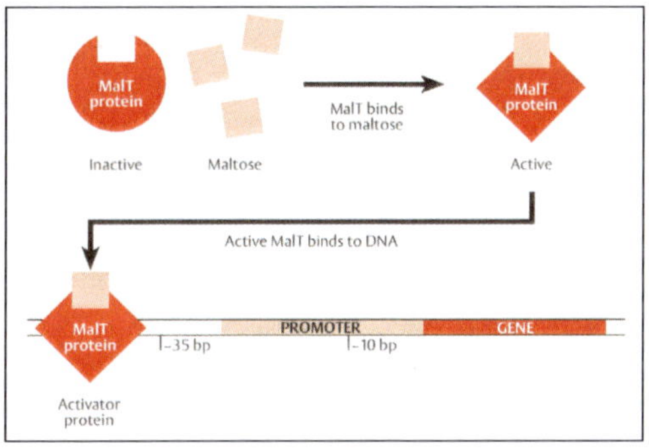

In E. coli, an activator protein, MalT, helps RNA polymerase to bind to the promoter upstream of the genes that encode enzymes enabling maltose to be metabolised. The activator protein can only bind to DNA in the presence of maltose, ensuring that the genes are only switched on when necessary.

Negative Regulation—Repressors

In addition to activator proteins that switch genes on, there are also repressor proteins that switch genes off. The most famous of these is the Lac repressor protein encoded by the gene LacI. It regulates the genes that are needed by E. coli to digest the sugar lactose.

When there is no lactose in the growth medium, the Lac repressor protein binds to the DNA just before the genes that are involved in digesting lactose. By so doing, it blocks part of the promoter sequence, so that RNA polymerase cannot bind to the DNA. Consequently, the genes downstream of the promoter are not transcribed into mRNA.

When lactose is present in the medium, however, it interacts with the Lac repressor protein, forcing it to change shape. With its new shape, the protein cannot bind to DNA, so the RNA polymerase binding site is not blocked and the lactose-digesting proteins are made The tryptophan and L-arabinose operons referred to above are controlled in a similar way by tryptophan and L-arabinose repressor proteins respectively.

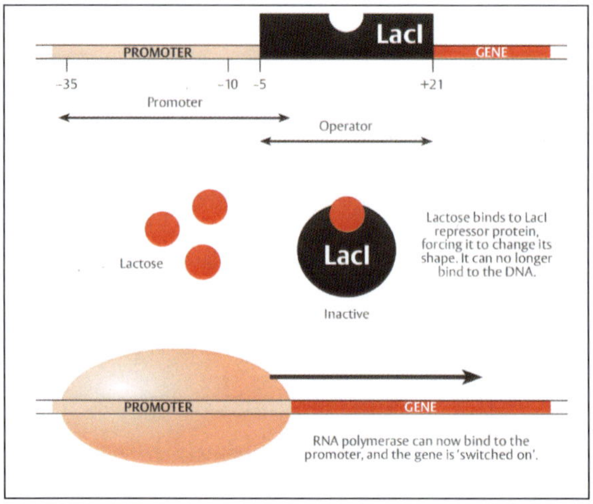

The lactase gene in E. coli is only expressed when lactose is present in the surrounding growth medium.

Positive and Negative Regulation—The Lac Operon: A Logic Gate

In fact, the Lac operon, like many regulatory mechanisms, is slightly more complex than suggested above and the promoter is controlled by two different regulatory molecules.

In response to low levels of its preferred energy source (glucose), E. coli produces a molecule called cyclic AMP (cAMP). This signal molecule interacts with the Cyclic AMP Receptor Protein (CRP), forcing that protein to change shape so that it can bind to DNA. When it does so (in several places throughout the E. coli genome), it activates genes that would allow E. coli to utilise alternative sugars such as lactose, maltose or fructose. CRP is thus a general-purpose activator which operates when glucose is not available.

It would, however, be wasteful of E. coli to start producing for example, the three enzymes needed to take up and utilise lactose if there was none of that sugar available either. The process of switching on the genes that encode the lactose-utilising enzymes is therefore as follows:

- A general signal is sent by cAMP ('Help! The glucose has run out').
- cAMP binds to the activator protein, CRP.
- CRP guides the RNA polymerase so that it can bind to DNA at the promoter site.
- But it can only do so if another signal molecule, lactose, is present and interacts with the Lac repressor protein, unblocking the promoter site.

This can be compared to an AND logic gate: only when glucose is absent AND lactose is present are the relevant genes switched on. Many genes are regulated in this way.

CRP protein	Represso protein	RNA polymerase	Switch ON or OFF
Present	Absent	Binds	On
Absent	Present	Cannot bind	Off
Present	Present	Cannot bind	Off
Absent	Absent	Cannot bind	Off

Protein Synthesis

The synthesis of proteins consumes more of a cell's energy than any other metabolic process. In turn, proteins account for more mass than any other macromolecule of living organisms. They perform virtually every function of a cell, serving as both functional (e.g., enzymes) and structural elements. The process of translation, or protein synthesis, the second part of gene expression, involves the decoding by a ribosome of an mRNA message into a polypeptide product.

The Genetic Code

Translation of the mRNA template converts nucleotide-based genetic information into the "language" of amino acids to create a protein product. A protein sequence consists of 20 commonly occurring amino acids. Each amino acid is defined within the mRNA by a triplet of nucleotides called a codon. The relationship between an mRNA codon and its corresponding amino acid is called the genetic code.

The three-nucleotide code means that there is a total of 64 possible combinations (4^3, with four different nucleotides possible at each of the three different positions within the codon). This number is greater than the number of amino acids and a given amino acid is encoded by more than one codon. This redundancy in the genetic code is called degeneracy. Typically, whereas the first two positions in a codon are important for determining which amino acid will be incorporated into a growing polypeptide, the third position, called the wobble position, is less critical. In some cases, if the nucleotide in the third position is changed, the same amino acid is still incorporated.

Whereas 61 of the 64 possible triplets code for amino acids, three of the 64 codons do not code for an amino acid; they terminate protein synthesis, releasing the polypeptide from the translation machinery. These are called stop codons or nonsense codons. Another codon, AUG, also has a special function. In addition to specifying the amino acid methionine, it also typically serves as the start codon to initiate translation. The reading frame, the way nucleotides in mRNA are grouped into codons, for translation is set by the AUG start codon near the 5′ end of the mRNA. Each set of three nucleotides following this start codon is a codon in the mRNA message.

first letter	second letter: U	second letter: C	second letter: A	second letter: G	third letter
U	UUU Phe	UCU Ser	UAU Tyr	UGU Cys	U
	UUC Phe	UCC Ser	UAC Tyr	UGC Cys	C
	UUA Leu	UCA Ser	UAA stop	UGA stop	A
	UUG Leu	UCG Ser	UAG stop	UGG Trp	G
C	CUU Leu	CCU Pro	CAU His	CGU Arg	U
	CUC Leu	CCC Pro	CAC His	CGC Arg	C
	CUA Leu	CCA Pro	CAA Gln	CGA Arg	A
	CUG Leu	CCG Pro	CAG Gln	CGG Arg	G
A	AUU Ile	ACU Thr	AAU Asn	AGU Ser	U
	AUC Ile	ACC Thr	AAC Asn	AGC Ser	C
	AUA Ile	ACA Thr	AAA Lys	AGA Arg	A
	AUG Met	ACG Thr	AAG Lys	AGG Arg	G
G	GUU Val	GCU Ala	GAU Asp	GGU Gly	U
	GUC Val	GCC Ala	GAC Asp	GGC Gly	C
	GUA Val	GCA Ala	GAA Glu	GGA Gly	A
	GUG Val	GCG Ala	GAG Glu	GGG Gly	G

Above figure shows the genetic code for translating each nucleotide triplet in mRNA into an amino acid or a termination signal in a nascent protein. The first letter of a codon is shown vertically on the left, the second letter of a codon is shown horizontally across the top, and the third letter of a codon is shown vertically on the right.

The genetic code is nearly universal. With a few exceptions, virtually all species use the same genetic code for protein synthesis, which is powerful evidence that all extant life on earth shares a common origin. However, unusual amino acids such as selenocysteine and pyrrolysine have been observed in archaea and bacteria. In the case of selenocysteine, the codon used is UGA (normally a stop codon). However, UGA can encode for selenocysteine using a stem-loop structure (known as the selenocysteine insertion sequence, or SECIS element), which is found at the 3′ untranslated region of the mRNA. Pyrrolysine uses a different stop codon, UAG. The incorporation of pyrrolysine requires the pylS gene and a unique transfer RNA (tRNA) with a CUA anticodon.

The Protein Synthesis Machinery

In addition to the mRNA template, many molecules and macromolecules contribute to the process of translation. The composition of each component varies across taxa; for instance, ribosomes may consist of different numbers of ribosomal RNAs (rRNAs) and polypeptides depending on the organism. However, the general structures and functions of the protein synthesis machinery are comparable from bacteria to human cells. Translation requires the input of an mRNA template, ribosomes, tRNAs, and various enzymatic factors.

Ribosomes

A ribosome is a complex macromolecule composed of catalytic rRNAs (called ribozymes) and structural rRNAs, as well as many distinct polypeptides. Mature rRNAs make up approximately 50% of each ribosome. Prokaryotes have 70S ribosomes, whereas eukaryotes have 80S ribosomes in the cytoplasm and rough endoplasmic reticulum, and 70S ribosomes in mitochondria and chloroplasts. Ribosomes dissociate into large and small subunits when they are not synthesizing proteins and reassociate during the initiation of translation. In E. coli, the small subunit is described as 30S (which contains the 16S rRNA subunit), and the large subunit is 50S (which contains the 5S and 23S rRNA subunits), for a total of 70S (Svedberg units are not additive). Eukaryote ribosomes have a small 40S subunit (which contains the 18S rRNA subunit) and a large 60S subunit (which contains the 5S, 5.8S and 28S rRNA subunits), for a total of 80S. The small subunit is responsible for binding the mRNA template, whereas the large subunit binds tRNAs.

Each mRNA molecule is simultaneously translated by many ribosomes, all synthesizing protein in the same direction: reading the mRNA from 5′ to 3′ and synthesizing the polypeptide from the N terminus to the C terminus. The complete structure containing an mRNA with multiple associated ribosomes is called a polyribosome (or polysome). In both bacteria and archaea, before transcriptional termination occurs, each protein-encoding transcript is already being used to begin synthesis of numerous copies of the encoded polypeptide(s) because the processes of transcription and translation can occur concurrently, forming polyribosomes. The reason why transcription and translation can occur simultaneously is because both of these processes occur in the same 5′ to 3′ direction, they both occur in the cytoplasm of the cell, and because the RNA transcript is not processed once it is transcribed. This allows a prokaryotic cell to respond to an environmental signal requiring new proteins very quickly. In contrast, in eukaryotic cells, simultaneous transcription and translation is not possible. Although polyribosomes also form in eukaryotes, they cannot do so until RNA synthesis is complete and the RNA molecule has been modified and transported out of the nucleus.

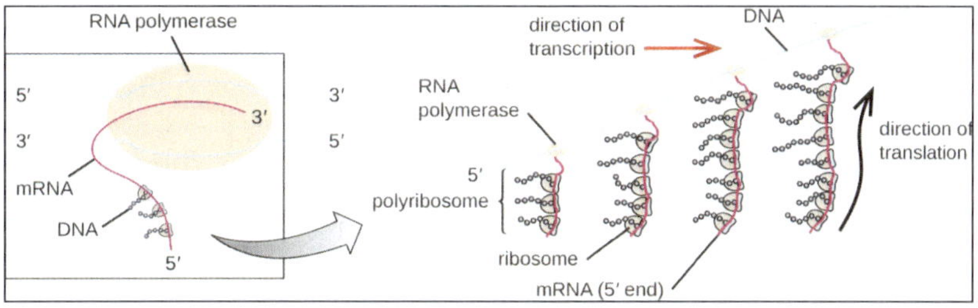

In prokaryotes, multiple RNA polymerases can transcribe a single bacterial gene while numerous ribosomes concurrently translate the mRNA transcripts into polypeptides. In this way, a specific protein can rapidly reach a high concentration in the bacterial cell.

Transfer RNAs

Transfer RNAs (tRNAs) are structural RNA molecules and, depending on the species, many different types of tRNAs exist in the cytoplasm. Bacterial species typically have between 60 and 90 types. Serving as adaptors, each tRNA type binds to a specific codon on the mRNA template and adds the corresponding amino acid to the polypeptide chain. Therefore, tRNAs are the molecules that actually "translate" the language of RNA into the language of proteins. As the adaptor molecules of translation, it is surprising that tRNAs can fit so much specificity into such a small package. The tRNA molecule interacts with three factors: aminoacyl tRNA synthetases, ribosomes, and mRNA.

Mature tRNAs take on a three-dimensional structure when complementary bases exposed in the single-stranded RNA molecule hydrogen bond with each other. This shape positions the amino-acid binding site, called the CCA amino acid binding end, which is a cytosine-cytosine-adenine sequence at the 3′ end of the tRNA, and the anticodon at the other end. The anticodon is a three-nucleotide sequence that bonds with an mRNA codon through complementary base pairing.

An amino acid is added to the end of a tRNA molecule through the process of tRNA "charging," during which each tRNA molecule is linked to its correct or cognate amino acid by a group of enzymes called aminoacyl tRNA synthetases. At least one type of aminoacyl tRNA synthetase exists for each of the 20 amino acids. During this process, the amino acid is first activated by the addition of adenosine monophosphate (AMP) and then transferred to the tRNA, making it a charged tRNA, and AMP is released.

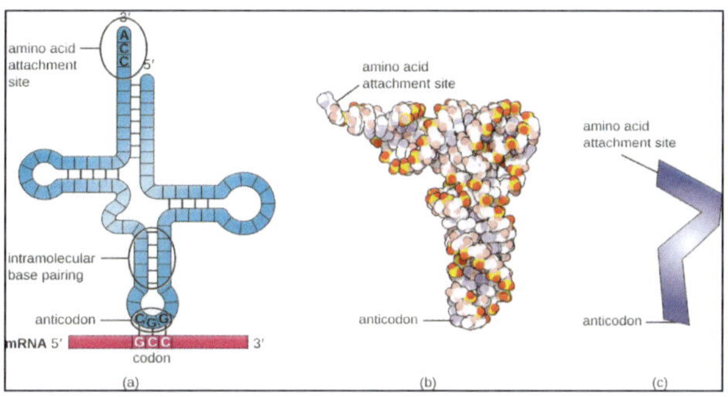

In figure, (a) After folding caused by intramolecular base pairing, a tRNA molecule has one end that contains the anticodon, which interacts with the mRNA codon, and the CCA amino acid binding end. (b) A space-filling model is helpful for visualizing the three-dimensional shape of tRNA. (c) Simplified models are useful when drawing complex processes such as protein synthesis.

The Mechanism of Protein Synthesis

Translation is similar in prokaryotes and eukaryotes. Here we will explore how translation occurs in E. coli, a representative prokaryote, and specify any differences between bacterial and eukaryotic translation.

Initiation

The initiation of protein synthesis begins with the formation of an initiation complex. In E. coli, this complex involves the small 30S ribosome, the mRNA template, three initiation factors that help the ribosome assemble correctly, guanosine triphosphate (GTP) that acts as an energy source, and a special initiator tRNA carrying N-formyl-methionine (fMet-tRNAfMet). The initiator tRNA interacts with the start codon AUG of the mRNA and carries a formylated methionine (fMet). Because of its involvement in initiation, fMet is inserted at the beginning (N terminus) of every polypeptide chain synthesized by E. coli. In E. coli mRNA, a leader sequence upstream of the first AUG codon, called the Shine-Dalgarno sequence (also known as the ribosomal binding site AGGAGG), interacts through complementary base pairing with the rRNA molecules that compose the ribosome. This interaction anchors the 30S ribosomal subunit at the correct location on the mRNA template. At this point, the 50S ribosomal subunit then binds to the initiation complex, forming an intact ribosome.

In eukaryotes, initiation complex formation is similar, with the following differences:

- The initiator tRNA is a different specialized tRNA carrying methionine, called Met-tRNAi.
- Instead of binding to the mRNA at the Shine-Dalgarno sequence, the eukaryotic initiation complex recognizes the 5′ cap of the eukaryotic mRNA, then tracks along the mRNA in the 5′ to 3′ direction until the AUG start codon is recognized. At this point, the 60S subunit binds to the complex of Met-tRNAi, mRNA, and the 40S subunit.

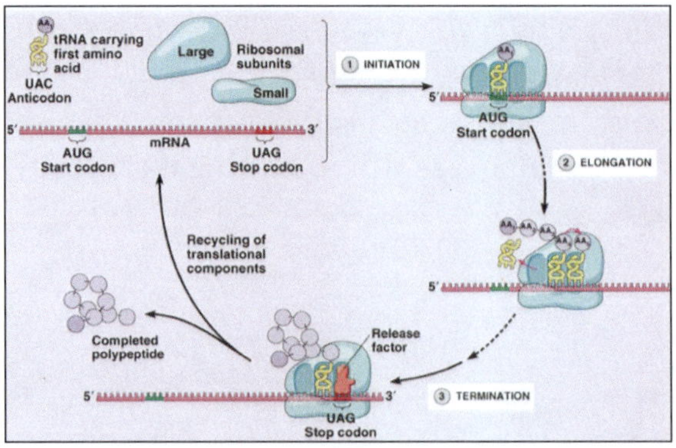

Translation in bacteria begins with the formation of the initiation complex, which includes the small ribosomal subunit, the mRNA, the initiator tRNA carrying N-formyl-methionine, and initiation factors. Then the 50S subunit binds, forming an intact ribosome.

Elongation

In prokaryotes and eukaryotes, the basics of elongation of translation are the same. In E. coli, the binding of the 50S ribosomal subunit to produce the intact ribosome forms three functionally important ribosomal sites: The A (aminoacyl) site binds incoming charged aminoacyl tRNAs. The P (peptidyl) site binds charged tRNAs carrying amino acids that have formed peptide bonds with the growing polypeptide chain but have not yet dissociated from their corresponding tRNA. The E (exit) site releases dissociated tRNAs so that they can be recharged with free amino acids. There is one notable exception to this assembly line of tRNAs: During initiation complex formation, bacterial fMet–tRNAfMet or eukaryotic Met-tRNAi enters the P site directly without first entering the A site, providing a free A site ready to accept the tRNA corresponding to the first codon after the AUG.

Elongation proceeds with single-codon movements of the ribosome each called a translocation event. During each translocation event, the charged tRNAs enter at the A site, then shift to the P site, and then finally to the E site for removal. Ribosomal movements, or steps, are induced by conformational changes that advance the ribosome by three bases in the 3′ direction. Peptide bonds form between the amino group of the amino acid attached to the A-site tRNA and the carboxyl group of the amino acid attached to the P-site tRNA. The formation of each peptide bond is catalyzed by peptidyl transferase, an RNA-based ribozyme that is integrated into the 50S ribosomal subunit. The amino acid bound to the P-site tRNA is also linked to the growing polypeptide chain. As the ribosome steps across the mRNA, the former P-site tRNA enters the E site, detaches from the amino acid, and is expelled. Several of the steps during elongation, including binding of a charged aminoacyl tRNA to the A site and translocation, require energy derived from GTP hydrolysis, which is catalyzed by specific elongation factors. Amazingly, the E. coli translation apparatus takes only 0.05 seconds to add each amino acid, meaning that a 200 amino-acid protein can be translated in just 10 seconds.

Termination

The termination of translation occurs when a nonsense codon (UAA, UAG, or UGA) is encountered for which there is no complementary tRNA. On aligning with the A site, these nonsense codons are recognized by release factors in prokaryotes and eukaryotes that result in the P-site amino acid detaching from its tRNA, releasing the newly made polypeptide. The small and large ribosomal subunits dissociate from the mRNA and from each other; they are recruited almost immediately into another translation initiation complex.

In summary, there are several key features that distinguish prokaryotic gene expression from that seen in eukaryotes. These are illustrated in and listed in table.

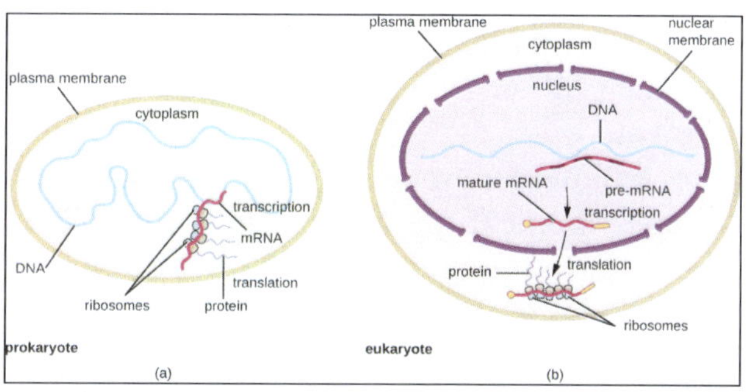

(a) In prokaryotes, the processes of transcription and translation occur simultaneously in the cytoplasm, allowing for a rapid cellular response to an environmental cue. (b) In eukaryotes, transcription is localized to the nucleus and translation is localized to the cytoplasm, separating these processes and necessitating RNA processing for stability.

Comparison of Translation in Bacteria versus Eukaryotes		
Property	**Bacteria**	**Eukaryotes**
Ribosomes	70S 30S (small subunit) with 16S rRNA subunit 50S (large subunit) with 5S and 23S rRNA subunits	80S 40S (small subunit) with 18S rRNA subunit 60S (large subunit) with 5S, 5.8S, and 28S rRNA subunits
Amino acid carried by initiator tRNA	fMet	Met
Shine-Dalgarno sequence in mRNA	Present	Absent
Simultaneous transcription and translation	Yes	No

Protein Targeting, Folding and Modification

During and after translation, polypeptides may need to be modified before they are biologically active. Post-translational modifications include:

1. Removal of translated signal sequences—short tails of amino acids that aid in directing a protein to a specific cellular compartment,
2. Proper "folding" of the polypeptide and association of multiple polypeptide subunits, often facilitated by chaperone proteins, into a distinct three-dimensional structure,
3. Proteolytic processing of an inactive polypeptide to release an active protein component,
4. Various chemical modifications (e.g., phosphorylation, methylation, or glycosylation) of individual amino acids.

DNA Replication in Bacteria, Archaea and Eukarya

The genetic information within the cells of our body is stored in the double helix of DNA, a long cylinderlike structure with a radius that is only 10 Å or one billionth of a meter but can be of considerable length. A single DNA molecule within a bacterium that grows in our gut flora is approximately 5 million base pairs in length and when stretched out, is about 1.6 mm in length, roughly the diameter of a pinhead. In contrast, the single DNA molecule in the largest human chromosome is 245,203,898 base pairs or about 8.33 cm long. The entire human genome, consisting of its 24 different chromosomes in a male is about 3 billion base pairs or 1 m long. Each cell in our body, with rare exceptions, contains two copies of the genome and thus 2 m of total DNA. Thus the scale and complexity of duplicating genomes is remarkable. For example, ~2200 human cells can sit on the top of a 1.5 mm pinhead and when extracted and laid out in a line, the DNA from these cells would be ~4.5 km (2.8 miles) long. In our body, about 500–700 million new blood cells are born every minute in the bone marrow containing a total of about 1 million km of DNA, or enough DNA to wrap around the equator of the earth 25 times. Thus DNA replication is a serious business in our body, occurring from the time that a fertilized egg first begins duplicating DNA to yield the many trillions of cells that make up an adult body and continuing in all tissues of the adult body throughout our life. The amount of DNA duplicated in an entire human body represents an unimaginable amount of information transfer. Moreover, each round of duplication needs to be highly accurate, making one mistake in less than 100 million bases copied per cell division. How copying of the double helix occurs and how it is so highly accurate is the topic of this collection. Inevitably the processes of accurate copying of the genome can go awry, yielding mutations that affect our lives, and thus the collection outlines the disorders that accelerate human disease.

However, the problem of copying DNA is much more complicated than indicated above. The 2 m of DNA in each human cell is wrapped up with histone proteins within the cell's nucleus that is only about 5 μm wide, presenting a compaction in DNA length of about 2 million-fold. How can the copying process deal with the fact that the DNA is wrapped around proteins and scrunched into a volume that creates a spatial organization problem of enormous magnitude? Not only is the DNA copied, but the proteins associated with the DNA need to be duplicated, along with all the chemical modifications attached to DNA and histones that greatly influence developmental patterning of gene expression. The protein machineries that replicate DNA and duplicate proteins within the chromosomes are some of the most complex and intriguing machineries known. Furthermore, the regulations of the processes are some of the most complex because they need to ensure that each DNA molecule in each chromosome is copied once, and only once each time before a cell divides. Errors in the regulation of DNA replication lead to accelerated mutation rates, often associated with increased rates of cancer and other diseases.

The process of accurately copying a genome can be broken down into various subprocesses that combine to provide efficient genome duplication. Central to the entire process is the machinery that actually copies the DNA with high fidelity, including proteins that start the entire process and the proteins that actually copy one helix to produce two. Superimposed on this fundamental process are mechanisms that detect and repair errors and damage to the DNA. Also associated with the DNA replication apparatus are the proteins that ensure that the histone proteins and their modifications in chromatin are inherited along with the DNA. Finally, other machineries cooperate with the DNA replication apparatus to ensure that the resulting two DNA molecules, the sister chromatids, are tethered together until the cell completes duplicating all of its DNA and segregates the sister chromatids evenly to the two daughter cells. Only by combining all of these processes can genetic inheritance ensure that each cell has a faithful copy of its parent's genome.

Where to Start

Replication begins at particular positions in chromosomes called "origins" where designated initiator proteins bind to DNA to start the process of replication. There are important differences among bacteria, archaea, and eukaryotes in this process, but there are also many striking similarities that suggest the process dates back to the last universal cellular ancestor. Bacteria often contain only one chromosome with one origin at which two replication forks assemble and move in opposite directions. Although not all bacteria follow this paradigm, this is the case for the Escherichia coli circular 4.4 Mb genome, forming a single replicon or unit of replication from a single origin. At a rate of 1 Kb/s for each fork, this genome is replicated within 30 min. In contrast, eukaryotes typically have multiple linear chromosomes, each with many origins. Multiple origins are a necessity for eukaryotes as they have much larger genomes than bacteria and eukaryotic replication forks move about 20 times more slowly than bacterial replication forks. As an example, the largest human chromosome (chromosome #1) is 250 Mb and if it had only one origin, it would require more than 50 days to replicate compared to the typical 24 h division time of a eukaryotic cell and approximately 8 h for copying DNA in S phase. Initiation at each origin produces two divergent DNA replication forks along the chromosome to create a replicon that is duplicated only once per cell division. The duplication of many replicons eventually yields two daughter chromosomes called sister chromatids that are tethered together until they separate during mitosis. Although few archaea species have been characterized, they appear to be evolutionary hybrids between bacteria and eukaryotes, because some species have a single chromosome with a single origin, whereas other species have multiple origins per chromosome. Moreover, the ploidy of the genome in archaea varies considerably, with some species having a 1C–2C distribution throughout their cell cycle, whereas others have up to 25 copies of their genome in proliferating cells. The rate of DNA replication fork progression also appears to be in between that in bacteria and eukaryotes, at about 20 kB/min, ~10 times faster than that in eukaryotes. Although some bacteria like E. coli replicate their genomes much faster, others such as Caulobacter crescentus replicate at roughly the same rate as some archaea, such as Pyrococcus abyssi.

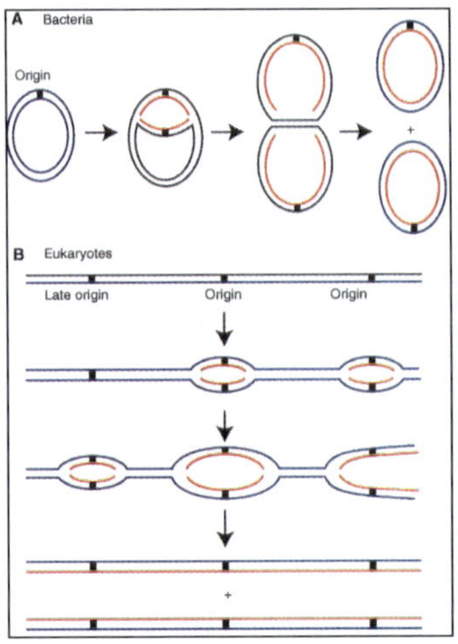

Replication initiation in bacteria and eukaryotes.

In figure, (A) Most bacteria have a circular chromosome with one origin, although there are exceptions to this. Illustrated here is the E. coli chromosome that has one origin from which two replication forks proceed in opposite directions. (B) Eukaryotes have long linear chromosomes. Bidirectional replication is initiated at multiple origins along each chromosome.

Bacterial origins are well-defined sequences to which the replication initiator proteins bind. In contrast, eukaryotic origins are not typically defined at the level of DNA sequence (with the important exception of the budding yeast Saccharomyces cerevisiae). Significant recent progress has indicated that eukaryotic origins are defined less by DNA sequence than by chromatin organization, with many origins corresponding to regions of DNA with transcriptional activity or other features that allow access to origin-binding proteins. Many human cell origins occur in sequences that are evolutionarily conserved among mammals, suggesting that they are far from arbitrary. In most eukaryotes, a small subset of potential origins is used in a typical cell cycle in individual cells, but origin utilization can be greatly increased to facilitate astonishingly rapid cell division, as seen in the fertilized eggs of many animals. Whether an origin is used or not is a stochastic process that depends on the chromatin context and in some cases the developmental state of cells in multicellular organisms. Furthermore the multiple origins in eukaryotic chromosomes are organized into clusters that are activated at specific times during S phase of the cell cycle, and the temporal patterning varies again with developmental patterning of cells.

To begin the process of activating an origin for replication, bacterial, archaeal, and eukaryotic cells use origin-binding proteins composed of AAA^+ family subunit(s). AAA^+ proteins generally function as multimeric machines. For example in bacteria, multiple copies of the DnaA origin-binding protein form a helical filament that binds the origin. The DnaA filament binds ATP to unwind an A/T-rich region of the origin, resulting in a single-strand DNA (ssDNA) "bubble" onto which the replicative helicase loads.

Eukaryotes contain a six subunit origin-binding protein referred to as ORC (origin recognition complex). Five of the ORC subunits are related to AAA^+ proteins and together with another AAA^+ protein called Cdc6 that is highly related in sequence to the largest ORC subunit, Orc1, they form a ring-shaped hexamer that binds DNA. However, unlike bacterial DnaA, ORC does not unwind DNA at regions to which it binds. Archaeal cells also use AAA^+ proteins that are related to the largest subunit of ORC, Orc1 and to Cdc6, but the number of these subunits varies depending on the particular type of archaeal cell. Both DnaA and ORC are used in other processes besides replication.

Loading the Helicase

The objective of origin-binding proteins in bacteria, archaea, and eukaryotes is the loading of two helicases onto DNA, which eventually give rise to two DNA replication forks that move in opposite directions from each origin. In all three domains of life, the helicase is a six subunit complex that unwinds DNA by encircling one strand of the parental duplex. Each helicase uses ATP hydrolysis to translocate along the single strand, acting as a moving wedge to force the parental duplex apart. The cellular DNA helicases are similar to hexameric enzymes present in several eukaryotic cell viruses such as the simian virus 40 T antigen and the human papillomavirus E1 helicase. Beyond these important similarities lie many differences among the replicative helicases of bacteria, archaea, and eukaryotic cells and their viruses. The bacterial helicase is a homohexamer that is placed around ssDNA generated by the DnaA protein at the origin; it travels 5′–3′ along the strand onto which it is bound. This directionality places the bacterial helicase around the lagging-strand template at a replication fork. In contrast, the eukaryotic helicase is a heterohexamer known as the MCM2-7 complex. Each of the six MCM subunits is encoded by a separate gene but they are related in sequence and are AAA^+ protein ATPases, whereas the bacterial helicase ATPase architecture is based on a RecA-like fold. Eukaryotic MCM encircles the leading strand template at a replication fork and tracks along ssDNA 3′–5′, the opposite polarity of the bacterial helicase. Another distinctive feature of the eukaryotic MCM2-7 helicase is that it is initially loaded onto the origin as a head-to-head double hexamer with the double-strand DNA (dsDNA) passing through the hexamer channel and therefore must transition to ssDNA to function as a helicase. This transition, although not well understood, is an important feature regulating replication initiation and involves addition of the Cdc45 and GINS proteins to form an active helicase called the CMG (Cdc45-Mcm2-7-GINS). Without these accessory proteins, the MCM2-7 is inactive as a helicase. Interestingly, the archaeal helicase is also a double hexamer, but in this case made up from a single protein called MCM that is related to the eukaryotic cell helicase. It also travels on ssDNA in the 3′–5′ direction and hence on the leading strand template and does not require accessory proteins for its helicase activity.

All cells require other factors in addition to the origin-binding protein to load the helicase onto DNA. Before loading, bacterial DnaB and eukaryotic MCM are bound by DnaC and Cdt1, respectively, which facilitate delivery of the helicase complexes to the origin. DnaC is an AAA^+ protein that uses ATP to bind the DnaB helicase in an inactive form and it cooperates with DnaA to load DnaB onto the ssDNA bubble formed at the origin by DnaA. ATP hydrolysis ejects DnaC after the loading step, enabling the helicase to become active in DNA unwinding. In eukaryotes, Cdt1 brings the MCM2-7 helicase to the ORC-Cdc6 complex that is bound to the origin DNA. MCM loading triggers ATP hydrolysis by Cdc6, ejecting it from the DNA and promoting release of Cdt1. Archaea have the AAA^+ Orc1/Cdc6 origin-binding protein, but to date no archaeal Cdt1 homolog has been identified, so the MCM hexamer may bind directly to the initiator protein. The precise mechanism by which these proteins load the helicase is unknown in any system. MCM2-7 loading by ORC, Cdc6, and Cdt1 forms a prereplicative complex (the Pre-RC) in which MCM2-7 surrounds duplex DNA, but it remains inactive for DNA unwinding until cells commit to enter S phase of the cell cycle.

Loading the helicase and activating it to unwind DNA are central replication control points in all cell types, but the way bacteria and eukaryotes regulate this process is fundamentally different. E. coli DnaA binding at the origin is regulated by SeqA, which sequesters the origin and prevents access to DnaA. Sequestration is dependent on the methylation state of the origin DNA (SeqA can only bind newly replicated, hemimethylated DNA). Under optimal growth conditions, E. coli reinitiates DNA synthesis at the origin before completing the previous round of replication, yielding multiple chromosomes in one cell; the chromosomes eventually segregate into individual cells. Eukaryotes cannot afford the luxury of rereplication because of their requirement for multiple origins on each chromosome. Reinitiation at some origins and not others would lead to copy number differences within regions of a chromosome and problems with chromosome segregation during mitosis. Hence, under most circumstances, eukaryotic origin initiation is tightly regulated so that origins initiate once, and only once, per cell division. Eukaryotes achieve this exquisite level of control by separating initiation events into different phases of the cell cycle and imposing multiple regulatory processes on the mechanism of initiation of DNA replication, whereas bacteria lack a well-defined cell cycle. Progression from one eukaryotic cell phase to the next is driven by many regulated events including the synthesis of new proteins, the destruction of others, and protein modification such as phosphorylation by kinases, a modification that is largely absent among bacterial replication proteins. An additional level of control, also distinct from bacteria, is that eukaryotic replication occurs in the nucleus and this compartmentalization allows for tight regulation by excluding key proteins from the nucleus when their activity is not required or when it might be detrimental.

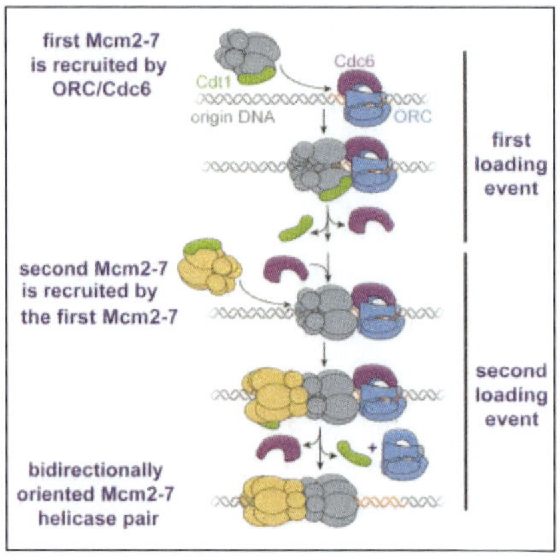

Origin activation and replisome assembly in bacteria and eukaryotes. (A) Origin activation in eukaryotes is regulated by DDK (Dbf4-Cdc7 kinase) and CDK (cyclin-dependent protein kinase) kinases that have low activity in G1 phase, and high activity in S phase. (B) Steps in origin activation and replisome assembly in bacteria and eukaryotes. The relatively more complex process of DNA replication in eukaryotes is reflected in the larger number of proteins required to initiate and elongate DNA synthesis from each origin.

Formation of the pre-RC occurs during mitotic exit in rapidly proliferating eukaryotic cells or during the G1 phase of the cell cycle, but the MCM2-7 helicase remains inactive after it is loaded onto the dsDNA. The establishment of active replication forks is regulated by kinases that drive the cell into S phase by directing the chromatin association of many factors, most importantly Cdc45 and GINS, which are now known to activate the MCM2-7 helicase activity. During S phase, the Cdc6 and Cdt1 proteins are eliminated by either selective protein degradation or nuclear exclusion, thereby preventing further MCM2-7 loading and reinitiation in S phase. These events underlie the phenomenon known as "licensing" of origins. Origins are licensed by MCM2-7 loading during mitotic exit or during G1, and then activated in S phase. By separating the helicase loading and DNA synthesis steps into two different phases of the cell cycle, eukaryotes prevent origins from initiating more than once. After replication, cells enter G2 phase and then M (mitosis) phase, in which the duplicated chromosomes are segregated into new daughter cells.

Reactions Leading to Priming

In all cell types studied so far, DNA polymerases cannot initiate new chains of nucleic acids and thus the synthesis of a primer by a DNA-dependent RNA polymerase is needed to begin cellular DNA replication. Priming occurs only on ssDNA, which requires prior helicase loading and unwinding activity. In E. coli, the primase is a single-subunit enzyme, DnaG, which transiently binds DnaB helicase to synthesize an RNA primer of ~12 nucleotides (nt). Binding of DnaG primase to DnaB also stimulates release of the regulatory protein DnaC from DnaB, indicating that initial priming and unwinding are tightly coordinated . Priming in eukaryotes is performed by the four-subunit DNA polymerase α-primase complex (Pol α/primase) that synthesizes an RNA of ~12 nt and extends this primer RNA with ~25 nt of DNA to form an RNA/DNA hybrid primer. Priming in eukaryotes occurs only after G1 phase transits to the S phase. How the Pol α/primase is recruited is not known, but in the case of SV40 DNA replication it binds directly to the helicase. In eukaryotic cells, recruitment of Pol α/primase may be mediated by the MCM10 protein that is required for initiation of DNA replication, but MCM10 then dissociates so that it cannot keep Pol α/primase on the lagging strand. The G1/S transition in eukaryotes, and eventual priming of DNA, involves activation of the S-phase CDK (cyclin-dependent protein kinase) and DDK (Dbf4-Cdc7 kinase) kinases. In all cells, elongation of the first primer becomes the leading strand. Subsequent priming events occur on the lagging strand to form Okazaki fragments. Okazaki fragments are 1–2 kb in bacteria, and only 100–200 bp in eukaryotes.

The reactions leading to priming in eukaryotes are still rather mysterious, and several proteins are involved that have no clear homologs in bacteria including Sld2, Sld3, Sld7, Cdc45, the four-subunit GINS complex, and Dpb11/TopBP1. An S-phase-specific CDK phosphorylates Sld2 and Sld3 that allows them to bind to Dbp11, whereas the DDK phosphorylation of the MCM4 subunit is necessary for initiation of DNA replication. Current studies indicate that DNA polymerase ε (Pol ε), a replicative DNA polymerase, associates with the origin before Pol α/primase in a complex with GINS, Dpb11, and CDK-phosphorylated Sld2. Thus the initial priming event may be performed by some protein other than Pol α/primase. Alternatively, Pol ε may act as a structural component that helps recruit Pol α/primase. At an undefined point in the process, the MCM2-7 complex transitions from encircling dsDNA to encircling ssDNA. Only when this transition is complete can DNA be unwound by the CMG and provide an ssDNA template for primase activity to initiate replication.

Replication Machine

Replication is performed by a multiprotein replication machine that synthesizes both daughter duplexes simultaneously. Replication machines have the same core components in all cells: DNA polymerases, circular sliding clamps, a pentameric clamp loader, helicase, primase, and SSB (single-strand binding protein). The way in which these proteins are arranged and connect to one another differs among cell types. The replication machine is often referred to as a "replisome." The bacterial replisome, illustrated in, is organized by the clamp loader, which contains three identical τ subunits that bind three C-family DNA polymerase III (Pol III) polymerases. The τ subunit also binds the homohexameric helicase (i.e., E. coli DnaB). As primase forms RNA primers, the clamp loader repeatedly loads new circular β clamps onto the primer/template for use by the lagging-strand Pol III(s). This primase/polymerase switch is facilitated by SSB, which binds ssDNA and enables the clamp loader to dislodge primase from the primed site. SSB protects the ssDNA from nucleases and facilitates elongation by Pol III-β on ssDNA. The bacterial replisome is highly processive, meaning that in E. coli it can synthesize ~86 kb on the leading strand without dissociating from the template. However, high processivity is a disadvantage on the lagging strand, which is synthesized as numerous short ~1000 nt Okazaki fragments and requires the polymerase to dissociate from the template DNA and reassociate at a new primed site for each Okazaki fragment. To overcome this "processivity barrier," specific mechanisms have evolved that pry the polymerase from the clamp to release an Okazaki fragment, on which the polymerase reassociates with a new clamp at the next RNA primed site. The three Pol IIIs in the bacterial replisome aid the production of multiple Okazaki fragments.

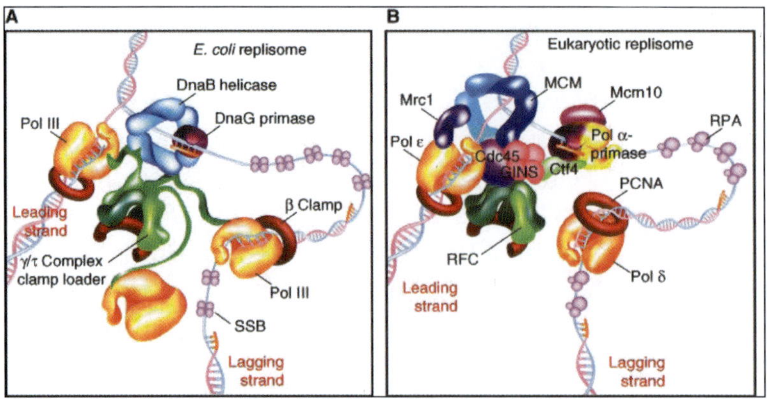

In figure, organization of bacterial and eukaryotic replisome machines. (A) Replisome architecture in E. coli. The helicase (DnaB) encircles the lagging strand. Three molecules of Pol III are attached to one clamp loader. The clamp loader binds the helicase and repeatedly assembles β clamps onto primed sites as they are formed by primase. (B) Proposed architecture of a eukaryotic replisome.

The MCM2-7 helicase encircles the leading strand; unwinding is aided by association of GINS and Cdc45 with MCM2-7 to form the CMG complex. The RFC clamp loader repeatedly loads PCNA (proliferating cell nuclear antigen) clamps onto lagging-strand primers formed by Pol α-primase. Unlike E. coli, the clamp loader may not form stabile attachments to the replisome. The leading-strand polymerase (Pol ε) is stabilized on DNA by Mrc1. Pol δ replicates the lagging strand. Contacts between Pol δ and other components of the replisome are not yet defined. Mcm10 and Ctf4 contact Pol α-primase.

Although eukaryotic and archaeal replisomes have similar components, the connections among the components are quite different from bacterial replisomes. The hexameric MCM2-7 encircles the leading strand, not the lagging strand and needs the Cdc45 and GINS proteins for it to function in unwinding DNA. Also, the eukaryotic replisome contains two different B-family polymerases that function separately for the leading and lagging strands, Pol ε and Pol δ, respectively, in addition to Pol α/primase, which starts each Okazaki fragment. The lagging strand is coated by replication protein A (RPA), a heterotrimer SSB that is structurally and functionally analogous to the bacterial SSB tetramer. As in bacteria, a primase/polymerase switch is mediated by RPA. Both Pol ε and Pol δ function with a ring-shaped proliferating cell nuclear antigen (PCNA) clamp of similar structure to E. coli β. PCNA is assembled on DNA by the pentameric replication factor C (RFC) clamp loader, composed of subunits with sequence homology and similar structure to the bacterial clamp loader. Unlike bacteria, RFC does not appear to contact the polymerases or the helicase, and the connections among replisome components remain unclear. There are numerous candidates among the many proteins known to be required for efficient eukaryotic replication that have no homolog in bacteria. For example, the GINS heterotetramer and the Cdc45 protein that form a complex with the MCM2-7 helicase to yield the CMG complex may bind Pol α/primase. GINS and Cdc45 are required for efficient DNA unwinding activity and may bind other factors in addition to the MCM2-7 complex to organize the replisome. In budding yeast, a replication progression complex has been identified containing potential links between the CMG and Pol α/primase via Ctf4/AND-1, and between CMG and Pol ε via Mrc1/Claspin.

The oligomeric structures of sliding clamps enable multiple proteins to bind one clamp simultaneously, referred to as the "toolbelt" hypothesis. An extreme case is an archaeal PCNA heterotrimer in which each different subunit binds a different partner protein (i.e., polymerase, ligase, and Fen1) that is involved in the synthesis and maturation of Okazaki fragments. These features of sliding clamps are discussed in, but one aspect will be mentioned here as it has important implications for replisome structure and function. In particular, all cells contain a variety of low-fidelity DNA polymerases that can bypass lesions in the DNA, referred to as "TLS" Pols (translesion synthesis polymerases). Different Pols can bind the clamp simultaneously and trade places with one another, making the replisome a much more dynamic machine than originally thought. E. coli contains three "translesion" DNA polymerases (TLS Pols), which are induced on DNA damage. Studies have shown that at high concentrations such as those induced on DNA damage, the TLS Pols bind the β clamp and trade places with Pol III at a moving fork, yet retain the helicase to form a "TLS replisome." TLS replisomes move much more slowly than Pol III and dictate the speed of the helicase. An obvious advantage of forming TLS replisomes on DNA damage is to slow the replication fork, giving time for DNA repair before a replication fork encounters a lesion. In the event a lesion is encountered, it can be bypassed by the TLS polymerase.

Dealing with DNA Damage during Replication

In bacteria and eukaryotes there are mechanisms to activate a stalled DNA replication fork that may be caused by the replisome running into DNA damage, a double-strand break, or a protein block. There are many repair mechanisms, such as error-free and error-prone DNA synthesis at the DNA replication fork, or post- replicative repair by nucleotide excision repair or base excision repair. The multiple DNA polymerases involved in DNA repair are discussed in and. Alternatively, branch migration followed by DNA synthesis or intersister chromatid recombination can be used to allow replicative bypass of DNA lesions. If these repair events occur while DNA replication is still occurring, stalled DNA replication forks must either be reactivated or bypassed, events that are discussed in. Unlike bacteria, eukaryotes have multiple origins and thus failure to replicate a complete replicon can be compensated by activation of adjacent origins of DNA replication, another way of bypassing a stalled DNA replication fork.

If significant DNA damage occurs either before or during DNA replication in eukaryotes, so-called checkpoint mechanisms signal to the cell-cycle regulatory machinery, principally the CDK and DDK protein kinases, that subsequent events in the cell cycle should wait until DNA damage is repaired. The biochemistry of these varied signaling events is still being worked out, but one common signal is the stable presence of RPA-coated ssDNA. Normally ssDNA should not exist in a cell, but its presence signals that a stalled replication fork exists, that resection of damaged DNA has occurred, or that recombination is taking place. In any event, the cell does not want to progress until the damage has been repaired. In both bacteria and eukaryotes, restarting a stalled DNA replication fork is a key process that involves DNA helicases and priming events that are unique to stalled fork recovery. For example, in eukaryotes cancer-prone disorders called Bloom's syndrome and Warner's syndrome have defects in specialized DNA helicases that are involved in dealing with DNA damage at stalled replication forks. Another cancer-prone syndrome called Fanconi anemia has revealed defects in a family of proteins that handle cross-links in DNA.

Spatial and Temporal Organization of DNA Replication

In bacteria, the DNA replication machinery is assembled at the single origin of DNA replication in a characteristic location. In Caulobacter, the replisome is located at one end of the rod-shaped bacterium but in E. coli, it is located in the middle of the cell. After initiation of DNA replication, the new replicated origins spool out of the replisome, which stays in place, and the origins move to predetermined locations. In E. coli, the origins move to the quarters of the cell but in Caulobacter, the origins move to the other end of the cell. Thus the origins of DNA replication are associated with DNA elements that can move the DNA and prepare the cell for separating the two daughter chromosomes during cell division, thereby remotely resembling centromeres in eukaryotic cells. Interestingly, pre-RC proteins in eukaryotes play a role at centromeres and centrosomes, perhaps reflecting an evolutionary link between chromosome replication and segregation.

As the two DNA bacterial forks move in opposite directions away from the origin, they eventually meet at a termination site that locates to the center of the cell before cell division. In rapidly growing bacteria, it is possible to reinitiate DNA replication from the origin before cell division actually takes place and this is made possible because of the spatial separation of the origin and termination regions of the genome and their placement relative to the plane of cell division. Such reinitiation is lethal in eukaryotic cells because it causes aneuploidy and genome instability.

In contrast to bacteria, eukaryotes have to deal with multiple chromosomes and numerous origins. The fastest way to replicate the multiple chromosomes in eukaryotes is for all origins to fire at the same time, but this is rare and occurs in the early and rapid cell divisions of Xenopus embryos and during the replication of the syncytial nuclei in the Drosophila embryo. In somatic cells origins of DNA replication fire at specific times during S phase, with some firing early and others late in a specific temporal pattern that is characteristic for each cell type. Within a chromosome, clusters of origins of DNA replication that are located adjacent to each other initiate DNA replication at the same time, creating megabase-pair-sized domains of chromosomes that are copied synchronously. Remarkably, whole genome analysis of these domains of chromosomes that are replicated at the same time has revealed that they correspond to regions of the genome that are spatially associated with each other in the nucleus. Thus, the genome within the nucleus is organized into domains that are physically associated with each other and these domains replicate at characteristic times during S phase. Such an arrangement explains why replication proteins such as the DNA polymerase clamp PCNA form temporally regulated patterns in the nucleus during S phase that correspond to sites of DNA synthesis of large regions of the genome.

Pre-RCs are assembled at all origins of DNA replication before the start of S phase and the temporal patterning of actual DNA synthesis from each origin is governed by the chromatin context, including the nature of histone modifications, and by rate-limiting proteins that establish the preinitiation complex. The activation of pre-RCs is regulated by at least two independent protein kinase signaling systems, the CDK and the DDK. Once the pre-RC is used or once passive DNA replication passes through a pre-RC, the pre-RC is destroyed and cannot be re-formed during the time in the cell cycle that CDKs are active. Thus pre-RCs cannot be established until the cells pass through the metaphase to anaphase transition when the cyclin moieties of CDKs are destroyed. Such a mechanism limits DNA replication from all origins to once per cell division cycle. So, unlike bacterial chromosome replication, it is clear that the complex genomes of eukaryotes have demanded more complex regulatory systems to maintain genome integrity.

What Needs to be Done?

Although the broad outline and many important details of DNA replication have been identified, many important aspects of this central process remain to be discovered. In large part, we still do not know how origins function. How do origin-binding proteins organize the DNA? Exactly how do helicase loaders function? How and when does the MCM helicase transition from encircling dsDNA to encircling ssDNA? The functions of several proteins required for origin activation and priming in eukaryotes are still shrouded in mystery. Priming and replisome assembly require numerous proteins that lack homologs in bacteria. What are the functions of Sld2, Sld3, Dbp11, Mcm10, GINS, and Cdc45 and how is their function influenced by phosphorylation? We lack an understanding of how the multiple origins in eukaryotes are coordinated and how the domain structure is established and maintained through multiple cell divisions. For example, just what are nuclear foci and how are replication foci organized within them? Are origins within one replicon clustered into one focus? Once replication forks are established, we know little about how they are regulated. If one replication fork in a focus were to stop, would it halt the other forks within that focus? How do replisomes move through nucleosomes, especially in highly condensed DNA and how are the parental nucleosomes inherited to the sister chromatids? How do replisomes deal with cohesin rings and how are these loaded? We have barely scratched the surface on questions surrounding the interface of replication with repair and recombination. For example, how can replication forks form during break-induced replication in S and G2 phase when the MCMs are thought to be loaded only in G1? How do checkpoint mechanisms act on moving replication forks? The newly revealed coordination of DNA metabolism with chromatin establishment, gene silencing, and epigenetic control is only beginning to be explored. Most of what we know about DNA replication has been learned in organisms with stable karyotypes and ploidy. However some organisms, particularly microbial eukaryotes, have extreme variations in ploidy and variable numbers of chromosomes. What mechanisms exist to facilitate this yet maintain order in this apparent chaos? Finally, and perhaps most important, some types of human disease, including certain cancers, have their basis in replication. Clearly many important questions remain, despite the enormous progress of recent years. We hold hope that understanding the mechanistic details of these processes may lead to cures, or at least treatment of human disease in the future.

Mutations

A mutation is a heritable change in the DNA sequence of an organism. The resulting organism, called a mutant, may have a recognizable change in phenotype compared to the wild type, which is the phenotype most commonly observed in nature. A change in the DNA sequence is conferred to mRNA through transcription, and may lead to an altered amino acid sequence in a protein on translation. Because proteins carry out the vast majority of cellular functions, a change in amino acid sequence in a protein may lead to an altered phenotype for the cell and organism.

Effects of Mutations on DNA Sequence

There are several types of mutations that are classified according to how the DNA molecule is altered. One type, called a point mutation, affects a single base and most commonly occurs when one base is substituted or replaced by another. Mutations also result from the addition of one or more bases, known as an insertion, or the removal of one or more bases, known as a deletion.

Effects of Mutations on Protein Structure and Function

Point mutations may have a wide range of effects on protein function. As a consequence of the degeneracy of the genetic code, a point mutation will commonly result in the same amino acid being incorporated into the resulting polypeptide despite the sequence change. This change would have no effect on the protein's structure, and is thus called a silent mutation. A missense mutation results in a different amino acid being incorporated into the resulting polypeptide. The effect of a missense mutation depends on how chemically different the new amino acid is from the wild-type amino acid. The location of the changed amino acid within the protein also is important. For example, if the changed amino acid is part of the enzyme's active site, then the effect of the missense mutation may be significant. Many missense mutations result in proteins that are still functional, at least to some degree. Sometimes the effects of missense mutations may be only apparent under certain environmental conditions; such missense mutations are called conditional mutations. Rarely, a missense mutation may be beneficial. Under the right environmental conditions, this type of mutation may give the organism that harbors it a selective advantage. Yet another type of point mutation, called a nonsense mutation, converts a codon encoding an amino acid (a sense codon) into a stop codon (a nonsense codon). Nonsense mutations result in the synthesis of proteins that are shorter than the wild type and typically not functional.

Deletions and insertions also cause various effects. Because codons are triplets of nucleotides, insertions or deletions in groups of three nucleotides may lead to the insertion or deletion of one or more amino acids and may not cause significant effects on the resulting protein's functionality. However, frameshift mutations, caused by insertions or deletions of a number of nucleotides that are not a multiple of three are extremely problematic because a shift in the reading frame results. Because ribosomes read the mRNA in triplet codons, frameshift mutations can change every amino acid after the point of the mutation. The new reading frame may also include a stop codon before the end of the coding sequence. Consequently, proteins made from genes containing frameshift mutations are nearly always nonfunctional.

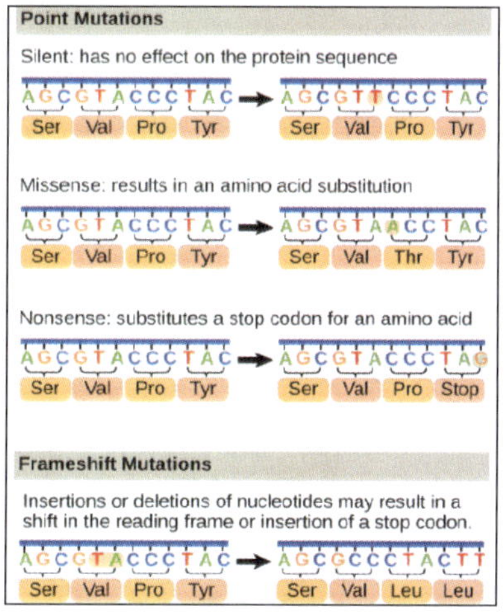

Mutations can lead to changes in the protein sequence encoded by the DNA.

Causes of Mutations

Mistakes in the process of DNA replication can cause spontaneous mutations to occur. The error rate of DNA polymerase is one incorrect base per billion base pairs replicated. Exposure to mutagens can cause induced mutations, which are various types of chemical agents or radiation. Exposure to a mutagen can increase the rate of mutation more than 1000-fold. Mutagens are often also carcinogens, agents that cause cancer. However, whereas nearly all carcinogens are mutagenic, not all mutagens are necessarily carcinogens.

A Summary of Mutagenic Agents			
Mutagenic Agents	**Mode of Action**	**Effect on DNA**	**Resulting Type of Mutation**
Nucleoside analogs			
2-aminopurine	Is inserted in place of A but base pairs with C	Converts AT to GC base pair	Point
5-bromouracil	Is inserted in place of T but base pairs with G	Converts AT to GC base pair	Point
Nucleotide-modifying agent			
Nitrous oxide	Deaminates C to U	Converts GC to AT base pair	Point
Intercalating agents			

Acridine orange, ethidium bromide, polycyclic aromatic hydrocarbons	Distorts double helix, creates unusual spacing between nucleotides	Introduces small deletions and insertions	Frameshift
Ionizing radiation			
X-rays, γ-rays	Forms hydroxyl radicals	Causes single- and double-strand DNA breaks	Repair mechanisms may introduce mutations
X-rays, γ-rays	Modifies bases (e.g., deaminating C to U)	Converts GC to AT base pair	Point
Nonionizing radiation			
Ultraviolet	Forms pyrimidine (usually thymine) dimers	Causes DNA replication errors	Frameshift or point

Chemical Mutagens

Various types of chemical mutagens interact directly with DNA either by acting as nucleoside analogs or by modifying nucleotide bases. Chemicals called nucleoside analogs are structurally similar to normal nucleotide bases and can be incorporated into DNA during replication. These base analogs induce mutations because they often have different base-pairing rules than the bases they replace. Other chemical mutagens can modify normal DNA bases, resulting in different base-pairing rules. For example, nitrous acid deaminates cytosine, converting it to uracil. Uracil then pairs with adenine in a subsequent round of replication, resulting in the conversion of a GC base pair to an AT base pair. Nitrous acid also deaminates adenine to hypoxanthine, which base pairs with cytosine instead of thymine, resulting in the conversion of a TA base pair to a CG base pair.

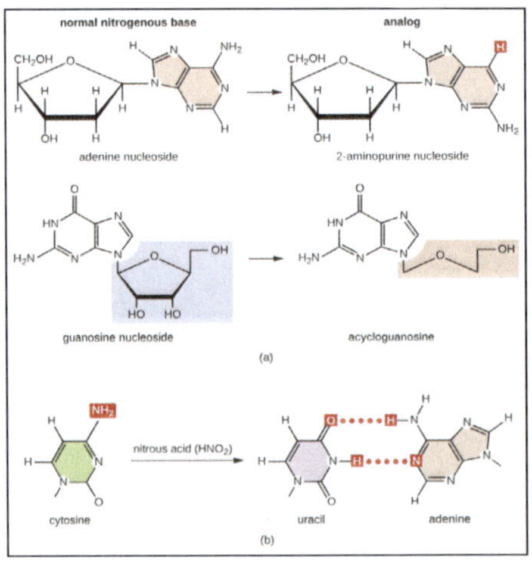

In figure, (a) 2-aminopurine nucleoside (2AP) structurally is a nucleoside analog to adenine nucleoside, whereas 5-bromouracil (5BU) is a nucleoside analog to thymine nucleoside. 2AP base pairs with C, converting an AT base pair to a GC base pair after several rounds of replication. 5BU pairs with G, converting an AT base pair to a GC base pair after several rounds of replication. (b) Nitrous acid is a different type of chemical mutagen that modifies already existing nucleoside bases like C to produce U, which base pairs with A. This chemical modification, as shown here, results in converting a CG base pair to a TA base pair.

Chemical mutagens known as intercalating agents work differently. These molecules slide between the stacked nitrogenous bases of the DNA double helix, distorting the molecule and creating atypical spacing between nucleotide base pairs. As a result, during DNA replication, DNA polymerase may either skip replicating several nucleotides (creating a deletion) or insert extra nucleotides (creating an insertion). Either outcome may lead to a frameshift mutation. Combustion products like polycyclic aromatic hydrocarbons are particularly dangerous intercalating agents that can lead to mutation-caused cancers. The intercalating agents ethidium bromide and acridine orange are commonly used in the laboratory to stain DNA for visualization and are potential mutagens.

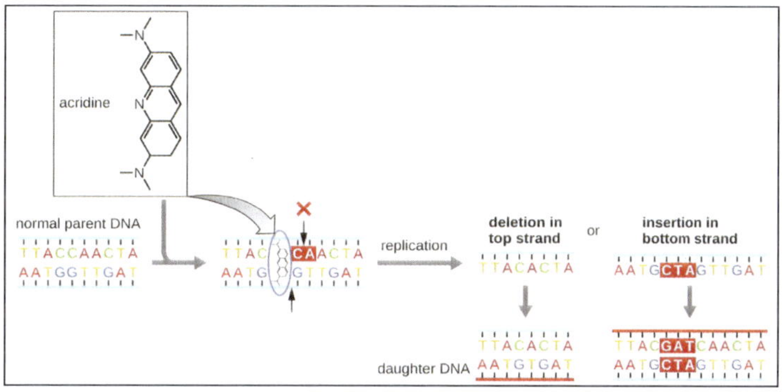

Intercalating agents, such as acridine, introduce atypical spacing between base pairs, resulting in DNA polymerase introducing either a deletion or an insertion, leading to a potential frameshift mutation.

Radiation

Exposure to either ionizing or nonionizing radiation can each induce mutations in DNA, although by different mechanisms. Strong ionizing radiation like X-rays and gamma rays can cause single- and double-stranded breaks in the DNA backbone through the formation of hydroxyl radicals on radiation exposure. Ionizing radiation can also modify bases; for example, the deamination of cytosine to uracil, analogous to the action of nitrous acid. Ionizing radiation exposure is used to kill microbes to sterilize medical devices and foods, because of its dramatic nonspecific effect in damaging DNA, proteins, and other cellular components.

Nonionizing radiation, like ultraviolet light, is not energetic enough to initiate these types of chemical changes. However, nonionizing radiation can induce dimer formation between two adjacent pyrimidine bases, commonly two thymines, within a nucleotide strand. During thymine dimer formation, the two adjacent thymines become covalently linked and, if left unrepaired, both DNA replication and transcription are stalled at this point. DNA polymerase may proceed and replicate the dimer incorrectly, potentially leading to frameshift or point mutations.

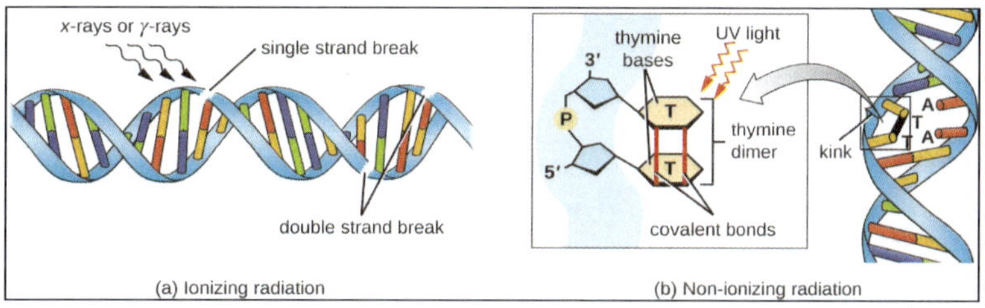

In figure, (a) Ionizing radiation may lead to the formation of single-stranded and double-stranded breaks in the sugar-phosphate backbone of DNA, as well as to the modification of bases (not shown). (b) Nonionizing radiation like ultraviolet light can lead to the formation of thymine dimers, which can stall replication and transcription and introduce frameshift or point mutations.

DNA Repair

The process of DNA replication is highly accurate, but mistakes can occur spontaneously or be induced by mutagens. Uncorrected mistakes can lead to serious consequences for the phenotype. Cells have developed several repair mechanisms to minimize the number of mutations that persist.

Proofreading

Most of the mistakes introduced during DNA replication are promptly corrected by most DNA polymerases through a function called proofreading. In proofreading, the DNA polymerase reads the newly added base, ensuring that it is complementary to the corresponding base in the template strand before adding the next one. If an incorrect base has been added, the enzyme makes a cut to release the wrong nucleotide and a new base is added.

Mismatch Repair

Some errors introduced during replication are corrected shortly after the replication machinery has moved. This mechanism is called mismatch repair. The enzymes involved in this mechanism recognize the incorrectly added nucleotide, excise it, and replace it with the correct base. One example is the methyl-directed mismatch repair in E. coli. The DNA is hemimethylated. This means that the parental strand is methylated while the newly synthesized daughter strand is not. It takes several minutes before the new strand is methylated. Proteins MutS, MutL, and MutH bind to the hemimethylated site where the incorrect nucleotide is found. MutH cuts the nonmethylated strand (the new strand). An exonuclease removes a portion of the strand (including the incorrect nucleotide). The gap formed is then filled in by DNA pol III and ligase.

Repair of Thymine Dimers

Because the production of thymine dimers is common (many organisms cannot avoid ultraviolet light), mechanisms have evolved to repair these lesions. In nucleotide excision repair (also called dark repair), enzymes remove the pyrimidine dimer and replace it with the correct nucleotides. In E. coli, the DNA is scanned by an enzyme complex. If a distortion in the double helix is found that was introduced by the pyrimidine dimer, the enzyme complex cuts the sugar-phosphate backbone several bases upstream and downstream of the dimer, and the segment of DNA between these two cuts is then enzymatically removed. DNA pol I replaces the missing nucleotides with the correct ones and DNA ligase seals the gap in the sugar-phosphate backbone.

The direct repair (also called light repair) of thymine dimers occurs through the process of photoreactivation in the presence of visible light. An enzyme called photolyase recognizes the distortion in the DNA helix caused by the thymine dimer and binds to the dimer. Then, in the presence of visible light, the photolyase enzyme changes conformation and breaks apart the thymine dimer, allowing the thymines to again correctly base pair with the adenines on the complementary strand. Photoreactivation appears to be present in all organisms, with the exception of placental mammals, including humans. Photoreactivation is particularly important for organisms chronically exposed to ultraviolet radiation, like plants, photosynthetic bacteria, algae, and corals, to prevent the accumulation of mutations caused by thymine dimer formation.

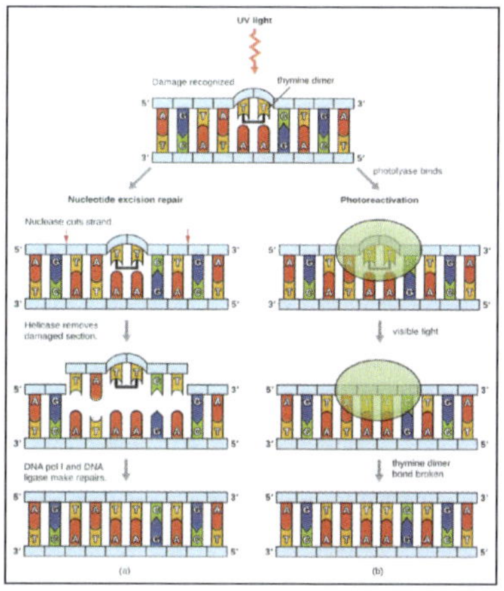

In figure, bacteria have two mechanisms for repairing thymine dimers. (a) In nucleotide excision repair, an enzyme complex recognizes the distortion in the DNA complex around the thymine dimer and cuts and removes the damaged DNA strand. The correct nucleotides are replaced by DNA pol I and the nucleotide strand is sealed by DNA ligase. (b) In photoreactivation, the enzyme photolyase binds to the thymine dimer and, in the presence of visible light, breaks apart the dimer, restoring the base pairing of the thymines with complementary adenines on the opposite DNA strand.

Identifying Bacterial Mutants

One common technique used to identify bacterial mutants is called replica plating. This technique is used to detect nutritional mutants, called auxotrophs, which have a mutation in a gene encoding an enzyme in the biosynthesis pathway of a specific nutrient, such as an amino acid. As a result, whereas wild-type cells retain the ability to grow normally on a medium lacking the specific nutrient, auxotrophs are unable to grow on such a medium. During replica plating, a population of bacterial cells is mutagenized and then plated as individual cells on a complex nutritionally complete plate and allowed to grow into colonies. Cells from these colonies are removed from this master plate, often using sterile velvet. This velvet, containing cells, is then pressed in the same orientation onto plates of various media. At least one plate should also be nutritionally complete to ensure that cells are being properly transferred between the plates. The other plates lack specific nutrients, allowing the researcher to discover various auxotrophic mutants unable to produce specific nutrients. Cells from the corresponding colony on the nutritionally complete plate can be used to recover the mutant for further study.

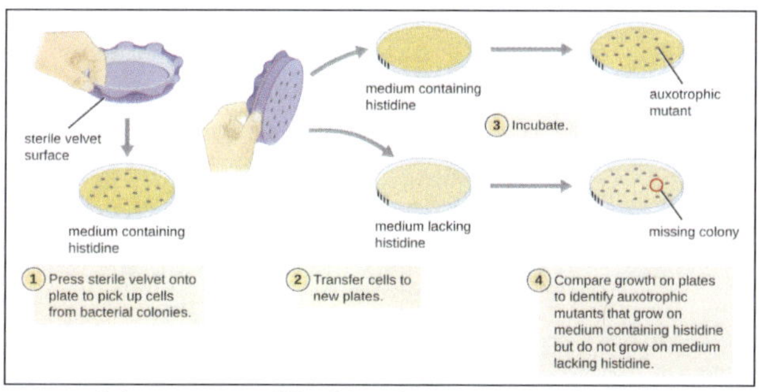

Identification of auxotrophic mutants, like histidine auxotrophs, is done using replica plating. After mutagenesis, colonies that grow on nutritionally complete medium but not on medium lacking histidine are identified as histidine auxotrophs.

The Ames Test

The Ames test, developed by Bruce Ames in the 1970s, is a method that uses bacteria for rapid, inexpensive screening of the carcinogenic potential of new chemical compounds. The test measures the mutation rate associated with exposure to the compound, which, if elevated, may indicate that exposure to this compound is associated with greater cancer risk. The Ames test uses as the test organism a strain of Salmonella typhimurium that is a histidine auxotroph, unable to synthesize its own histidine because of a mutation in an essential gene required for its synthesis. After exposure to a potential mutagen, these bacteria are plated onto a medium lacking histidine, and the number of mutants regaining the ability to synthesize histidine is recorded and compared with the number of such mutants that arise in the absence of the potential mutagen. Chemicals that are more mutagenic will bring about more mutants with restored histidine synthesis in the Ames test. Because many chemicals are not directly mutagenic but are metabolized to mutagenic forms by liver enzymes, rat liver extract is commonly included at the start of this experiment to mimic liver metabolism. After the Ames test is conducted, compounds identified as mutagenic are further tested for their potential carcinogenic properties by using other models, including animal models like mice and rats.

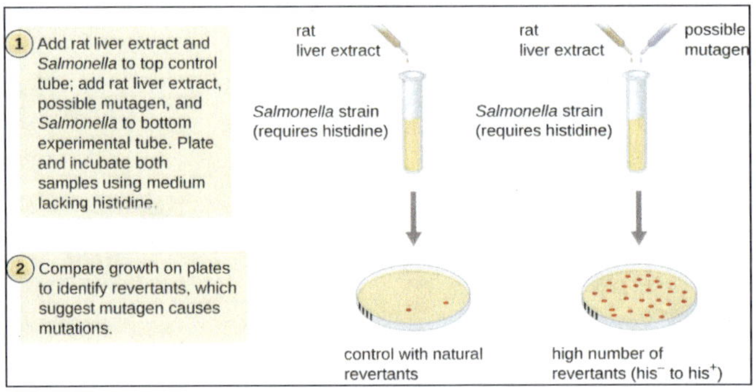

Above figure show the Ames test is used to identify mutagenic, potentially carcinogenic chemicals. A Salmonella histidine auxotroph is used as the test strain, exposed to a potential mutagen/carcinogen. The number of reversion mutants capable of growing in the absence of supplied histidine is counted and compared with the number of natural reversion mutants that arise in the absence of the potential mutagen.

Viral Replication

Viral replication is the term used indicate the formation of biological viruses during the infection process in the target host cells. Viruses must first penetrate and enter the cell before viral replication can occur. From the perspective of the virus, the purpose of viral replication is to allow reproduction and survival of its kind. By generating abundant copies of its genome and packaging these copies into viruses, the virus is able to continue infecting new hosts.

Replication between viruses is varied and depends on the type of genes involved. Most DNA viruses assemble in the nucleus; most RNA viruses develop solely in cytoplasm. Viral populations do not grow through cell division, because they are acellular. Instead, they hijack the machinery and metabolism of a host cell to produce multiple copies of themselves, and they assemble inside the cell.

The life cycle of viruses differs greatly between species but there are six common basic stages:

Attachment is a specific binding between viral capsid proteins and specific receptors on the host cellular surface. This specificity determines the host range of a virus. For example, HIV can infect only a limited range of human leukocytes. Its surface protein, gp120, specifically interacts only with the CD4 molecule – a chemokine receptor – which is most commonly found on the surface of CD4+ T-Cells. This mechanism has evolved to favor those viruses that infect only cells within which they are capable of replication. Attachment to the receptor can fore the viral envelope protein to undergo either changes that result in the fusion of viral and cellular membranes, or changes of non-enveloped virus surface proteins that allow the virus to enter.

Penetration follows attachment: Virions enter the host cell through receptor-mediated endocytosis or membrane fusion. This is often called viral entry. The infection of plant and fungal cells is different from that of animal cells. Plants have a rigid cell wall made of cellulose, and fungi one of chitin, so most viruses can get inside these cells only after trauma to the cell wall. However, nearly all plant viruses (such as tobacco mosaic virus) can also move directly from cell to cell, in the form of single-stranded nucleoprotein complexes, through pores called plasmodesmata. Bacteria, like plants, have strong cell walls that a virus must breach to infect the cell. However, since bacterial cell walls are much less thick than plant cell walls due to their much smaller size, some viruses have evolved mechanisms that inject their genome into the bacterial cell across the cell wall, while the viral capsid remains outside.

Uncoating is a process in which the viral capsid is removed: This may be by degradation by viral or host enzymes or by simple dissociation. In either case the end-result is the release of the viral genomic nucleic acid.

Replication of viruses depends on the multiplication of the genome. This is accomplished through synthesis of viral messenger RNA (mRNA) from "early" genes (with exceptions for positive sense RNA viruses), viral protein synthesis, possible assembly of viral proteins, then viral genome replication mediated by early or regulatory protein expression. This may be followed, for complex viruses with larger genomes, by one or more further rounds of mRNA synthesis: "late" gene expression is, in general, of structural or virion proteins.

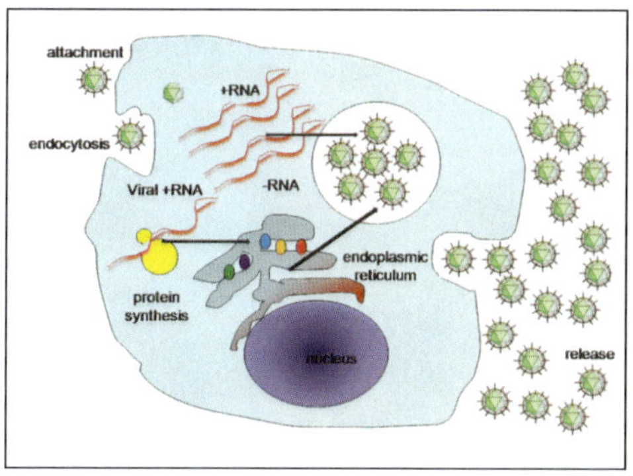

Hepatitis C virus: A simplified diagram of the Hepatitis C virus replication cycle.

Following the structure-mediated self-assembly of the virus particles, some modification of the proteins often occurs. In viruses such as HIV, this modification (sometimes called maturation) occurs after the virus has been released from the host cell.

Viruses can be released from the host cell by lysis, a process that kills the cell by bursting its membrane and cell wall if present. This is a feature of many bacterial and some animal viruses. Some viruses undergo a lysogenic cycle where the viral genome is incorporated by genetic recombination into a specific place in the host's chromosome. The viral genome is then known as a provirus or, in the case of bacteriophages a prophage. Whenever the host divides, the viral genome is also replicated. The viral genome is mostly silent within the host; however, at some point the provirus or prophage may give rise to active virus, which may lyse the host cells. Enveloped viruses (e.g., HIV) typically are released from the host cell by budding. During this process the virus acquires its envelope, which is a modified piece of the host's plasma or other internal membrane. The genetic material within virus particles, and the method by which the material is replicated, varies considerably between different types of viruses.

Steps of Virus Infections

Viral infection involves the incorporation of viral DNA into a host cell, replication of that material, and the release of the new viruses.

A virus must use cell processes to replicate. The viral replication cycle can produce dramatic biochemical and structural changes in the host cell, which may cause cell damage. These changes, called cytopathic (causing cell damage) effects, can change cell functions or even destroy the cell. Some infected cells, such as those infected by the common cold virus known as rhinovirus, die through lysis (bursting) or apoptosis (programmed cell death or "cell suicide"), releasing all progeny virions at once. The symptoms of viral diseases result from the immune response to the virus, which attempts to control and eliminate the virus from the body and from cell damage caused by the virus. Many animal viruses, such as HIV (Human Immunodeficiency Virus), leave the infected cells of the immune system by a process known as budding, where virions leave the cell individually. During the budding process, the cell does not undergo lysis and is not immediately killed. However, the damage to the cells that the virus infects may make it impossible for the cells to function normally, even though the cells remain alive for a period of time. Most productive viral infections follow similar steps in the virus replication cycle: attachment, penetration, uncoating, replication, assembly, and release.

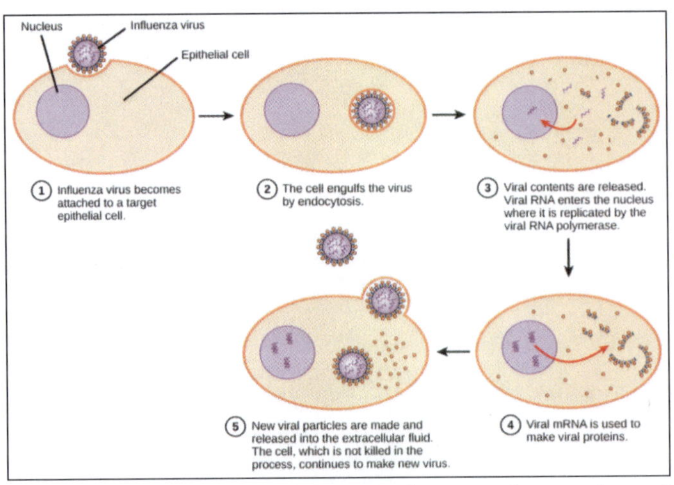

Pathway to viral infection: In influenza virus infection, glycoproteins attach to a host epithelial cell. As a result, the virus is engulfed. RNA and proteins are made and assembled into new virions.

Attachment

A virus attaches to a specific receptor site on the host cell membrane through attachment proteins in the capsid or via glycoproteins embedded in the viral envelope. The specificity of this interaction determines the host (and the cells within the host) that can be infected by a particular virus. This can be illustrated by thinking of several keys and several locks where each key will fit only one specific lock.

Entry

The nucleic acid of bacteriophages enters the host cell naked, leaving the capsid outside the cell. Plant and animal viruses can enter through endocytosis, in which the cell membrane surrounds and engulfs the entire virus. Some enveloped viruses enter the cell when the viral envelope fuses directly with the cell membrane. Once inside the cell, the viral capsid is degraded and the viral nucleic acid is released, which then becomes available for replication and transcription.

Replication and Assembly

The replication mechanism depends on the viral genome. DNA viruses usually use host cell proteins and enzymes to make additional DNA that is transcribed to messenger RNA (mRNA), which is then used to direct protein synthesis. RNA viruses usually use the RNA core as a template for synthesis of viral genomic RNA and mRNA. The viral mRNA directs the host cell to synthesize viral enzymes and capsid proteins, and to assemble new virions. Of course, there are exceptions to this pattern. If a host cell does not provide the enzymes necessary for viral replication, viral genes supply the information to direct synthesis of the missing proteins. Retroviruses, such as HIV, have an RNA genome that must be reverse transcribed into DNA, which then is incorporated into the host cell genome.

To convert RNA into DNA, retroviruses must contain genes that encode the virus-specific enzyme reverse transcriptase, which transcribes an RNA template to DNA. Reverse transcription never occurs in uninfected host cells; the needed enzyme, reverse transcriptase, is only derived from the expression of viral genes within the infected host cells. The fact that HIV produces some of its own enzymes not found in the host has allowed researchers to develop drugs that inhibit these enzymes. These drugs, including the reverse transcriptase inhibitor AZT, inhibit HIV replication by reducing the activity of the enzyme without affecting the host's metabolism. This approach has led to the development of a variety of drugs used to treat HIV and has been effective at reducing the number of infectious virions (copies of viral RNA) in the blood to non-detectable levels in many HIV-infected individuals.

Egress

The last stage of viral replication is the release of the new virions produced in the host organism. They are then able to infect adjacent cells and repeat the replication cycle. As you have learned, some viruses are released when the host cell dies, while other viruses can leave infected cells by budding through the membrane without directly killing the cell.

Tissue Tropism in Animal Viruses

Host tropism refers to the way in which viruses/pathogens determine which cells become infected by a given pathogen.

A tropism is a biological phenomenon, indicating growth or turning movement of a biological organism in response to an environmental stimulus. In tropisms, this response is dependent on the direction of the stimulus (as opposed to nastic movements which are non-directional responses). Viruses and other pathogens also affect what is called "host tropism" or "cell tropism. " Case tropism refers to the way in which different viruses/pathogens have evolved to preferentially target specific host species or specific cell types within those species.

Host tropism is the name given to a process of tropism that determines which cells can become infected by a given pathogen. Host tropism is determined by the biochemical receptor complexes on cell surfaces that are permissive or non-permissive to the docking or attachment of various viruses.

Various factors determine the ability of a pathogen to infect a particular cell. For example, viruses must bind to specific cell surface receptors to enter a cell. If a cell does not express these receptors then the virus cannot normally infect it. Viral tropism is determined by a combination of susceptibility and permissiveness: a host cell must be both permissive (allow viral entry) and susceptible (possess the receptor complement needed for viral entry) for a virus to establish infection. An example of this is the HIV virus, which exhibits tropism for CD4 related immune cells (e.g. T helper cells, macrophages or dendritic cells). These cells express a CD4 receptor, to which the HIV virus can bind, through the gp120 and gp41 proteins on its surface.

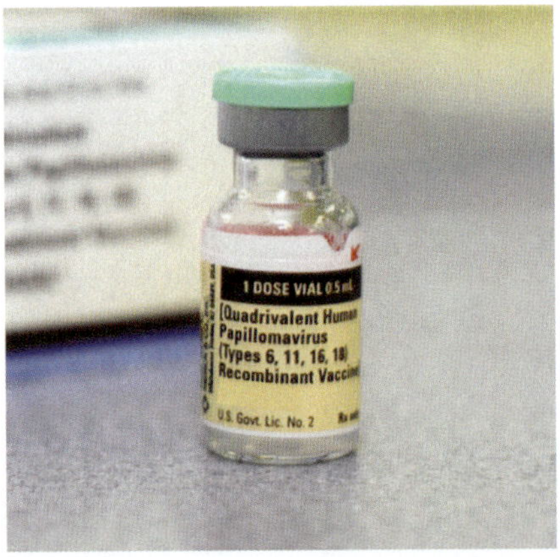

Human papillomavirus vaccine: Gardasil is a human papillomavirus vaccine on the market and it protects against HPV-16 and HPV-18 which cause 70% of cervical cancers, 80% of anal cancers, 60% of vaginal cancers, and 40% of vulvar cancers.

In virology, Tissue tropism is the cells and tissues of a host that support growth of a particular virus or bacteria. Some viruses have a broad tissue tropism and can infect many types of cells and tissues. Other viruses may infect primarily a single tissue.

Factors influencing viral tissue tropism include: 1) the presence of cellular receptors permitting viral entry, 2) availability of transcription factors involved in viral replication, 3) the molecular nature of the viral tropogen, and 4) the cellular receptors are the proteins found on a cell or viral surface.

These receptors are like keys allowing the viral cell to fuse with a cell or attach itself to a cell. The way that these proteins are acquired is through similar process to that of an infection cycle.

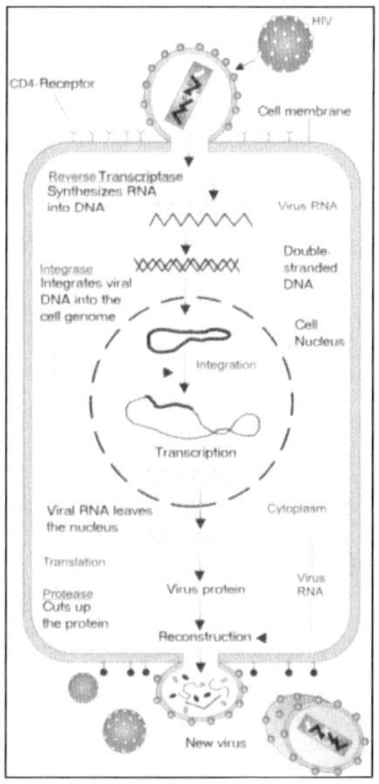

HIV life cycle: HIV has a gp120 which is precisely what the CD4 marker is on the surface of the macrophages and T cells. Therefore, HIV can enter T cells and macrophages.

Animal Viruses

Animal viruses have their genetic material copied by a host cell after which they are released into the environment to cause disease.

Animal viruses, unlike the viruses of plants and bacteria, do not have to penetrate a cell wall to gain access to the host cell. Non-enveloped or "naked" animal viruses may enter cells in two different ways. When a protein in the viral capsid binds to its receptor on the host cell, the virus may be taken inside the cell via a vesicle during the normal cell process of receptor-mediated endocytosis. An alternative method of cell penetration used by non-enveloped viruses is for capsid proteins to undergo shape changes after binding to the receptor, creating channels in the host cell membrane. The viral genome is then "injected" into the host cell through these channels in a manner analogous to that used by many bacteriophages. Enveloped viruses also have two ways of entering cells after binding to their receptors: receptor-mediated endocytosis and fusion. Many enveloped viruses enter the cell by receptor-mediated endocytosis in a fashion similar to some non-enveloped viruses. On the other hand, fusion only occurs with enveloped virions. These viruses, which include HIV among others, use special fusion proteins in their envelopes to cause the envelope to fuse with the plasma membrane of the cell, thus releasing the genome and capsid of the virus into the cell cytoplasm.

After making their proteins and copying their genomes, animal viruses complete the assembly of new virions and exit the cell. Using the example of HIV, enveloped animal viruses may bud from the cell membrane as they assemble themselves, taking a piece of the cell's plasma membrane in the process. On the other hand, non-enveloped viral progeny, such as rhinoviruses, accumulate in infected cells until there is a signal for lysis or apoptosis, and all virions are released together.

Animal viruses are associated with a variety of human diseases. Some of them follow the classic pattern of acute disease, where symptoms worsen for a short period followed by the elimination of the virus from the body by the immune system with eventual recovery from the infection. Examples of acute viral diseases are the common cold and influenza. Other viruses cause long-term chronic infections, such as the virus causing hepatitis C, whereas others, like herpes simplex virus, cause only intermittent symptoms. Still other viruses, such as human herpes viruses 6 and 7, which in some cases can cause the minor childhood disease roseola, often successfully cause productive infections without causing any symptoms at all in the host; these patients have an asymptomatic infection.

In hepatitis C infections, the virus grows and reproduces in liver cells, causing low levels of liver damage. The damage is so low that infected individuals are often unaware that they are infected, with many infections only detected by routine blood work on patients with risk factors such as intravenous drug use. Since many of the symptoms of viral diseases are caused by immune responses, a lack of symptoms is an indication of a weak immune response to the virus. This allows the virus to escape elimination by the immune system and persist in individuals for years, while continuing to produce low levels of progeny virions in what is known as a chronic viral disease. Chronic infection of the liver by this virus leads to a much greater chance of developing liver cancer, sometimes as much as 30 years after the initial infection.

As mentioned, herpes simplex virus can remain in a state of latency in nervous tissue for months, even years. As the virus "hides" in the tissue and makes few if any viral proteins, there is nothing for the immune response to act against; immunity to the virus slowly declines. Under certain conditions, including various types of physical and psychological stress, the latent herpes simplex virus may be reactivated and undergo a lytic replication cycle in the skin, causing the lesions associated with the disease. Once virions are produced in the skin and viral proteins are synthesized, the immune response is again stimulated and resolves the skin lesions in a few days by destroying viruses in the skin. As a result of this type of replicative cycle, appearances of cold sores and genital herpes outbreaks only occur intermittently, even though the viruses remain in the nervous tissue for life. Latent infections are common with other herpes viruses as well, including the varicella-zoster virus that causes chickenpox. After having a chickenpox infection in childhood, the varicella-zoster virus can remain latent for many years and reactivate in adults to cause the painful condition known as "shingles".

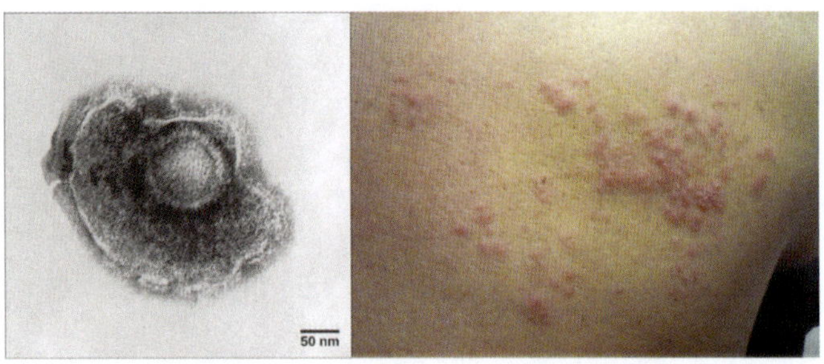

(a) (b)

Chicken pox virus: (a) Varicella-zoster, the virus that causes chickenpox, has an enveloped icosahedral capsid visible in this transmission electron micrograph. Its double-stranded DNA genome incorporates into the host DNA and reactivates after latency in the form of (b) shingles, often exhibiting a rash.

Plant Virus Life Cycles

Plant viruses are often spread from plant to plant by organisms known as vectors.

Plant viruses are viruses that affect plants. Like all other viruses, plant viruses are obligate intracellular parasites that do not have the molecular machinery to replicate without a host. Plant viruses are pathogenic to higher plants.

There are many types of plant virus, but often they only cause a loss of yield, and it is not economically viable to try to control them. Plant viruses are often spread from plant to plant by organisms (vectors). These are normally insects, but some fungi, nematode worms and single-celled organisms have been shown to be vectors. When control of plant virus infections is considered economical, (for perennial fruits for example), efforts are concentrated on killing the vectors and removing alternate hosts such as weeds. Plant viruses are harmless to humans and other animals because they can only reproduce in living plant cells.

Viral Life Cycle

For the virus to reproduce and thereby establish infection, it must enter cells of the host organism and use those cells' materials. To enter the cells, proteins on the surface of the virus interact with proteins of the cell. Attachment, or adsorption, occurs between the viral particle and the host cell membrane. A hole forms in the cell membrane, then the virus particle or its genetic contents are released into the host cell, where viral reproduction may commence. Next, a virus must take control of the host cell's replication mechanisms. At this stage, a distinction between susceptibility and permissibility of a host cell is made. Permissibility determines the outcome of the infection. After control is established and the environment is set for the virus to begin making copies of itself, replication occurs quickly by the millions. After a virus has made many copies of itself, it usually has exhausted the cell of its resources. The host cell is now no longer useful to the virus, therefore the cell often dies and the newly produced viruses must find a new host. The process by which virus progeny are released to find new hosts, is called shedding. This is the final stage in the viral life cycle. Some viruses can "hide" within a cell, either to evade the host cell defenses or immune system, or simply because it is not in the best interest of the virus to continually replicate. This hiding is deemed latency. During this time, the virus does not produce any progeny, it remains inactive until external stimuli—such as light or stress—prompts it to activate.

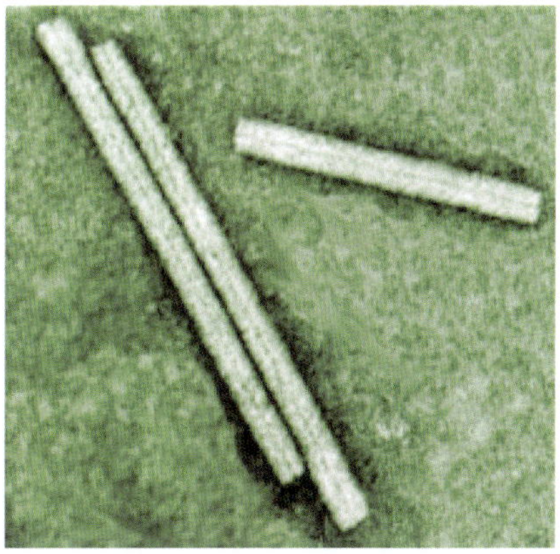

Viruses can be spread by direct transfer of sap, and by contact of a wounded plant with a healthy one. Such contact may occur during agricultural practices, when damage is caused by tools or hands, or naturally, when an animal feeds on the plant. Generally Tobacco mosaic virus (TMV), potato viruses, and cucumber mosaic viruses are transmitted via sap.

Tobacco mosaic virus and Cauliflower mosaic virus (CaMV) are frequently used in plant molecular biology. Of special interest is the CaMV 35S promoter, which is a very strong promoter most frequently used in plant transformations.

Tobacco Mosaic Virus: Tobacco mosaic virus (TMV) is a positive-sense single stranded RNA virus that infects plants, especially tobacco and other members of the family Solanaceae. The infection causes characteristic patterns (mottling and discoloration) on the leaves (hence the name). TMV was the first virus to be discovered. Although it was known from the late 19th century that an infectious disease was damaging tobacco crops, it was not until 1930 that the infectious agent was determined to be a virus.

Microbial Genomes

Microbial genomes are widely variable and reflect the enormous diversity of bacteria, archaea and lower eukaryotes. Bacterial genomes usually consist of a single circular chromosome, but species with more than one chromosome (eg. Deinococcus radiodurans), linear chromosomes (eg. some Bacillis subtilis strains) and combinations of linear and circular chromosomes (eg. Agrobacterium tumefaciens) also exist.

One major difference between the genomes of microorganisms and higher eukaryotes, is the presence of circular, extra-chromosomal DNA called plasmids. Plasmids can be transferred via horizontal DNA transfer from on cell of the same generation to another, mediating the rapid evolution of many different organisms.

The study of microbial genomes helps us to better understand the broader biology of bacteria, and how their genetic composition contributes to their tangible characteristics. The study of genomics is also important to infer the evolution of bacteria. Bacteria often evolve not just through small, single nucleotide level changes but through quantum evolutionary events. These include through the transfer of plasmids between species and also the transposition of large genetic elements within single cells. Understanding these processes allows us to determine the origins of bacteria and map the transfer of genes such as those conferring antibiotic resistance.

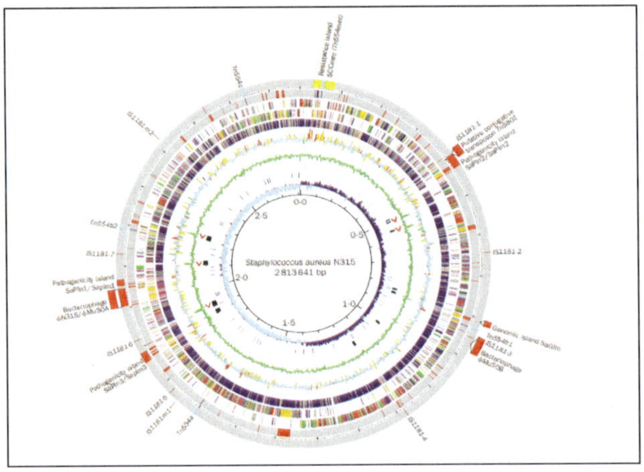

The figure above shows an annotated diagram of the methicillin Staphylococcus aureus strain N315 published by Kuroda et al in 2001. Genomics approaches allowed the authors to identify many different genes (particularly antibiotic resistance genes) that seem to have been acquired horizontally and which have driven the evolution of this superbug.

In the years preceding the development of full genome sequencing techniques we were restricted to studying microbial genomes in the lab using techniques such as PFGE.

Pulsed Field Gel Electrophoresis (PFGE)

The principle underlying PFGE is very similar to normal gel electrophoresis. PFGE however allows us to resolve far large, 'genome scale' pieces of DNA (greater than 20 kilobases in size). PFGE is still an important technique used to estimate the size of microbial genomes and in epidemiology studies.

DNA Sequencing

DNA sequencing has been an enormous advancement in the field of microbial genomics and indeed genetics as a whole, allowing us to amass vast amounts of genetic data form our organisms of choice. The first method for sequencing DNA was developed by Frederick Sanger and his group in 1977. Their method, termed Sanger Sequencing was a platform for innovation in the field of DNA sequencing, and we now have methods for sequencing entire bacterial genomes with relative ease. Whole genome sequencing produces immense amounts of data from which we can derive a catalogue of important information. From the need to analyse this data, the field of bioinformatics has flourished and become an integral part of genetics research.

Bioinformatics

Bioinformatics is the use of computing technologies to manage and analyse biological data. The huge amounts of genomic data produced through DNA sequencing can often be very confusing and difficult to analyse. Therefore bioinformatics is often required to derive useful information from this data.

Microbial Whole-Genome Sequencing

Microbial whole-genome sequencing is an important tool for mapping genomes of novel organisms, finishing genomes of known organisms, or comparing genomes across multiple samples. Sequencing the entire microbial genome is important for generating accurate reference genomes, for microbial identification, and other comparative genomic studies.

Unlike capillary sequencing or PCR-based approaches, next-generation sequencing (NGS) allows microbiology researchers to sequence hundreds of organisms with the power of multiplexing. Unlike traditional methods, NGS-based microbial genome sequencing doesn't rely on labor-intensive cloning steps, saving time and simplifying the workflow. NGS can identify low frequency variants and genome rearrangements that may be missed or are too expensive to identify using other methods.

De Novo Microbial Genome Sequencing

De novo whole-genome sequencing involves assembling a genome without the use of a genomic reference and is often used to sequence novel microbial genomes. Illumina sequencers provide unparalleled raw read accuracy, read length and read depth for high-quality draft and complete microbial genome assemblies.

References

- Microbial-genetics, anatomy-and-physiology, medicine: encyclopedia.com, Retrieved 14 June, 2019
- Protein-synthesis-translation: lumenlearning.com, Retrieved 30 March, 2019
- Mutations: lumenlearning.com, Retrieved 3 January, 2019
- Viral-replication, microbiology: lumenlearning.com, Retrieved 23 May, 2019
- Microbial-genomes, microbial-genetics: le.ac.uk, Retrieved 8 February, 2019
- Microbial-whole-genome-sequencing, microbiology: illumina.com, Retrieved 21 April, 2019

6

Microbial Biotechnology

Biotechnology is the field of biology that is concerned with living systems and organisms to develop and make products. Microbial biotechnology is a subset of biotechnology that makes use of microorganisms for industrial purposes. All the diverse principles of microbial biotechnology have been carefully analyzed in this chapter.

Microbial biotechnology is that aspect of biotechnology which involves the use of microorganisms or their products.

Microbial biotechnology is sometimes also referred to as industrial microbiology which is an old field that has been given new dimensions because of the discoveries made in the field of genetic engineering in vitro manipulation of DNA molecules to generate new combinations of genes or sequences, to place the gene under the control of different regulatory systems, to introduce a specific mutation in a molecule, etc.

Industrial microbiology came into existence initially by establishment of alcoholic fermentation processes to produce wine and beer. Thereafter, came the microbial production of antibiotics, and food additives such as amino acids, enzymes, butanol and citric acid.

Genetic engineering has enabled us to use microorganisms for production of new substances which the microorganisms could not have produced normally, such as the production of hormone insulin a pancreatic hormone which stimulates the transportation of glucose into cells. The production of insulin by bacterium has been possible due to genetic engineering techniques of inserting human insulin gene into bacterium.

Microbial Biotechnology can be Divided under two Subheadings:

1. Traditional microbial technology which is the large scale manufacture of products which are normally produced by microorganisms.
2. Microbial technology with genetically engineered microorganisms in which new genes have been inserted.

Industrial Microorganisms

Industrial microorganisms are those microorganisms which have been selected carefully to make one or more specific products. Industrial microorganisms are selected for their metabolic activities which are capable of specific products and give high yield of particular metabolites.

To achieve the desired objectives of high metabolic specialization the industrial strains are genetically modified by mutation or recombination using techniques of genetic engineering. The minor metabolic pathways are either brought down or eliminated. Though the industrial strains may grow well under highly specialized artificial conditions of the fermenter they may show poor growth characteristics in natural conditions of competitive environments.

Source of Industrial Strains

The initial and ultimate source of industrial strains has no doubt been nature but through the experience of years of large scale microbial processes perfection has been achieved for greater yield. The strains so developed have been deposited in culture collections. To get a new industrial process patented, the applicant is required to deposit a strain capable of carrying on the process, to a recognized culture collection.

Although most industrial companies would be reluctant to deposit their best cultures with any recognized culture collection, yet these collections serve as a ready source of cultures.

Strain Improvement

Microorganisms from original sources are highly modified in the laboratory. The modification is done to achieve a target of higher yield. One of the interesting examples of the progressive improvement is the antibiotic penicillin produced by the fungus Penicillium chrysogenum. The production of penicillin on industrial scale was for the first time 1 to 10 pg/ml.

However, the programmes of strain development accompanied by changes in medium and the growth conditions after hard work of many years, increased the yield of antibiotic penicillin to about 50,000 pg/ml. Thus, almost 50,000 times increase was possible by mutation and selection without involving genetic engineering techniques. The addition of new genetic engineering techniques has brightened the chances for yet greater yield.

Requisites for an Industrial Microorganism

Not all but only a few selected microorganisms are suitable for application in industrial production. An industrial microorganism must grow readily in large-scale manufacturing equipment. No doubt an industrial microorganism's first requisite is to manufacture the product of interest but it must also grow rapidly and on relatively cheap culture media.

An industrial microorganism should have the possibility of undergoing manipulations genetically for its strain improvement. An industrial microorganism should neither be pathogenic nor it should produce any toxic products which may be harmful to humans, animals or plants.

Microbes and Biotechnology

Microbes / micro-organisms are mostly micropsic small creatures are placed in different groups such bacteria, fungi, protozoa, micro-algae and viruses. These organisms live in soil, water, food, animal intestines and other different environments. Various microbial habitats reflect an enormous diversity of biochemical and metabolic traits that have arisen by genetic variation and natural selection in microbial populations.

Men used some of microbial diversity in the production of fermented foods such as bread, yogurt, and cheese. Some soil microbes release nitrogen that plants need for growth and emit gases that maintain the critical composition of the Earth's atmosphere.

Other microbes challenge the food supply by causing yield-reducing diseases in food-producing plants and animals. In our bodies, different microbes help to digest food, ward off invasive organisms, and engage in skirmishes and pitched battles with the human immune system in the give-and-take of the natural disease process.

A genome is the totality of genetic material in the DNA of a particular organism. Genomes differ greatly in size and sequence across different organisms. Obtaining the complete genome sequence of a microbe provides crucial information about its biology, but it is only the first step toward understanding a microbe's biological capabilities and modifying them, if needed, for agricultural purposes.

Microbial biotechnology, enabled by genome studies, will lead to breakthroughs such as improved vaccines and better disease-diagnostic tools, improved microbial agents for biological control of plant and animal pests, modifications of plant and animal pathogens for reduced virulence, development of new industrial catalysts and fermentation organisms, and development of new microbial agents for bioremediation of soil and water contaminated by agricultural runoff.

Microbial biotechnology is an important area that promotes for advances in food safety, food security, value-added products, human nutrition and functional foods, plant and animal protection, and overall fundamental research in the agricultural sciences.

Genetic Manipulation

Microbial biotechnology involves the exploitation, genetic manipulation and alterations of micro-organisms to make commercial valuable products and that also involves fermentationand various upstream and downstream processes.

Microorganisms produce an amazing array of valuable products such as macromolecules (e.g. proteins, nucleic acids, carbohydrate polymers, even cells) or smaller molecules and are usually divided into metabolites that are essential for vegetative growth (primary metabolites) and those which give advantages over adverse environment (secondary metabolites). They usually produce these compounds in small amounts that are needed for their own benefit.

Genetic Engineering of Microorganisms for Biotechnology

Molecular genetics can be used to manipulate genes in order to alter the expression and production of microbial products, including the expression of novel recombinant proteins.

The compounds that are isolated from plants or animals can be synthesized by genetic manipulation of different micro-organisms to enhance the production and by environmental and other manipulations, even up to 1000-fold for small metabolites can be increased.

The advent of recombinant DNA technology (also referred to as gene cloning or in vitro genetic manipulation) has dramatically broadened the spectrum of microbial genetic manipulations. With the advancement of recombinant DNA technology, many novel host systems have been explored to produce commercially important products like therapeutic proteins, antibiotics, small molecules, biosimilars etc.

The basis of this technology is the use of restriction endonucleases, polymerases and DNA ligases as a means to specifically cut and paste fragments of DNA. Similarly, foreign DNA fragments can be introduced into a vector molecule (a plasmid or a bacteriophage), which enables the DNA to replicate after introduction into a bacterial cell.

The ability to modify and clone genes accelerated the rate of discovery and the development in biotech industries.

The basic steps in DNA cloning involves the following:

- A fragment of DNA is inserted into a carrier DNA molecule, called a vector, to produce a recombinant DNA.
- The recombinant DNA is then introduced into a host cell, where it can multiply and produce numerous copies of itself within the host. The most commonly used host is the bacteria, although other hosts can also be used to propagate the recombinant DNA.
- Further amplification of the recombinant DNA is achieved when the host cell divides, carrying the recombinant DNA in their progenies, where further vector replication can occur.
- After a large number of divisions and replications, a colony or clone of identical host cell is produced, carrying one or more copies of the recombinant DNA.
- The colony carrying the recombinant DNA of interest is then identified, isolated, analyzed sub-cultured and maintained as a recombinant strain.

Expression of a Foreign Protein in a Microbe

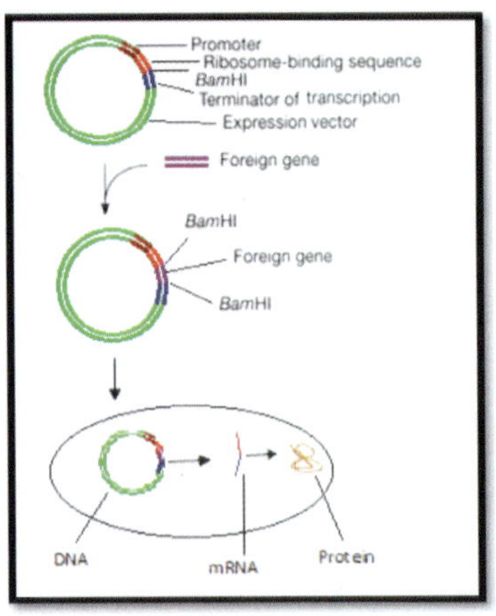

Foreign protein expression in a microbe

There are several methods by which genetic alterations of producer microbial strains can be done for maximization of products or metabolites.

Traditional Method of Strain Improvement

- The remarkable increases in antibiotic productivity and the resulting decreases in costs have resulted due to mutation and screening for higher producing microbial strains. In recent years, efforts have been devoted to miniaturize and automate the screening procedures and to enhance the frequency of improved strains by selection procedures, e.g., the isolation of anti-metabolite-resistant mutants in cases where the natural metabolite is a precursor, an inhibitor or a co-repressor of a biosynthetic pathway.
- Mutation has also served to shift the proportion of metabolites production in a fermentation broth to a more favorable distribution, to elucidate the pathways of secondary metabolism, and to yield new antibiotics.
- Targeted mutagenesis: It involves introduction of mutations at a specific location in DNA. As many antibiotics, growth hormones, regulatory factors production genes have now been cloned, targeted mutagenesis of the cloned DNA can be performed in vitro, followed by transformation of the recipient organism.
- The major problems of the classic strain improvement procedure based on random mutagenesis were the very low probability of introducing mutations into relevant genes and high rate of unwanted mutations in other, unrelated genes.

Classical Genetics

The most effective use of classical genetics in the past was the backcrossing of overproducing strains with parent strains to improve the vigor of mutant strains.

- After backcrossing such a strain with the wild type, progeny cells are produced that have inherited the overproducing traits from the mutant parent and the wild-type hardiness and vigor from the wild-type parent.
- Priorly this was not possible with Penicillium, which lacks a true sexual cycle. However, a parasexual cycle resulting in the production of heterokaryons was discovered in Penicillium in 1958 and was used to improve the strains.
- Protoplast fusion is relatively a new versatile technique to induce or promote genetic recombination in a variety of prokaryotic and eukaryotic cell especially, industrially useful microorganisms such as Streptomycete ambofaciens, Micromonsporae chinospora, because it breaks the barriers to genetic exchange. Another use of protoplast fusion has been the recombination of different strains from the same or different species to yield new antibiotics such as anthracyclines, aminoglycosides and rifamycins. Protoplast fusion has also been useful in elimination of an undesirable component from penicillin broths imposed by conventional mating systems.

Rational Selection

It involves selecting an improved producer out of a very large population of progeny. For example, some antibiotics, notably penicillin and tetracyclines, are chelators of heavy metal ions. The more of these antibiotics an organism produces, the more resistant it will be to heavy metals in the medium. Thus, selection for mutants resistant to heavy metals was used in the improvement of the penicillin producers.

Cloning of the Candidate Genes

Recombinant DNA technology can be used to introduce genes coding for antibiotic synthetases into producers of other antibiotics or into non-producing strains to obtain modified or hybrid antibiotics. The use of recombinant DNA technology in antibiotic improvement and discovery has been enhanced by the finding that some streptomycetes antibiotic biosynthetic pathways are coded by plasmid genes, eg. methylenomycin A. Even when the antibiotic biosynthetic pathway genes of streptomycetes are chromosomal, they appear to be clustered into operons which facilitate transfer of an entire pathway in a single manipulation.

The genes encoding individual enzymes of antibiotic biosynthesis which have already been cloned include those of the cephalosporin, clavulanic acid, prodigiosin, undecylprodigiosin, actinomycin, and candicidin pathways. The isopenicillin N synthetase ("CyClaSe") gene of Cephalosporium acremonium has been cloned in Escherichia coli and expressed at a level of 20% of total cell protein. Cyclase gene of Penicillium chrysogenum and Streptomyces clavuligerus has also been cloned in E.coli system. The expandase/ hydroxylase gene of C. acremonium has been cloned in E. coli. The protein accumulated as inclusion bodies in E. coli near to 15% of total cell protein.

Similar to E.coli, now a days Bacillus subtilis, Pichia pastoris, Saccharomyces cerevisiae have also emerged as a promising heterologous expression system for prokaryotic and eukaryotic candidate genes.

There are several factors which govern the production of recombinant therapeutic proteins in B. subtilis. The factors are as follows:

a) Well understood transcription and translation machinery including different regulatory factors responsible for extracellular product.

b) Generation of stable recombinant plasmids.

c) Development of novel recombinant B. subtilis strains with reduced nuclease and protease contents.

d) Better systematic understanding of the protein secretion method to elucidate the factors responsible for the secretion of intracellular proteins.

To get the soluble extracellular protein product from the gene, it can be linked to a DNA fragment coding for B. subtilis signal peptide for extra-cellular secretion. The signal sequence may be preceded by a efficient translation initiator sequence or ribosome binding site (RBS) which is followed by mRNA stability-enhancing sequences (SES) at 5' end of mRNA. The 'strong' promoter for this gene consist a cluster of other efficient promoters which would be regulated temporally by the growth condition or growth medium components. Thus genes possess suitable properties not only for efficient transcription, translation but would be under temporal control avoiding other exogenous induction. Thus generated mRNA would be stabilized by 3'SES which protects mRNA from degradation by 3' exonucleases. The protein product consists of a typical B. subtilis signal peptide linked to N-terminus and have to be further processed with proteases to recover only the desired portion.

Microbial Cell-Surface Display

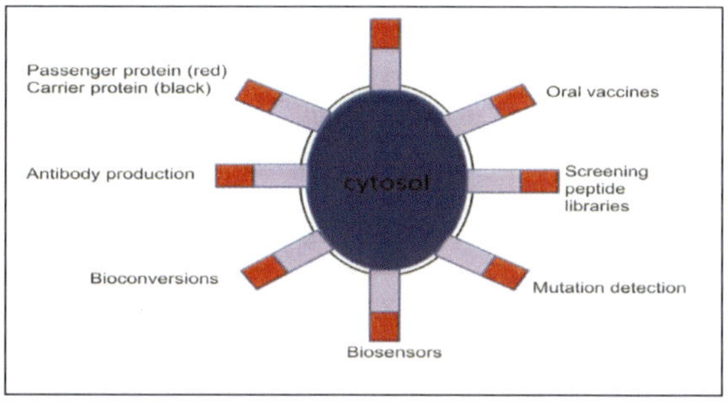

Different applications of various bio macro molecules generated in microbes

Cell-surface display allows proteins, peptides and other bio macro molecules to be displayed on the surface of microbial cells by fusing them with the anchoring motifs. The protein to be displayed -passenger protein - can be fused to an anchoring motif - the carrier protein - by N-terminal fusion, C-terminal fusion or sandwich fusion. The specific features of carrier protein, passenger protein and host cell, and various fusion methods affect the efficiency of surface display of bio-macromolecules. Microbial cell-surface display has many potential applications, including live vaccine development, peptide library screening, bioconversion using whole cell biocatalyst and bio-adsorption.

Potential Applications of Genetic Manipulation of Micro-organisms

One of the potential applications of microbial biotechnology is the production of pharmaceuticals, neutraceuticals by bacteria or other micro-organisms that produce economically, clinically important products like human insulin for diabetics or human growth hormone for dwarf individuals. Techniques are being perfected to transfer human genes into cows, sheep, and goats to obtain medically significant products from the milk of these animals.

Development of diagnostics is to detect disease-causing organisms and monitor the safety of food and water quality. Investigators are developing systems for identifying pathogens that may be used as biological weapons by rogue nations or even terrorist groups in future.

Bacteria can be genetically altered to emit a green fluorescent protein visible in ultraviolet light when they metabolize the explosive TNT leaking from land mines. Researchers envision a day when bacteria can be applied to a tract of land with a crop duster and then be analyzed from a helicopter. Genetically modified microorganisms can be used a living sensor to detect any particular chemicals in soil, air or other inorganic or biological specimens.

In Microbial Genome Program, alterations in the genome of the bacterium Deinococcus radiodurans are performed to increase its potential in cleaning up toxic-waste sites. The microbe's extraordinary DNA-repair processes enable it to thrive in high-radiation exposed environments.

Using various biotechnological processes, genes can be added from other organisms that will confer the ability to degrade toxinogenic chemicals such as toluene, commonly found in chemical and radiation waste sites.

Gene Manipulation in Bacteria

There is no meiosis in bacteria so special techniques have been worked out for manipulating genes in bacteria so that mapping experiments, strain construction, and complementation tests can be done.

First, we need a way of getting chromosomal DNA from one cell into another. There are several ways to do this. All of the methods have in common the use of special extra chromosomal elements for mobilizing chromosomal genes; the methods differ according to which extra chromosomal element is used.

We will consider a method that uses phage and is known as Transduction.

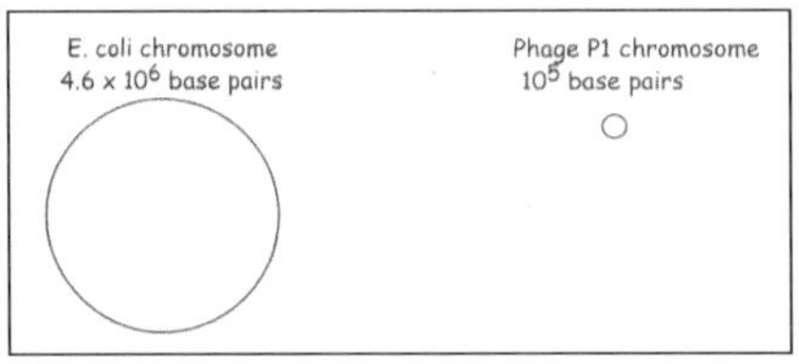

After infection of E. coli, the phage DNA is replicated by a mechanism known as a "rolling circle" and the phage is packaged into phage particles one headfull at a time.

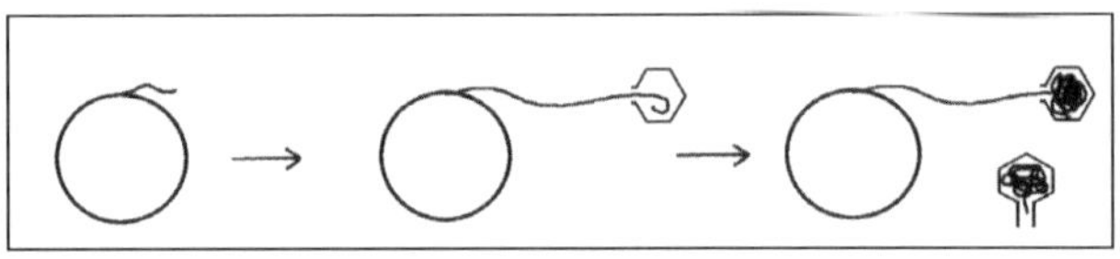

1/300 phage mistakenly packages E. coli chromosome DNA instead of phage DNA.

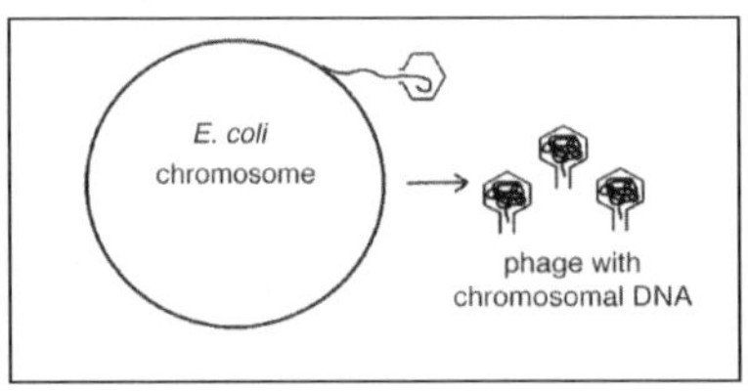

Each phage particle will package about 1/50 of the E. coli chromosome. By combining probabilities we see that about 1/15,000 phage will carry a particular E. coli gene.

A basic transduction experiment to measure the linkage between markers A and B is done as follows:

1. Grow P1 on A^+B^+
2. Infect A^-B^-
3. Select for A^+ and then screen for B^+

The idea is that we are looking for the rare cases where some chromosomal DNA carrying gene A is moved into the recipient. To find these recombinants, we select for A^+. Then we screen for B+ to see how often gene B comes along with gene A.

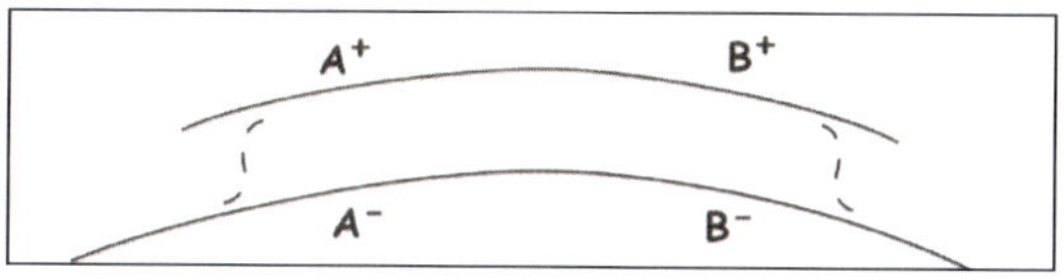

The measured frequency of cotransduction of B with A gives a measure of distance according to the following rules:

- If distance between A and B is greater than one headfull (10^5 bp) then there will be no cotransduction.
- If A and B are very close together then there will be 100% cotransduction.
- Cotransduction frequency is an inverse measure of distance.

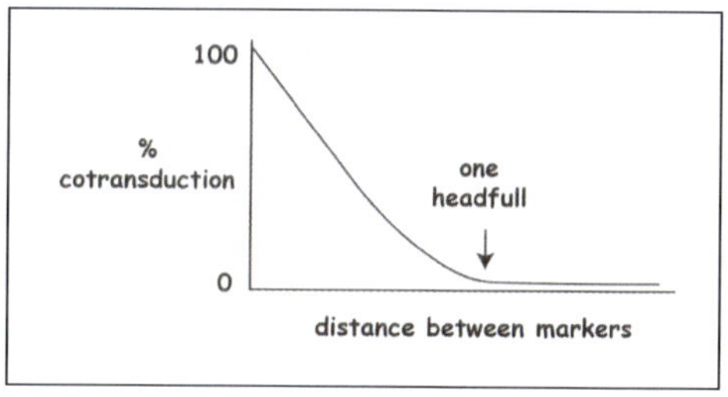

The experiment just described is the bacterial equivalent of a 2-factor cross and will give us relative distances between genes.

We can also do a 3-factor cross to determine gene order.

- Grow P1 on $A^+B^+C^+$
- Infect $A^-B^-C^-$
- Select for A^+ and then screen for B^+ and/or C^+

Genotypes

$A^+B^+C^+$	2 crossovers ← (A to C distance)
$A^+B^+C^-$	2 crossovers ← (A to B distance)
$A^+B^-C^-$	2 crossovers
$A^+B^-C^+$	4 crossovers (very rare)

There are only four possible genotypes because we select A^+.

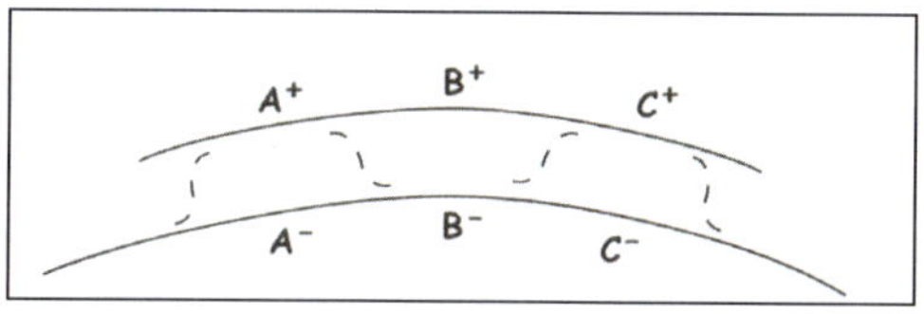

A limitation of transduction experiments is the need for a good selectable marker. Tn5 insertions provide a way to extend the utility of transduction for mapping and strain construction.

For example, let's say that we have isolated a new mutation in the MotA gene. MotA is a component of the bacterial flagellar motor and MotA⁻ mutants are nonmotile, a phenotype easily detected by the inability of MotA⁻ colonies to "swarm" outward on soft agar plates. Imagine that we want to map the MotA⁻ mutation or to move this mutation into an E. coli strain with a new genetic background. Clearly direct transduction of MotA⁻ would not be possible since we have no way to select for rare (1/15,000) transductants with the nonmotile MotA⁻ phenotype. One solution would be to use a nearby marker for which we can select to move MotA⁻ by its cotransduction with the selectable marker. Unfortunately, good selectable markers are not common and we are unlikely to have a good selectable marker placed within cotransduction distance of MotA⁻ readily available. A powerful alternative approach would be to isolate a random Tn5 insertion that is close to MotA⁻ and to use the Kanr trait conferred by Tn5 as the selectable marker for cotransduction.

The steps for finding a linked Tn5 insertion are as follows:

1. Start with a collection of random Tn5 insertions into wild type E. coli. Grow phage P1 on the mixture of 2×10^4 different Tn5 insertion mutants. Note that this donor strain is $MotA^+$.
2. Use the resulting P1 phage to infect a $MotA^-$ recipient strain. Select for transduction of the Tn5 insertions by selecting for growth of the transductants on kanamycin plates. Screen for cotransduction of $MotA^+$ by testing each of the Kan^r transductants for motility on soft agar. The desired cotransductant will be Kanr and will be motile. Given that one P1 phage headfull corresponds to about 1/50 of the E. coli chromosome, about 1 in 500 Tn5 insertions will be close enough to the MotA gene to show 90% cotransduction. Thus if we test about 10^3 Kan^r transductants for motility, we are likely to find at least one that has cotransduced the $MotA^+$ marker.
3. Once a Tn5 (Kan^r) MotA+ transductant has been identified, grow P1 on Tn5 (Kanr) $MotA^+$.
4. Use the P1 phage from step 3) to infect a $MotA^-$ recipient strain. Select for transduction of the Tn5 insertions by selecting for growth of the transductants on kanamycin plates.

Test the resulting Kan^r transductants for their motility. The transductants that have cotransduced the $MotA^+$ marker will be motile, whereas the transductants that still contain the $MotA^-$ allele will be nonmotile. The fraction of the total transductants that are motile will give the distance between $MotA^-$ and the Tn5 insertion as a cotransduction frequency.

A Tn5 (Kanr) $MotA^-$ transductant isolated in step 4) can then be used to transduce the $MotA^-$ marker into a new recipient strain by cotransduction with Tn5. Note that if we had isolated a second $MotA^-$ mutant, transduction into this strain would amount to a 3-factor cross and would provide a way to determine the order of the two different $MotA^-$ alleles.

Applications of Microbial Biotechnology

Chemical and Fuel Technological Microbiology

Obtaining chemicals such as organic acids via microbial activity is very promising, especially if it is thought to occur from renewable carbon sources. Most organic acids are natural products or intermediates of the microbial metabolism present in important metabolic pathways. Due to their functional groups, these acids, such as acetic, citric, lactic, and succinic acid, are extremely useful as raw materials for the chemical or food industry. Citric acid, for example, has been required on the market for use as a food additive, and all annual worldwide industrial-scale production occurs via the fermentation of glucose, beet molasses, cane molasses, or corn starch using A. niger. On the other hand, all the annual world production of lactic acid also comes from the fermentation performed by microorganisms. This acid and its derivatives are widely used in the food, pharmaceutical, leather, and textile industries. In addition, lactic acid fermentation processes have recently received more attention because of the growing demand for new biomaterials, such as biodegradable products and biocompatible polylactics. The most common method for obtaining this acid is using Lactobacillus spp. cultivated in whey. However, it may also be obtained via the activity of Rhizopus sp. under aerobic conditions in glucose-rich medium and with limited amounts of nitrogen and even via the fermentation of Saccharomyces cerevisiae in glucose- and cane juice-based medium In the future, the microbiological processes for obtaining a variety of organic acids are expected to will be competitive, become established in the market, and allow for an annual increase in the production of these compounds.

The microbial production of acetone and butanol, efficiently performed by the genus *Clostridium*, was one of the first large-scale industrial fermentation processes to gain global importance, but this production has been losing ground to chemical synthesis. Similarly, the centennial microbial synthesis of glycerol was impacted by the inability to compete with chemical synthesis from petrochemical feedstocks. However, in a scenario where the cost of propylene increased because its availability decreased, especially in developing countries, glycerol became an important raw material for the production of various chemicals, which made its alternative synthesis by fermentation more attractive.

Likewise, the production of 1,3-propanediol (1,3-PDO), which occurs through the fermentation of glycerol by bacteria of the genus *Clostridium* or *Enterobacteriaceae*, a technique described in 1881, lost status front to chemical synthesis through petroleum products, remaining forgotten for more than a century. In the last decade, however, research related to the synthesis of microbial 1,3-PDO expanded considerably because this diol began to be used in the synthesis of biodegradable polymers and for obtaining solvents, films, adhesives, antifreezes, and polyesters. Currently, a potentially viable alternative for the synthesis of 1,3-PDO is the use of genetically modified microorganisms. Genes from pathogenic bacteria, such as *Citrobacter freundii* and *Klebsiella pneumoniae*, were recently introduced in *E. coli* to allow the efficient synthesis of 1,3-PDO from waste glycerol.

This synthesis of chemicals through microbial metabolic processes meets an urgent need to reduce dependence on fossil fuels for energy generation. In modern biorefineries, renewable resources such as biomass or waste products are converted into substrates susceptible to microbial action and thus, interest in bio-based chemicals has recently been renewed because increasing climate change and environmental problems have pushed the industry, moving it away from fossil fuel consumption and toward renewable raw materials. Microorganisms have also been potentially explored for the production of a new generation of biofuels. The production of second-generation ethanol, for example, obtained from lignocellulosic biomass, already occurs in some countries, although improvements are still needed to make the technology economically competitive. Recent developments such as the discovery of functional xylose isomerases resulted in the creation of new yeasts capable of fermenting 5-carbon (C5) sugars, as well as 6-carbon (C6) sugars. Co-fermentation of C5 sugars with cane juice can produce up to 37% more ethanol in first-generation fermenters. Another problem to be overcome for the effective production of second-generation ethanol is the tolerance to acetic acid. This acid is one of the main inhibitors of lignocellulose hydrolysates. The polygenic basis of the high acetic acid tolerance present in some strains of *S. cerevisiae* is still unknown, but its identification may lead to greater efficacy in improving acetic acid tolerance in strains without negatively affecting other industrially important yeast properties. However, in addition to yeast genetic improvements, the prospection of new cellulose sources, such as forestry and crop residues (eucalyptus bark, corn, and rice husks), and the development of pretreatment techniques can leverage the production of second-generation ethanol.

In addition to bioethanol, other energy molecules such as biogas can be obtained from the microbial conversion of biomass. Biogas is a combination of methane, CO_2, nitrogen, H_2S, and traces of other gasses produced by anaerobic digestion (AD). Although AD processes have been carried out for several decades, knowledge about the microbial consortia involved in this process is limited due to the lack of phylogenetic and metabolic data on these predominantly unculturable microorganisms Studies carried out to isolate and identify the microbial community associated with the production of biogas revealed the presence of Proteobacteria, Chloroflexi, Firmicutes, Bacteroidetes, Actinobacteria, Bacteroides, Acidobacteria, and Spirochetes . Methanogenic Archaea, such as Methanosarcina barkeri, M. frisius, and Methanobacterium formicicum, were also identified in anaerobic digestions. However, the performance of AD for biogas production is dependent not only on the maintenance of a high density of these bioconversion microorganisms but also on the activity of several ion-specific transporters and enzyme systems not yet well-known, so that future production challenges comprise the knowledge of genes that control these systems with high efficiency.

The expectation is that, in the future, at least 25% of all bioenergy can originate from biogas and therefore, studies that seek to optimize the methanogenesis process or describe the structure of microbial communities have been encouraged. Metagenomic approaches associated with next-generation sequencing (NGS) techniques will help to unravel the diversity of natural communities and in biogas fermenters communities. However, studies have shown that most of the microorganisms isolated from the reactors are still unexplored and may be a source for new products and services in the future.

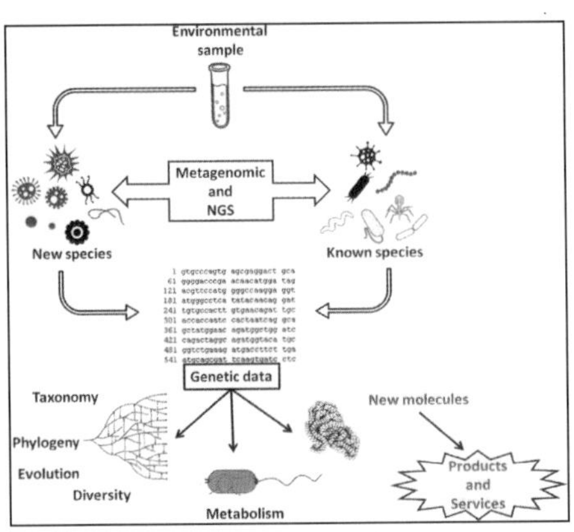

Use of metagenomics and next-generation sequencing in the study of microbial communities to obtain genetic data. These data can be used in studies of microbial diversity, group phylogeny, species diversification, and microbial metabolism. These data have also allowed the discovery of new or modified molecules used to obtain improved products, new products, or services.

Environmental Technological Microbiology

A large variety of microorganisms, including heterotrophic or autotrophic aerobic bacteria, actinomycetes, fecal coliforms, and thermophiles, as well as yeasts and other fungi, have been reported in solid waste composting processes. Many factors determine the microbial community present during composting, but under aerobic conditions, temperature is the main factor determining not only the microbial types but also the species diversity and metabolic rate. On the other hand, the direct use of microbial enzymes in the treatment of effluents, especially industrial effluents, has been encouraged because the enzymatic action is faster, dispensing with the conditions necessary for the fermentative process. Lipases, for example, are used in the treatment of wastewater containing mainly triglycerides. The presence of these enzymes in activated sludge and in other aerobic degradation processes is important for the continuous removal of the fat layers formed on the surface of aerated tanks to allow oxygen transport. Peroxidases, phenoloxidases, dioxygenases, and phenoloxidase-like compounds have also been used for the removal of contaminants present in wastewater. Peroxidases, polyphenol oxidases, and tyrosinases obtained from microorganisms such as P. syringae, Arthromyces ramosus, and Agaricus bisporus may be applied to the removal of phenols, biphenols, and chlorophenols. Laccases of P. cinnabarinus were found to be efficient for the degradation of benzopyrene, while manganese peroxidases of Phanerochaete chrysosporium, Nematoloma frowardii, and Phlebia radiata can be applied to the elimination of lignin in wastewater.

Currently, research efforts have focused on integrating the treatment of solid wastes or even wastewater with the use of microbial fuel cells (MFC), i.e., microbial cells that use electrons donated by low-value organic substrates, contained in the waste, to generate energy. This alternative technology can be carried out using mixed MFC cultures adaptable to a wide variety of substrates and offers the dual advantage of effluent treatment and electricity generation.

Research efforts have also been directed at improving the purification of drinking water. A recent biotechnological process called biologically active carbon (BAC) has been found to be very efficient in removing water contaminants. In this process, microbial cells colonize the surface of the granular activated carbon (GAC) used in the filtering mechanism. The biofilm formed is able to degrade significant amounts of dissolved organic matter and contaminants trapped in the GAC pores. In addition, the BAC biofilm can also biodegrade the cyanotoxins and organic substances that can change the taste and odor of potable water.

Waste treatment based on enzymatic processes tends to be less expensive; however, the enzymes are biodegradable, and further studies and prospection of microbial enzymes that are thermostable or resilient to large pH variations are needed. The use of enzymes in waste treatment has also been affected by the poor knowledge about the enzyme-producing species potentially applicable in the process, given that only approximately 2% of the world's microorganisms have been tested as enzyme sources. Genetic improvement, as well as the genetic manipulation of cells and the heterologous expression of genes, is expected to help increase the enzymatic biosynthesis in microorganisms of interest, or even to contribute to the development of microorganism biofactories for important enzymes not only for food or industry but also for environmental applications, thereby expanding alternatives for the elimination of the wastes that have historically accumulated in soils and watercourses.

The increasing market demand for biodegradable polymers has also stimulated the prospection of microorganisms that act in the synthesis of these compounds. The class of biodegradable biopolymers of major interest is the polyhydroxyalkanoates (PHAs), and the best known among these are poly(beta-hydroxybutyrate; PHB), poly(beta-hydroxyvalerate; PHV), and poly(hydroxybutyrate-co-valerate; PHB-V), the latter being commercially known as Biopol. These biopolymers are accumulated intracellularly by bacteria as a carbon and/or energy reserve under the limitation of a nutrient essential for their growth, such as nitrogen, phosphorus, sulfur, or oxygen. The species *Cupriavidus necator* is one of those responding more favorably to the conditions for industrial production. This bacterium can accumulate approximately 80% of its dry mass in polymer and uses different types of substrates, such as glucose, fructose, and crude glycerin. Biopolymers offer a possible solution for eliminating the problem of the residuals associated with petroleum-based plastic, but for researchers, the real challenge lies not in obtaining these molecules, which are the products of microbial metabolism, but in finding applications that consume large amounts of these materials, promoting price reductions and allowing biopolymers to compete economically in the market.

Another aspect of environmental technological microbiology is advances in the knowledge and use of the symbiotic relationship between plants and mycorrhizal fungi as a strategy to increase plant biomass or increase the yields of products of agricultural or pharmacological interest. Such benefits are the product of the positive and multifunctional roles of mycorrhizal fungi in plant nutrition, pathogen protection, stress tolerance, and soil structure supply. Due to their unique biological traits, which include obligatory biotrophy and intracellular development within plant tissues, and multiple ecological functions, arbuscular mycorrhizal fungi (AMFs; which are also thought to have helped plants conquer terrestrial environments) have aroused agronomic interests with the aim of sustainable production with low chemical consumption. In sustainable agriculture, AMFs are known as biofertilizers. In the plant-microorganism relationship, the mineral nutrients (mainly phosphorus, nitrogen and water) are extracted from the soil through the extensive network of hyphae and transferred to the plant, and organic compounds of carbon are transferred from the plant to the AMF. These microorganisms therefore reduce the need for application of chemical fertilizers in soils In the current context, "mycorrhizal technology" aims to increase the abundance and diversity of AMFs in soils in general to enhance the function of the mycorrhizal community and, consequently, crop production efficiency.

Medical Technological Microbiology

The participation of microorganisms in the generation of medical products or services involves four distinct aspects: (1) biocontrol of diseases, (2) production of vaccines, (3) production of antibiotics, and (4) production of biotherapeutics (hormones, biomaterials, and others). These aspects will be discussed throughout this session.

A problem commonly encountered in developing countries is the difficulty of implementing public policies to control the spread of parasitic vectors such as those of the genera Aedes and Anopheles. However, recent epidemic outbreaks of emerging and reemerging diseases have stimulated the development of biotechnological techniques that can not only assist diagnosis but also serve as alternatives for controlling transmission. An example is the potential presented by the introduction of the bacterium Wolbachia as an endosymbiont of the mosquito Aedes aegypti, which transmits diseases such as dengue, yellow fever, chikungunya, and the more recently detected Zika virus. The focus of this approach is on the reduction of mosquito longevity and not on abundance. The presence of the bacteria reduces the mosquito life span, thus decreasing the possibility of dengue virus transmission, since only adult females are able to transmit. Mosquitoes containing the Wolbachia wMelPop-CLA strain showed an approximately 50% reduction of the survival of females compared to mosquitoes without the strain. These bacteria are transmitted vertically from the female to the offspring. To guarantee this transmission, the bacterium manipulates its host in diverse ways such as feminization, death of males, parthenogenesis, and cytoplasmic incompatibility. In cytoplasmic incompatibility, fertilization of females not infected with Wolbachia by infected males results in embryonic mortality. By contrast, females infected with the bacterium will produce the highest number of viable offspring, increasing the number of infected individuals in the population. Cytoplasmic incompatibility facilitates the propagation of Wolbachia in natural populations and their persistence over time.

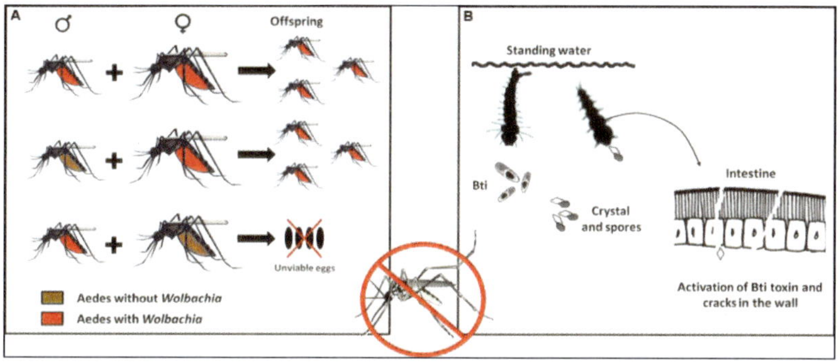

In figure, use of Technological Microbiology to prevent the proliferation of the Aedes aegypti vector and the DENV virus. (A) Mechanisms of transmission of the bacterium Wolbachia to the offspring of the vector. Cytoplasmic incompatibility causes females with Wolbachia to always breed offspring with Wolbachia, whether mating with males with or without the bacterium. When females without Wolbachia mate with males with Wolbachia, the fertilized eggs die. With successive generations, the number of male and female mosquitoes with the bacterium tends to increase until the entire mosquito population bears this characteristic. (B) Use of Bacillus thuringiensis israelensis bioinsecticide to fight dengue. This bacterium synthesizes protein crystals that, when consumed by Aedes larvae, solubilize in the mosquito's intestine and are transformed into efficient toxins that damage the intestinal wall, allowing the attack of pathogenic bacteria that cause the death of the larva.

Another alternative, also focused on preventing the proliferation of the dengue virus vector, is the use of a biological insecticide. The action of this insecticide is based on the activation of endotoxins (Cry and Cyt) produced naturally by the bacterium *B. thuringiensis* serotype *israelensis* (Bti). These toxins, previously mentioned in this text, are inactive crystals that, when ingested by *Aedes* larvae, are solubilized by intestinal proteases and converted into active toxins that interact with the cell membrane of the midgut, leading to pore formation, cell lysis, septicemia, and finally death. This alternative has been successfully applied in several regions of the world and many Bti-based commercial products have been introduced to the market, although some strains of *A. aegypti* have already shown resistance to Bt toxins.

At the same time, the technological race for the development of a vaccine against dengue continues. Recently, the first dengue vaccine, the recombinant yellow fever-17D-dengue virus, live, attenuated, and tetravalent (which induces antibodies against four DENV virus serotypes; CYD-TDV; Dengvaxia, Sanofi Pasteur, Singapore, Singapore), was licensed for use in individuals aged

9–45 years old in Mexico, Brazil, the Philippines, El Salvador, and Paraguay. This represents an important complement to the other techniques that focus mainly on vector control. This vaccine, however, is currently only available under private health care systems at very high prices.

The HPV (human papilloma virus) vaccine is another example of a newly developed vaccine that has been widely used in many countries. The focus of campaigns in developing countries has been on women who are not yet sexually active, although being sexually active does not contraindicate the vaccine, which can be tetravalent (protects against types 6, 11, 16, and 18) or bivalent (protects against types 16 and 18). This virus has been frequently linked to cases of cervical cancer, with types 16 and 18 accounting for approximately 70% of cases. However, these same viral types account for 86–95% of cases of non-cervical cancer, i.e., anal, oropharynx, vulval, and vaginal cancers in women, as well as anal, oropharyngeal, and penile cancers in men. Thus, prophylaxis should also extend to men as a way to prevent potential non-cervical cancers; however, affordable prices, funding mechanisms, and multidisciplinary partnerships are essential for the HPV vaccine to reach most populations in need, especially considering that cervical cancer is the second leading cause of death of cancer in women and is more worrying in populations that do not have screening programs to detect precursor lesions.

For the production of vaccines, microorganisms do not function as biofactories but are instead only used (whole or fractionated) to stimulate the synthesis of specific antibodies. Vaccines are classified according to the type of antigen they possess: (A) attenuated or live, (B) inactivated (subdivided into B1 – whole or fractionated, B2 – subunit vaccines, B3 – toxoids, B4 – carbohydrate vaccines, and B5 – conjugates), (C) DNA vaccines, and (D) recombinant vaccines. In attenuated vaccines, the pathogens (virus or bacteria) are alive and induce immune reactions similar to those resulting from a real infection. Attenuated vaccines have been efficiently developed for a range of diseases: mumps, polio (Sabin), rubella, measles, smallpox, chickenpox, tuberculosis, yellow fever, and dengue. These vaccines are considered highly immunogenic and efficiently stimulate humoral immunity, such that only one dose is capable of conferring immunity for decades. Recent studies have revealed strategies for the development of attenuated strains of the influenza virus that trigger robust immune responses. These studies not only provide perspectives for producing more effective vaccines against influenza but also suggest innovative approaches for the generation of live attenuated strains for viral pathogens, which remains a challenge. Studies have also shown that live vaccines induce non-specific immunity, conferring protection beyond the target pathogen and that revaccination may be indicated to reduce general mortality caused by all diseases.

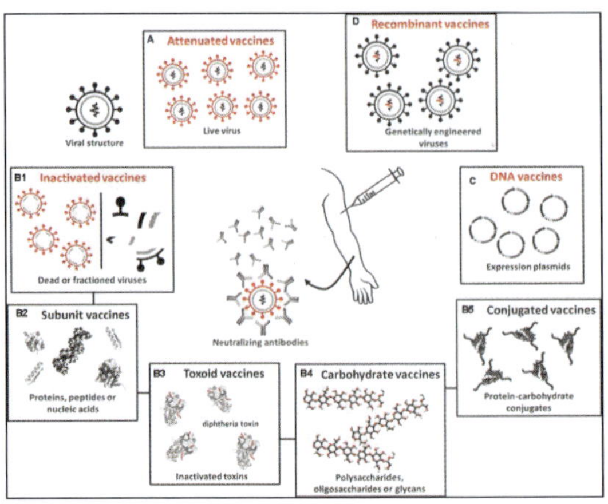

In figure, different biotechnological techniques used in the production of currently available vaccine types. (A) Attenuated or live vaccines, which use attenuated pathogens. (B) Inactivated vaccines containing completely inactivated or fractionated pathogens or only antigenic components of these pathogens, subdivided into (B1) whole or fractioned; (B2) subunit vaccines, which use proteins, peptides, or nucleic acids as antigens; (B3) toxoids, which use inactivated pathogen toxins as antigens; (B4) carbohydrate vaccines produced from polysaccharides, oligosaccharides, and glycans; and (B5) conjugate vaccines, which have polysaccharides combined with transport proteins. (C) DNA vaccines carrying plasmids containing genes encoding immunogenic antigens. (D) Recombinant vaccines containing viruses engineered to carry genes encoding antigens from other disease-causing viruses.

Inactivated vaccines, which are classified as whole or fractionated, contain completely inactivated or fractionated pathogens or only antigenic components of these pathogens (including subunit vaccines, toxoid vaccines, carbohydrate vaccines, and conjugate vaccines). In general, these vaccines are effective stimulators of humoral immune responses, but in many cases, multiple doses are required for long-term immunity since they do not stimulate the production of immunoglobulin A (IgA) or cytotoxic T cell responses because the virus does not replicate. The advantage of inactivated vaccines is that they are safer because the virulence of dead organisms cannot be reversed. Inactivated vaccines are currently available for hepatitis A, rabies, cholera, influenza, poliomyelitis (Salk), typhoid fever, and pertussis. Subunit vaccines use antigens, proteins, peptides or nucleic acids identified as immunogenic that can be rapidly manufactured in response to new outbreaks. These antigens are purified from microorganisms, produced by recombinant DNA techniques, or chemically synthesized. They are poorly reactogenic, which is an advantage in terms of adverse effects but a disadvantage in terms of stimulating potent and long-lasting immune responses. Thus, these vaccines often require the co-administration of efficient adjuvants to activate and modulate immune responses. Despite this poor reactogenicity, subunit vaccines have been developed against a variety of pathogens, including *Streptococcus pneumoniae*, hepatitis B virus, and HPV. By contrast, toxoid vaccines stimulate the immune response by using inactivated pathogen toxins as antigens. Vaccines of this type are available for tetanus, diphtheria, and anthrax. Carbohydrate vaccines are developed based on the knowledge that the vast majority of pathogens have dense distributions of polysaccharides, oligosaccharides, and complex glycans on their cell surface (glycocalyx) and that infected organisms detect the presence of pathogens through glycocalyx recognition and pattern recognition receptors (PRRs) that stimulate host defense responses. Low immunogenicity is a major obstacle to the manufacture of carbohydrate vaccines, and because of the lack of long-term immune protection by these antigens, few such vaccines are commercially available today. These include vaccines against *Salmonella typhi* and *Neisseria meningitides* . Traditional covalent linkages between carbohydrates and carrier proteins have been used to increase immune responses to polysaccharide antigens thereby obtaining conjugate vaccines. In this case, proteins function as carriers for capsular polysaccharides, which induce T-cell dependent antibody responses and B-cell differentiation in plasma cells and long-term memory. As a result, protein-polysaccharide vaccines induce immunological memory, a protection that is longer lasting than that induced by the use of simple polysaccharide antigens. Conjugate vaccines are already currently used in the control of *Haemophilus influenza* type B and *S. pneumoniae*.

DNA vaccines consist of an expression plasmid containing genes encoding one or more immunogenic antigens of interest. The use of viral promoters enhances gene expression and improves mRNA stability related to antigen synthesis. In addition, the incorporation of immunological adjuvants and new methods of insertion of this material into cells of the host organism beyond the intramuscular and dermal routes have increased the popularity as well as the immunogenicity of DNA vaccines, since the action of nucleases can inactivate plasmid DNA. The other insertion pathways with potential applications in humans include (1) bioballistics, which requires further technical enhancement for use in humans; (2) intradermal needleless administration of the DNA plasmid: in tests conducted with non-human primates, these intradermal needleless devices did not increase immunogenicity compared to conventional syringes; (3) intradermal tattoo, in which the DNA plasmid is delivered to the epidermal layer using thousands of injections; and (4) through the mucosa, which simulates the entry of pathogens by this route. One approach for this last route uses bacteria as carriers of a DNA plasmid to target specific cells and activate receptors of pathogen-associated molecular patterns. These vaccines are still under clinical testing, and microbiological research efforts have focused on HIV, hepatitis B, hepatitis C, influenza, and HPV.

Recombinant (gene) vaccines are prepared from viruses engineered to carry genes encoding antigens from other disease-causing viruses for expression in the host after inoculation. This expression induces antibody production and immunization. The immunity induced by recombinant vaccines is usually attributed to the ability of the recombinant virus to express the gene of interest at high levels within the host cells. The viral vectors used for this purpose are attenuated to the host and are therefore intrinsically safe. The viruses with the greatest potential for the production of this type of vaccine are those with an extensive genome, such as vaccinia virus In the manufacture of such vaccines, all genes that are not essential for replication arc first eliminated from the virus. Subsequently, the genes of the other virus are introduced. Recombinant vaccines are not yet available clinically but are quite promising.

Recently, the incorporation of biotechnological techniques has allowed wide access to numerous monoclonal antibodies (mAbs). Human virus-neutralizing MAbs have already been isolated from non-immune and immune sources using a range of newly developed antibody isolation technologies. One such technology employs microorganisms, such as phages, yeasts, bacteria, and viruses, to display repertoires of single-chain variable-domain antibody fragments (ScFvs), antigen-binding fragments (Fab), or domain antibodies (Dabs) on their surfaces. These antibodies can also be obtained directly from memory B cells of viral-infected patients or even from mouse lymphocyte cells. These antibodies have also been used for the treatment of infectious diseases. Recently, two antibodies were approved for this purpose: palivizumab, a human respiratory syncytial virus (RSV)-neutralizing monoclonal antibody that blocks virus replication; and raxibacumab, which prevents binding of the protective antigen of the anthrax toxin to its receptors in host cells. MAbs represent one of the largest classes of drugs in development, and between 2010 and 2014, 17 of the 54 protein drugs approved were mAbs (31.5%). These drugs therefore provide a new and promising way of thinking about the treatment of diseases caused by microorganisms.

In addition to increasing progress in the production of vaccines and MAbs, technological advances have also increased the availability of new drugs, such as antibiotics and hormones. The boom of antibiotic discovery occurred between 1950 and 1960. However, despite the need for new antibiotics, only two new classes of antibiotics have been introduced in medicine since 1963, both of which are based on nalidixic acid. Limited research results, inappropriate prescription of antibiotics, and misuse of antibiotics by the general population have threatened antibiotic potency and increased the occurrence of superbugs, i.e., microorganisms that appear at an alarming rate and are resistant to most or all clinical antibiotics in use.

In the United States, the Centers for Disease Control and Prevention estimate that antibiotic-resistant bacteria infect more than two million people annually. In this country alone, methicillin-resistant *Staphylococcus aureus* (MRSA) accounts for approximately 10,000 cases of hospital-acquired bloodstream infections, whereas *Clostridium difficile*, associated with diarrhea, is the most common infection in the United States, with more than 80,000 estimated annual cases. Many multi-antibiotic-resistant gram-negative bacilli also fit the description of superbugs, such as *P. aeruginosa, K. pneumoniae, E. coli, Acinetobacter baumannii*, and *Stenotrophomonas maltophilia*, and polymyxins have emerged as the major last-line of defense against these gram-negative superbugs. developed an important method that allowed the *in situ* growth of soil microorganisms not cultivable under laboratory conditions. Hence, chemicals produced naturally by the microorganisms could be tested, such as teixobactin, the first compound of a new important class of antibiotics. Teixobactin is capable of eliminating MRSA, and bacteria are believed not to be susceptible to developing resistance to teixobactin, which targets the lipids essential for the maintenance of the bacterial cell wall. Thus, biotechnological alternatives have been developed to circumvent the problem of the low rate of discovery of new antibiotic molecules. Selenium nanoparticles (SeNPs) combined with the synergistic properties of quercetin and acetylcholine showed inhibitory effect on MRSA. A study showed that SeNPs attach to the bacterial cell wall, causing irreversible damage to the membrane, thus achieving a remarkable synergistic antibacterial effect that inhibits MRSA.

The use of recombinant microbial cells has allowed large-scale production of a large number of products of pharmaceutical interest, such as hormones, anticoagulants, high-value proteins, antibodies or antigens, and others. This has been crucial in determining the structure-function relation of proteins, as well as for developing a better understanding of immune system reactions, cell biology, and signaling events. The major microorganisms explored as biofactories are the bacterium *E. coli*, followed by the yeast *S. cerevisiae*; both prokaryotic and eukaryotic systems are constantly evolving and competing to improve their properties and intensify as platforms of choice for the production of biopharmaceuticals In the early 1980s, the FDA approved the clinical use of human insulin, obtained by heterologous expression via *E. coli*, for the treatment of type I and type II diabetes and this was the first recombinant pharmaceutical product to be introduced into the market. Since then, the improvement of new heterologous protein production systems via *E. coli*enabled the commercial approval of several other products, including hormones (calcitonin, parathyroid hormone, human growth hormone, glucagon, and somatropin), interferons, and interleukins For example, approximately 30% of commercially available recombinant proteins are currently produced in prokaryotic systems. This production method is due to the unusual physiology of the cells as well as the ease of genetically manipulating them, but the understanding is that it is possible, by adding heterologous reactions, to synthesize 1,777 non-native products from *E. coli*, of which 279 have commercial applications. Among the latter are 4-hydroxybenzoate, tyrosine, and phenylalanine, which are precursors common to a large number of non-native commercial products.

By contrast, the products of heterologous proteins obtained from *S. cerevisiae* comprise hormones (insulin, insulin analogs, and glucagon), vaccines (hepatitis B virus surface antigen), and virus-like particles (VLPs) Prokaryotic production systems are required whenever the recombinant proteins are smaller or do not require post-translational modifications (PTMs), such as glycosylation, phosphorylation, or proteolytic cleavage. However, production in yeasts, such as *S. cerevisiae* and *Pichia pastoris*, is generally required when the target protein cannot be produced in a soluble form in a prokaryotic system, when it is rich in disulfide bonds, or when a specific PTM, essential for the biological activity of the protein, cannot be artificially created from the purified product.

Modified strains of *S. carnosus, Corynebacterium glutamicum, B. subtilis*, and *Lactococcus lactis*, for example, have been used in the controlled biological synthesis of calcitonin, amino acids (glutamate and lysine), proinsulin and protein nanoparticles respectively. Further manipulations of these species are expected to create strains capable of producing a wide variety of non-native commercial products.

Filamentous fungi, used for centuries in traditional Chinese medicine, have also been evaluated for the potential production of biopharmaceuticals. Polysaccharides of secondary metabolites can be obtained using *Ganoderma lucidum, Cordyceps sinensis*, and *C. militaris*. Endophytic fungi such as *Metarhizium anisopliae* and *C. gloeosporioides* were found to be efficient in the synthesis of Taxol, being a viable alternative to obtain this antineoplastic. The great diversity of molecules produced by filamentous fungi justifies the exploration of these microorganisms, and therefore, the development of production systems in bioreactors has been encouraged.

The metabolic engineering of molecules from microorganisms has been stimulated mainly by the need for new functional biomaterials in emerging drugs (nanostructured or not). A large number of substances with nanomedical application have emerged, including polymers, metallic nanoparticles, magnetic nanoparticles, VLPs, virions or virion components, and a growing diversity of self-organized protein materials, some with adjustable biomechanical properties such as stiffness, elasticity, adhesion or controllable disintegration or release of incorporated functional blocks or conventional chemical drugs. Obviously, the success of the use of microbial nanoparticles in nanotechnology and nanomedicine depends on the identification of new species of microorganisms and on the knowledge of the microbial interactions that occur in natural environments that can lead to the discovery of new molecules.

Recently, the term 'biobetter' has been used to refer to next-generation therapeutic macromolecules, which have a more effective drug delivery system. These macromolecules are modified by chemical and/or engineering methods using molecular biology techniques to display better pharmacological properties, such as higher activity, greater stability, fewer side effects, and lower immunogenicity. Because they require original research and development, and because they propose alternative methods of administration such as dermatological applications and inhaled formulations in order to minimize biological instability, biobetters still have significantly higher costs compared to reference biopharmaceutical versions. However, the future popularization of protein engineering techniques, especially site-directed mutagenesis (SDM), which allows the substitution, elimination or insertion of one or more amino acids in the sequence of a protein, is expected to enable the availability of less expensive biobetters, which are the main growing class of biopharmaceuticals.

Materials Technological Microbiology

The application of biotechnological techniques to microbiology has also made it possible to obtain a great diversity of biomaterials and biosensors. Biomaterials are artificial or natural products, usually synthesized by microorganisms in different environmental conditions, that can act in biological systems (tissues or organs). Biosensors integrate microorganisms with a physical transducer to generate a measurable signal proportional to the concentration of analytes, allowing rapid and accurate detection of analysis targets in diverse fields such as medicine, environmental monitoring, food processing, and others.

An important family of biomaterials includes the bioplastics. Bioplastics are polyesters that accumulate intracellularly in microorganisms in the form of storage granules, with physicochemical properties similar to petrochemical plastics. However, these properties, as well as the monomeric composition, can be altered according to the microbial origin of the bioplastic, and the main interest in these polymers lies in their biodegradability and biocompatibility. Bioplastic can also be produced as a byproduct of biorefinery using acidogenic fermentation or pyrolysis of lignocellulosic biomass, as well as a by-product of the biotreatment of solid or liquid wastes.

Bioplastics are being used in the manufacture of high-value-added medical materials, such as films that function as vehicles for drug delivery. For use as medical materials, the high purity of PHAs is enhanced by using bioceramics and/or bioactive glasses that improve their biomechanical properties and increase their bioactivity. Bioplastic formulations were recently tested for seed coating of agronomic species. These coatings, which contain spores of growth promoters such as *T. harzianum*, may help in the control of agricultural pests in the future. Bioplastics also have the potential to lead to the rise, within civil construction, of materials that have low incorporated energy, contributing to energy efficiency.

Polysaccharides of microbial origin, such as chitosan, alginate, xanthan gum, and cellulose, are another class of biomaterials that have gained considerable interest for medical use because of their properties, including that they are renewable, biodegradable, and mimic the components of the extracellular matrix, which make them key elements in biological processes. Chitosan can be easily recovered from the cell wall of fungi such as *A. niger, Rhizopus oryzae, Cunninghamella elegans*, and *Mucor indicus*, among others. Specific interactions with extracellular matrix components allow the use of chitosan in the field of tissue engineering to repair skin, bone, and cartilage. The potential of fungal chitosan, when present in bioactive filters, to chelate heavy metals and inhibit pathogenic microbial agents in contaminated water was recently evaluated for the development of a renewable, ecofriendly, and cost-effective polymer that can help overcome the current problems of chemical and microbial water pollution.

Alginate is a polysaccharide synthesized by several genera of brown algae and two genera of bacteria: *Pseudomonas* and *Azotobacter*. Its properties have broadened its use in the encapsulation or controlled release of drugs, enzymes, or cells, or as a matrix for tissue engineering, similarly to chitosan. When mixed in the aqueous phase, alginate and chitosan combine spontaneously by strong electrostatic attraction, forming a polyelectrolyte complex (PEC) that may be employed for the production of thin, transparent membranes that allow good absorption of physiological fluids, as well as the incorporation of several bioactive compounds. Xanthan gum is also a good alternative for combination with chitosan, forming a complex used in the immobilization of enzymes and in the production of microparticles and membranes. This exopolysaccharide is commercially synthesized by the bacterium *Xanthomonas campestris*, using different carbon sources.

Cellulose synthesized (in abundance) by bacteria such as *Gluconacetobacter xylinus* displays the same polymeric structure of cellulose from plants but is superior in its mechanical properties, purity, and uniformity, allowing the production of higher-quality devices. These devices include dialysis membranes and scaffolds for tissue engineering. Microbial cellulose has great potential for the treatment of skin lesions and replacement of small-diameter blood vessels.

The focus on microorganisms as an alternative in the production of biosensors is mainly due to the ability to produce them massively through cell culture. In addition, the recombinant DNA technique has facilitated the availability of microbial biosensors in the market, providing a new direction to manipulate their selectivity and sensitivity at the DNA level. This technique consists of the construction of recombinant microbial strains that contain a reporter gene (*lux*, GFP, or *lacZ*), i.e., a gene that generates a signal when the biological reaction between a microorganism and analyte occurs. An example of a microbial biosensor currently being used for pollutant monitoring purposes consists of immobilized recombinant *E. coli* cells expressing organophosphorus hydrolase (OPH). The OPH catalyzes the hydrolysis of organophosphorus pesticides releasing protons, whose concentration is proportional to the amount of substrate analyzed. Recently, *E. coli* cells were developed to function as CadC-T7 biosensors, which are based on synthetic genetic circuits that combine a fluorescence reporter gene and heavy metal-responsive proteins. These biosensors showed high specificity for the detection of heavy metals. For environmental monitoring of cadmium, a biosensor in which cells express β-galactosidase in the presence of this metal is also available. Bioluminescent *E. coli* have also been used to signal DNA damage, superoxide radical production, and membrane damage caused by potentially toxic liquids.

In addition to the use of *E. coli* in biosensors, other microorganisms have already been evaluated. *P. putida*has already been tested as a biosensor for catechol, nitrophenol, benzene, toluene, and others; *S. cerevisiae*, for Cu^{2++}; *Acidithiobacillus ferrooxidans* and *Leptospirillum ferrooxidans*, for Fe^{2+}, $S_2O_3^{2-}$, $Cr_2O_7^{2-}$, and others; and *Gluconobacter oxydans*, for propanediol and ethanol, among others. Despite the great leap forward made by biotechnology in the area of microbial biosensor development, many challenges still need to be overcome. New microorganisms still need to be evaluated for efficiency, more precise methods for immobilizing microbial cells still need to be developed, and the induction techniques need to be continuously evaluated because they may vary in terms of their efficiency depending on the analyte.

7

Applied Microbiology

Microbiology finds application in numerous different areas. A few of them are industrial microbiology, diagnostic microbiology, pharmaceutical microbiology and soil microbiology. This chapter closely examines these key applications of microbiology to provide an extensive understanding of the subject.

Applied microbiology is a scientific discipline that deals with the application of microorganisms for various purposes in science and engineering. Applications include biotechnology, agriculture, medicine, food microbiology and bioremediation.

Industrial Microbiology

Industrial microbiology includes the use of microorganisms to manufacture food or industrial products in large quantities. Numerous microorganisms are used within industrial microbiology; these include naturally occurring organisms, laboratory selected mutants, or even genetically modified organisms (GMOs). Currently, the debate in the use of genetically modified organisms (GMOs) in food sources is gaining both momentum, with more and more supporters on both sides. However, the use of microorganisms at an industrial level is deeply rooted into today's society. The following is a brief overview of the various microorganisms that have industrial uses, and of the roles they play.

Archaea are specific types of prokaryotic microbes that exhibit the ability to sustain populations in unusual and typically harsh environments. Those suriving in the most hostile and extreme settings are known as extremophile archaea. The isolation and identification of various types of Archaea, particularly the extremophile archaea, have allowed for analysis of their metabolic processes, which have then been manipulated and utilized for industrial purposes.

Extremophile archaea species are of particular interest due to the enzymes and molecules they produce that allow them to sustain life in extreme climates, including very high or low temperatures, extremely acid or base solutions, or when exposed to other harmful factors, including radiation. Specific enzymes which have been isolated and used for industrial purposes include thermostable DNA polymerases from the Pyrococcus furiosus. This type of polymerase isa common tool in molecular biology; it is capable of withstanding the high temperatures that are necessary to complete polymerase chain reactions. Additional enzymes isolated from Pyrococcus speciesinclude specific types of amylases and galactosidases which allow food processing to occur at high tempertatrues as well.

Corynebacteria are characterized by their diverse origins. They are found in numerous ecological niches and are most often used in industry for the mass production of amino acids and nutritional factors. In particular, the amino acids produced by Corynebacterium glutamicum include the amino acid glutamic acid. Glutamic acid is used as a common additive in food production, where it is known as monosodium glutamate (MSG). Corynebacterium can also be used in steroid conversion and in the degradation of hydrocarbons. Steroid conversion is an important process in the development of pharmaceuticals. Degradation of hydrocarbons is key in the breakdown and elimination of environmental toxins. Items such as plastics and oils are hydrocarbons; the use of microorganisms which exhibit the ability to breakdown these compounds is critical for environmental protection.

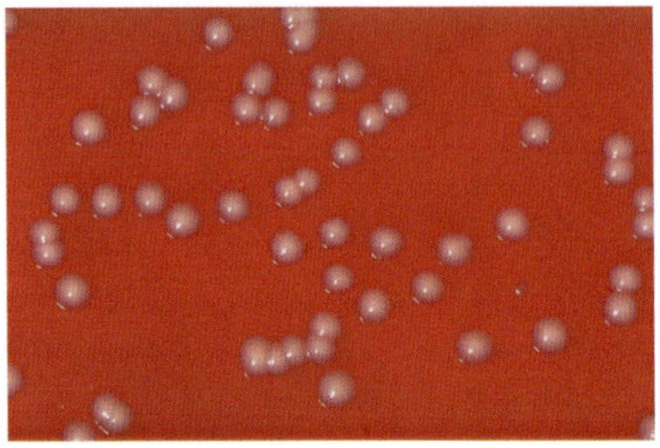

Corynebacterium: Corynebacterium species are often used to mass produce amino acids utilized in food processing.

Xanthomonas, a type of Proteobacteria, is known for its ability to cause disease in plants. The bacterial species which are classified under Xanthomonas exhibit the ability to produce the acidic exopolysaccharide commonly marketed as xanthan gum, used as a thickening and stabilizing agent in foods and in cosmetic ingredients to prevent separation.

Another type of microorganism utilized by industry includes various species of Aspergillus. Thisgenusincludes several hundred types of mold. Aspergillus has become a key component in industrial microbiology, where it is used in the production of alcoholic beverages and pharmaceutical development. Aspergillus niger is most commonly used to produce citric acid, which is used in numerous products ranging from household cleaners, pharmaceuticals, foods, cosmetics, photography and construction. Aspergillus is also commonly used in large-scale fermentation in the production of alcoholic beverages such as Japanese sake.

Molecular Products from Microbes

The isolation of molecular products from microbes is considered to be a key component of molecular biology research.

The expansion and growing popularity of the field of molecular biology has resulted in a higher demand for tools used to study molecular biology. The field of molecular biology specifically deals with the molecular mechanisms of a cell and focuses on the regulation of cellular interactions. Topics of particular interest within the field include gene expression (transcription and translation) and protein synthesis. Studying these mechanisms in the laboratory has been made possible by the use of molecules derived from microbes. The following is a brief overview of some of the molecular products derived from microbes that allow for the performance of popular molecular biology techniques.

Taq Polymerase

Taq polymerase is an enzyme that was first isolated from the microbe Thermus aquaticus. T.

aquaticus is a specific type of bacterial species, a DNA polymerase, that is thermostable – it can withstand extremely high temperatures. The isolation of this polymerase has resulted in the ability to perform polymerase chain reactions (PCR), a process used to amplify DNA segments, in a single step. Prior to the isolation of Taq polymerase, a new DNA polymerase had to be added to the reaction after every cycle because of thermal denaturation. With the addition of Taq polymerase to the reaction tube, the cycle can be performed much more quickly, and less enzyme needs to be used. Currently, Taq polymerase is manufactured and produced on a large scale and is available for commercial sale.

Restriction Enzymes

Restriction enzymes are a specific class of enzymes isolated from various bacteria and archaea, in which they grow naturally as a means of protection against viral infection. These enzymes have the ability to cut DNA at specific recognition sequences and have served as invaluable tools in DNA modification and manipulation. The enzymes have the ability to recognize foreign DNA and cut it up. The bacteria and archaea from which these enzymes are isolated from have innate mechanisms to protect their own DNA sequences from these enzymes, such as methylation. The isolation of approximately 3000 restriction enzymes has allowed molecular biologists to utilize them in processes such as cloning and the production of recombinant DNA.

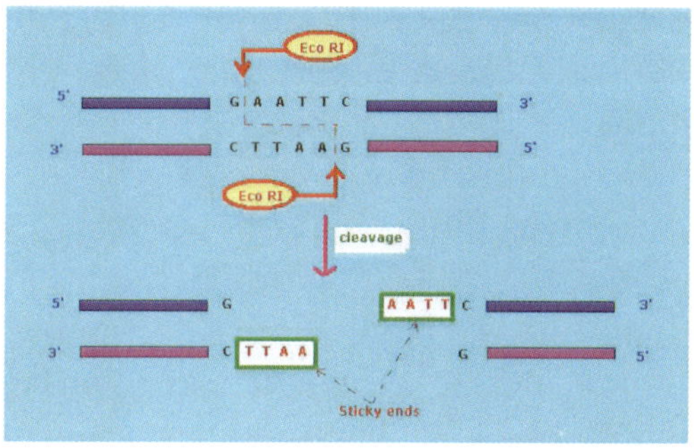

EcoRI Restriction Enzyme: An example of a specific restriction enzyme, EcoRI, which exhibits the ability to target specific sequences within DNA.

DNA Ligase

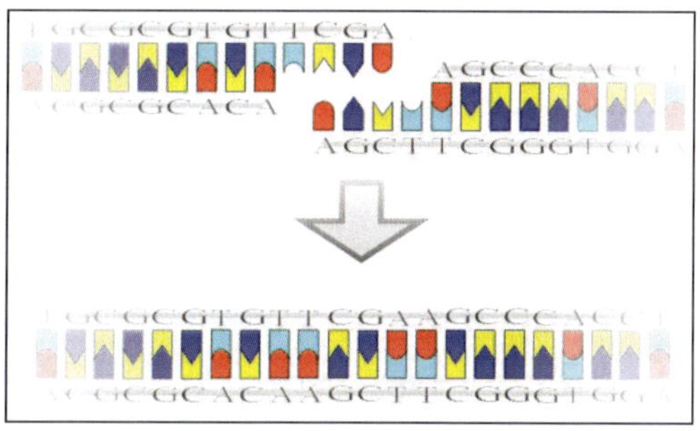

Diagram of a DNA ligation.

Another enzyme that was isolated from T. aquaticus and that has been undeniably important to the field of molecular biology is DNA ligase. DNA ligase plays a key role in molecular biology processes due to its ability to insert DNA fragments into plasmids. The process of DNA ligation is defined as the ability of DNA ligase to covalently link, or ligate, fragments of DNA together. In molecular biology — specifically, during the process of developing recombinant DNA — DNA ligase can be used to ligate a fragment of DNA into a plasmid vector. The most commonly used DNA ligase is derived from the T4 bacteriophage and is referred to as T4 DNA ligase.

Primary and Secondary Metabolites

Primary and secondary metabolites are often used in industrial microbiology for the production of food, amino acids, and antibiotics.

Bacterial metabolism can be classified into three major categories: the kind of energy used for growth, the carbon source, and the electron donors used for growth. Pathogenic bacteria are capable of exhibiting various types of metabolism. Metabolites, the intermediates and products of metabolism, are typically characterized by small molecules with various functions. Metabolites can be categorized into both primary and secondary metabolites. These metabolites can be used in industrial microbiology to obtain amino acids, develop vaccines and antibiotics, and isolate chemicals necessary for organic synthesis.

Primary Metabolites

Primary metabolites are involved in growth, development, and reproduction of the organism. The primary metabolite is typically a key component in maintaining normal physiological processes; thus, it is often referred to as a central metabolite. Primary metabolites are typically formed during the growth phase as a result of energy metabolism, and are deemed essential for proper growth. Examples of primary metabolites include alcohols such as ethanol, lactic acid, and certain amino acids. Within the field of industrial microbiology, alcohol is one of the most common primary metabolites used for large-scale production. Specifically, alcohol is used for processes involving fermentation which produce products like beer and wine. Additionally, primary metabolites such as amino acids– including L-glutamate and L-lysine, which are commonly used as supplements– are isolated via the mass production of a specific bacterial species, Corynebacteria glutamicum. Another example of a primary metabolite commonly used in industrial microbiology includes citric acid. Citric acid, produced by Aspergillus niger, is one of the most widely used ingredients in food production. It is commonly used in pharmaceutical and cosmetic industries as well.

Secondary Metabolites

Secondary metabolites are typically organic compounds produced through the modification of primary metabolite synthases. Secondary metabolites do not play a role in growth, development, and reproduction like primary metabolites do, and are typically formed during the end or near the stationary phase of growth. Many of the identified secondary metabolites have a role in ecological function, including defense mechanism(s), by serving as antibiotics and by producing pigments. Examples of secondary metabolites with importance in industrial microbiology include atropine and antibiotics such as erythromycin and bacitracin. Atropine, derived from various plants, is a secondary metabolite with important use in the clinic. Atropine is a competitive antagonist for acetycholine receptors, specifically those of the muscarinic type, which can be used in the treatment of bradycardia. Antibiotics such as erythromcyin and bacitracin are also considered to be secondary metabolites. Erythromycin, derived from Saccharopolyspora erythraea, is a commonly used antibiotic with a wide antimicrobial spectrum. It is mass produced and commonly administered orally.

Lastly, another example of an antibiotic which is classified as a secondary metabolite is bacitracin. Bacitracin, derived from organisms classified under Bacillus subtilis, is an antibiotic commonly used a topical drug. Bacitracin is synthesized in nature as a nonribosomal peptide synthetase that can synthesize peptides; however, it is used in the clinic as an antibiotic.

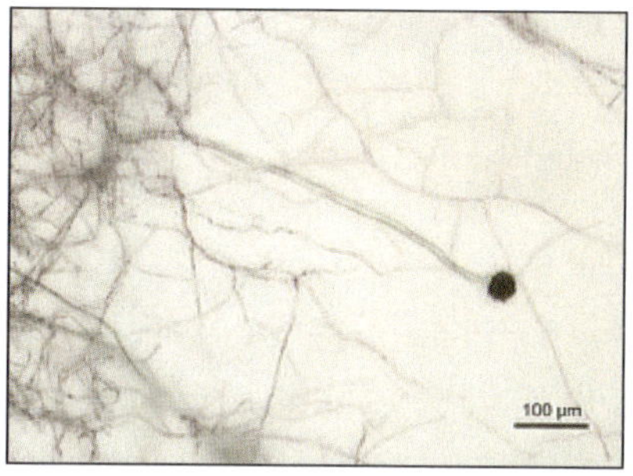

Aspergillus niger: Microorganisms such as Aspergillus niger are used in industrial microbiology for mass production of citric acid.

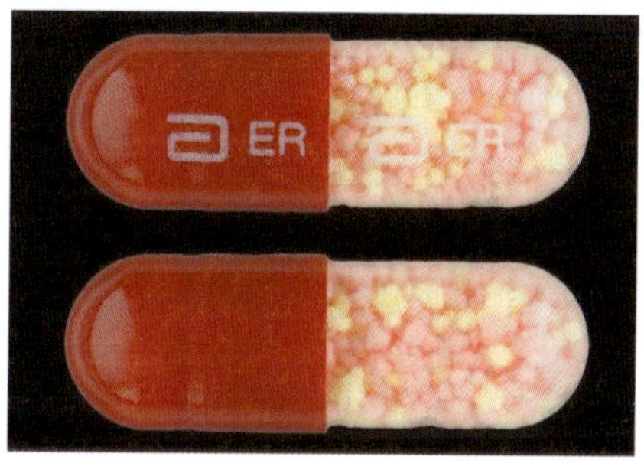

Erythromycin tablets: Erythromycin is an example of a secondary metabolite used as an antibiotic and mass produced within industrial microbiology.

Large-Scale Fermentations

Large-scale fermentations are key to the production of numerous products ranging from food to pharmaceutical items.

Fermentation includes the processes by which energy is extracted from the oxidation of organic compounds. The oxidation of organic compounds occurs by utilizing an endogenous electron acceptor to transfer electrons released from nutrients to molecules obtained from the breakdown of these same nutrients.

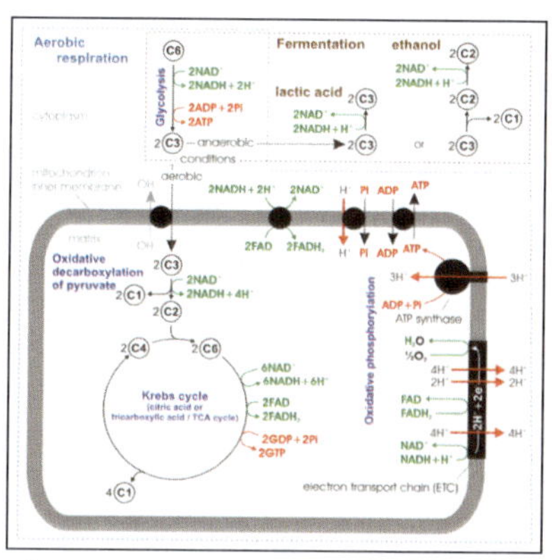

Common types of fermentation: These are common types of fermentation utilized in eukaryotic cells.

There are various types of fermentation which occur at the industrial level such as ethanol fermentation and fermentation processes used to produce food and wine. The ability to utilize the fermentation process in anaerobic conditions is critical to organisms which demand ATP production by glycolysis. Fermentation can be carried out in aerobic conditions as well, as in the case of yeast cells which prefer fermentation to oxidative phosphorylation. The following is a brief overview of a few types of the large-scale fermentations utilized by industries in production creation.

Ethanol Fermentation

Ethanol fermentation is used to produce ethanol for use in food, alcoholic beverages, and both fuel and industry. The process of ethanol fermentation occurs when sugars are converted into cellular energy. The sugars which are most often used include glucose, fructose, and sucrose. These sugars are converted into cellular energy and produce both ethanol and carbon dioxide as waste products. Yeast is the most commonly used organism to produce ethanol via the fermentation process for beer, wine, and alcoholic drink production. As stated previously, despite abundant amounts of oxygen which may be present, yeast prefer to utilize fermentation. Hence, the use of yeast on a large-scale to produce ethanol and carbon dioxide occurs in an anaerobic environment.

The ethanol which is produced can then be used in bread production. Yeast will convert the sugars present in the dough to cellular energy and produce both ethanol and carbon dioxide in the process. The ethanol will evaporate and the carbon dioxide will expand the dough. In regards to alcohol production, yeast will induce fermentation and produce ethanol. Specifically, in wine-making, the yeast will convert the sugars present in the grapes. In beer and additional alcohol such as vodka or whiskey, the yeast will convert the sugars produced as a result of the conversion of grain starches to sugar by amylase. Additionally, yeast fermentation is utilized to mass produce ethanol which is added to gasoline. The major source of sugar utilized for ethanol production in the US is currently corn; however, crops such as sugarcane or sugar beets can be used as well.

Fermentation in grapes: This is a photograph of grapes undergoing fermentation during the wine-making process.

Recombinant Products

Fermentation is also utilized in the mass production of various recombinant products. These recombinant products include numerous pharmaceuticals such as insulin and hepatitis B vaccine. Insulin, produced by the pancreas, serves as a central regulator of carbohydrate and fat metabolism and is responsible for the regulation of glucose levels in the blood. Insulin is used medically to treat individuals diagnosed with diabetes mellitus. Specifically, individuals with type 1 diabetes are unable to produce insulin and those with type 2 diabetes often develop insulin resistance where the hormone is no longer effective.

The increase in individuals diagnosed with diabetes mellitus has resulted in an increase in demand for external insulin. The mass production of insulin is performed by utilizing both recombinant DNA technology and fermentation processes. E. coli, which has been genetically altered to produce proinsulin, is grown to a large amount to produce sufficient amounts in a fermentation broth. The proinsulin is then isolated via disruption of the cell and purified. There is further enzymatic reactions that occur to then convert the proinsulin to crude insulin which can be further altered for use as a medicinal compound.

An additional recombinant product that utilizes the fermentation process to be produced is the hepatitis B vaccine. The hepatitis B vaccine is developed to specifically target the hepatitis B virus infection. The creation of this vaccine utilizes both recombinant DNA technology and fermentation. A gene, HBV, which is specific for hepatitis B virus, is inserted into the genome of the organism yeast. The yeast is used to grow the HBV gene in large amounts and then harvested and purified. The process of fermentation is utilized to grow the yeast, thus promoting the production of large amounts of the HBV protein which was genetically added to the genome.

Diagnostic Microbiology

Diagnostic microbiology concentrates on the laboratory analysis of clinical specimens in cases when an infectious disease is suspected. The diagnosis of staphylococcal infections may involve clinical specimens isolated from humans, animals, or food products, as well as samples collected from the environment. A presented overview of the diagnosis of staphylococcal infections includes the following topics: specimen collection and direct specimen Gram staining; microscopic examination of Gram stained material; inoculation into general-purpose media as well as into selective-differential media; incubation, colony morphology identification, Gram staining, and isolation of pure culture; catalase testing; furazolidone susceptibility testing; detection of free coagulase and bound coagulase; identification of bacterial species based on biochemical tests; detection of resistance mechanisms (beta-lactamases, methicillin resistance, macrolide, and lincozamine resistance, or, the MLSB resistance mechanism); and antibiogram execution. Automated methods applied to staphylococcal diagnostics are presented.

The purpose of diagnostic microbiology is to confirm the suspicion of infectious disease and to identify the etiologic agent, often by bacterial or fungal culture or virus isolation. When the pathologist suspects infectious disease, microbiologic assays are selected based on the differential diagnosis established from the history, postmortem examination, or histologic evaluation, and on the availability of validated tests. Staying abreast of emerging diseases and rapidly developing diagnostic methods requires continuing education.

Pharmaceutical Microbiology

When microbiological concepts, processes and techniques are applied to pharmaceutical operations, the subject is then called 'pharmaceutical microbiology'. It can be defined as the study of microorganisms that are pertinent to the production of antibiotics, enzymes, vitamins, vaccines, and other pharmaceutical products; it also incorporates the study of microorganisms that cause pharmaceutical contaminations, and degradation, deterioration and spoil of pharmaceutical raw materials and finished products.

Generally, pharmaceutical microbiology provides knowledge and understanding with regards to the significance of the presence of bacteria, yeasts, moulds, viruses and toxins in pharmaceutical raw materials, intermediates, products and pharmaceutical production environments, as well as the microbiological control of pharmaceutical products, production environments and people.

One of the key objectives of pharmaceutical microbiology is to ensure safety and efficacy of pharmaceutical products. It embraces the processes like the validation of disinfectants, evaluation of the efficacy of disinfectants in suspension, on surfaces, and through field trials. Pharmaceutical microbiology offers protocols and techniques associated with the operation and assurance of clean-room, aseptic-room and controlled environments for preventing any possible microbial contamination, and introduces risk assessment and practical contamination control strategies.

Since the introduction of this applied subject area of microbiology, over the years, pharmaceutical microbiology has evolved and expanded significantly to encompass various other aspects, e.g., research and development of new anti-infective agents, the use of microorganisms to detect mutagenic and carcinogenic potential in drugs, and the use of microorganisms in the manufacture of insulin and human growth hormone.

As pharmaceutical microbiology is crucial to the manufacture and production of drugs, biologics and devices, it is absolutely important that people working in these industries have a strong understanding and in depth knowledge of pharmaceutical microbiological principles, techniques, processes and strategies in order to avoid any potentially costly and life-threatening failures and consequences. Traditionally, in pharmacy and pharmaceutical sciences curricula, pharmaceutical microbiology is taught at various lengths and depths. Also, in many microbiology curricula, often pharmaceutical microbiology along with medical microbiology is taught. Pharmaceutical microbiology will continue to provide evidence-based and scientifically-proven approaches and procedures to ensure sterility, determine antimicrobial effectiveness, microbial contamination or bioburden, or to analyze endotoxins. With the emergence of 'super bugs' or 'multiple drug resistant strains' of pathogenic microorganisms, the role of pharmaceutical microbiology has become even more significant and relevant to quality healthcare provisions, and must be taught in great lengths, especially in healthcare and medicine related curricula world-wide.

Role of Microorganisms used in Wastewater Treatment

The role of microorganisms in wastewater treatment helps to treat and purify wastewater and make it less harmful to the environment.

While there are many different microbes used in sewage treatment, there are three well-known microbes that play an instrumental role in keeping sewage clean. Each of these types of bacteria help the treatment process in a unique way to ensure there is little to no impact on the surrounding environment.

Aerobic Bacteria

Aerobic bacteria are mostly used in new treatment plants in what is known as an aerated environment. This bacterium uses the free oxygen within the water to degrade the pollutants in the wastewater and then converts it into energy that it can use to grow and reproduce.

For this type of bacteria to be used correctly, it must have oxygen added mechanically. This will ensure the bacteria are able to do their job correctly and continue to grow and reproduce on its food source.

Anaerobic Bacteria

Anaerobic bacteria are used in wastewater treatment on a normal basis. The main role of these bacteria in sewage treatment is to reduce the volume of sludge and produce methane gas from it.

The great thing about this type of bacteria and why it's used more frequently than aerobic bacteria is that the methane gas, if cleaned and handled properly, can be used as an alternative energy source. This is a huge benefit considering the already high wastewater treatment energy consumption levels.

Unlike aerobic bacteria, this type of bacteria is able to get more than enough oxygen from its food source and will not require adding oxygen to help do its job. Phosphorus removal from wastewater is another benefit of anaerobic microbes used in sewage treatment.

Facultative

Facultative microorganisms in sewage treatment are bacteria that can change between aerobic and anaerobic depending on the environment they are in. Note that these bacteria normally prefer to be in an aerobic condition.

Soil Microbiology

Soil microbiology is the scientific discipline that is concerned with the study of all biological aspects of the microorganisms (bacteria, archaea, viruses, fungi, parasites and protozoa) that exist in the soil environment. This is a subdiscipline of environmental microbiology.

Contribution of Microbes to Nutrient Cycling

In soils, microbes play a pivotal role in cycling nutrients essential for life. For example, soil microbes play major roles in cycling carbon, nitrogen (N) and phosphorus, which are essential for producing biomolecules such as amino acids, proteins, DNA and RNA – the fundamental compounds of life. Many plant nutrients are ultimately derived from weathering of minerals. Silicate minerals such as feldspar, mica and hornblende provide calcium, magnesium and potassium, whereas apatite is the primary mineral source of phosphorus. Mineral weathering by soil bacteria and fungi plays a signifi cant role in ion cycling and plant nutrition.

Carbon Cycling

Microbes play major roles in the cycling of carbon – the key constituent of all living organisms. Primary producers fix carbon dioxide and convert it to organic material. In terrestrial ecosystems the primary producers of organic material are plants, although surface-dwelling algae and cyanobacteria, both free-living and symbiotic as lichens, can contribute signifi cantly to carbon fixation in some ecosystems. Within soil, autotrophic microbes can also fix carbon dioxide.

Organic materials resulting from primary production reside in living organisms and the non-living organic materials derived from them. Heterotrophic bacteria and fungi are the ultimate recyclers of non-living organic material. These soil saprotrophs complete the carbon cycle, converting organic material formed by primary producers back to carbon dioxide during respiration. They are sometimes aided in this process by higher animals (herbivores and carnivores) that digest particulate organic material with the help of microbes residing in their intestinal tracts. This process is known as decomposition and involves the degradation of nonliving organic material to obtain energy for growth. Mineralisation of the organic compound occurs when it is degraded completely into inorganic products such as carbon dioxide, ammonia, and water.

In soil ecosystems, the major agents of organic matter decomposition are fungi, which constitute the majority of soil biomass. However, both bacteria and fungi degrade complex organic molecules that higher organisms cannot break down. A wide variety of bacteria, especially those belonging to Actinobacteria and Proteobacteria, degrade soluble organic molecules such as organic acids, amino acids, and sugars. Likewise, some bacteria, such as Bacteroidetes, help degrade more recalcitrant carbon compounds such as cellulose, lignin and chitin. Bacteria that target these recalcitrant carbon compounds may require relatively high levels of available N to support the production of extracellular and transport enzymes In contrast, bacteria adapted to low N environments are more adept at metabolising organic N compounds such as amino acids. Net carbon mineralisation in soils was reported to be positively correlated with β- Proteobacteria and Bacteroidetes abundance and negatively correlated with Acidobacteria.

Microbes are unique in their capacity to carry out anaerobic (fermentative) degradation of organic matter, which results in the fermentation of organic compounds to organic acids, and generates gases such as hydrogen and carbon dioxide. Under strictly anaerobic conditions the hydrogen may be used by methanogens to reduce carbon dioxide to produce methane gas. Some methanogens can metabolise methanol, acetate or methylamine to methane and carbon dioxide.

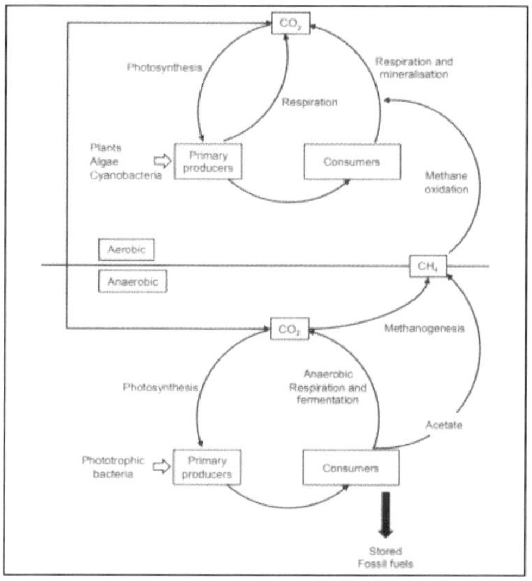

Microbial role in the global carbon cycle.

Nitrogen Cycling

All organisms require nitrogen, because it is an essential element in protein and nucleic acids. Animals derive nitrogen from organic sources while plants require inorganic nitrogen sources such as ammonium and nitrate, or relatively depolymerised nitrogen sources such as single amino acids (e.g. glycine). Most microbes can use ammonium or nitrate for growth.

Microbes play an important role in the nitrogen cycle. They carry out several processes not carried out by other organisms, namely nitrogen fixation, dissimilatory nitrate reduction to ammonia (DNRA), nitrification, anammox, and denitrification. Because nitrogen is often the major limiting nutrient for plant biomass production in terrestrial habitats, the rates of these microbial processes often limit ecosystem productivity. Some steps in the nitrogen cycle are mediated by few microbial groups (e.g. nitrogen fixation or nitrification) and are referred to as narrow processes, whereas others are mediated by many groups (e.g. DRNA) and are considered broad processes. The release of ammonium from soil organic matter during decomposition is known as ammonification.

Only bacteria and archaea carry out biological nitrogen fixation (N-fixation), the reduction of atmospheric nitrogen gas to ammonium. N-fixation is the only natural process through which new N enters the biosphere, so it is critically important for ecosystem function. N-fixation is catalysed by the enzyme nitrogenase. This enzyme is extremely sensitive to oxygen, requiring a low oxygen environment for activity. N-fixation is energetically expensive, consuming 16 moles of ATP per mole of N fixed. The ammonium produced through N-fixation is assimilated into amino acids and subsequently polymerised into proteins. Under nitrogen-limiting conditions, N-fixing microbes have an advantage. N-fixation is carried out by free-living microbes (e.g. Azotobacter, Burkholderia, Clostridium and some methanogens), some of which may be associated with the rhizosphere of plants, and bacteria that form symbiotic relationships with plants (e.g. Rhizobium, Mesorhizobium, Frankia). Exudates from plants may supply some of the energy required for N-fixation. In agricultural soils in New Zealand, rhizobia that form root nodules in symbiotic relationships with introduced legumes such as clover, lucerne or lotus are a signifi cant source of N. Similarly, native legumes (e.g. Sophora and Clianthus) form symbiotic relationships with Mesorhizobium or Rhizobium leguminosarum. Notably, the strains of rhizobia on native legumes differed from those on weed legumes like gorse. The rates of N-fixation by symbiotic rhizobia are often two or three orders of magnitude higher than by free-living bacteria in soil.

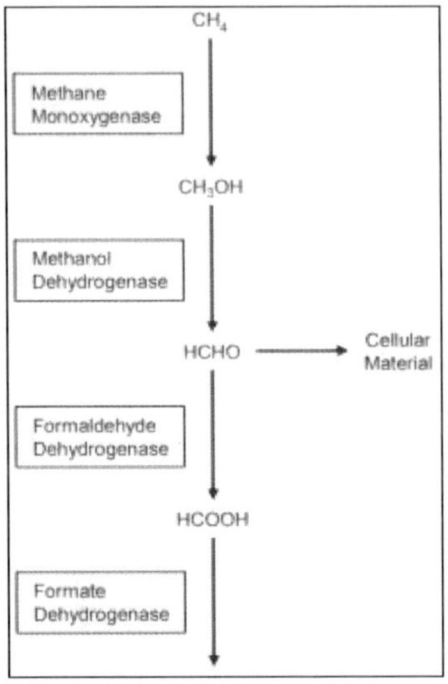

Illustration of single carbon compound (C1) metabolism used by methanotrophs for biomass production.

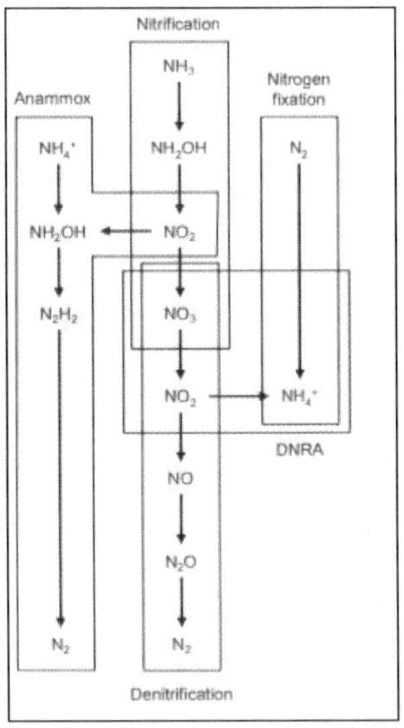

Microbial role in the nitrogen cycle.

During nitrification, ammonia or ammonium ions are oxidised to nitrite and then to nitrate. In soil, nitrification is aerobic and appears to be restricted to a few autotrophic bacteria and Crenarchaea. The two steps in nitrification – the formation of nitrite, then nitrate – are carried out by different microbial groups. In soils, oxidation of ammonia to nitrite is mediated by bacteria like Nitrosospira and Nitrosomonas or the crenarchaeum Nitrososphaera, whereas the oxidation of nitrite to nitrate is mediated by bacteria such as Nitrobacter and Nitrospira. Nitrifying microbes utilise the energy derived from nitrification to assimilate carbon dioxide. Nitrification is especially important in soils, because the oxidation of ammonium to nitrite and nitrate ions changes their charge from positive to negative. This leads to nitrate leaching, because the positively charged ammonium ions (NH_4^+) tend to be bound by negatively charged clay particles but the negatively charged nitrate ions (NO_3-) can be readily leached into groundwaters. To minimise this nitrate leaching, nitrification inhibitors have been applied to New Zealand soils. In New Zealand grassland soils, ammonia-oxidising bacteria (AOB) were more abundant in topsoils, whereas ammonia-oxidising archaea (AOA) were more abundant in one of the subsoils. Apparently, AOB and AOA may prefer different soil N concentrations to grow: AOB dominate under high ammonia substrate conditions, AOA dominate under low ammonia substrate concentrations.

Denitrification is a microbial respiratory process during which soluble nitrogen oxides are used as an alternative electron acceptor when oxygen is limiting. It consists of the sequential reduction of nitrate (NO_3-), nitrite (NO_2-) and nitric oxide (NO) to the greenhouse gas nitrous oxide (N_2O) or benign nitrogen gas (N_2). It occurs predominantly in waterlogged areas that have become anaerobic. Complete denitrification (to N_2) is the major biological mechanism by which fixed N returns to the atmosphere from soil and water, completing the nitrogen cycle. It results in considerable loss of fixed N from soil, and has important consequences because nitrogen is the limiting nutrient for crop production. The ability to denitrify has been identifi ed in a diverse range of phylogenetically unrelated soil bacteria including members of the Proteobacteria, Actinobacteria, and Firmicutes, as well as in fungi and other soil eukaryotes. However, many denitrifiers lack one or more of the enzymes involved in denitrification, and are thus often said to be 'incomplete'; for example, most fungi and approximately one-third of sequenced bacterial denitrifiers lack N_2O reductase, so their final denitrification product is $N\ O$. Incomplete denitrification is a major source of greenhouse gas emissions from pastoral agriculture in New Zealand.

In an alternative process called dissimilatory reduction of nitrate, a variety of facultative anaerobic bacteria including Alcaligenes or Escherichia reduce nitrate to nitrite under anaerobic conditions. The nitrite produced by these species is excreted or, under appropriate conditions, some microbes reduce nitrite via hydroxylamine to ammonia. These organisms do not produce gaseous nitrogen products: that is, they do not denitrify.

Anammox bacteria anaerobically oxidise ammonium to nitrogen gas (N_2). The anammox reaction depends on the concomitant presence of both oxidised and reduced inorganic nitrogen compounds under anaerobic conditions. In soil, anammox bacteria have been detected in permafrost and agricultural soil, and from bulk soils and soil associated with nitrogen-fixing plants. The bacteria that carry out this reaction form a deep-branching, monophyletic group within the Planctomycetes. Phylogenetic analysis has revealed that 16S rRNA gene sequences cloned from these bacteria in soils were closely related to sequences from the candidate genera 'Kuenenia' and 'Brocadia'.

Some bacteria can participate in multiple steps in the nitrogen cycle. For example, Rhizobium, Bradyrhizobium and Azospirillum have members that both fix nitrogen and denitrify. Moreover nitrifying bacteria such as Nitrosomonas can carry out denitrification: this process is called nitrifi er denitrification.

Phosphorus Cycling

Phosphorus (P) is not an abundant element in the environment, and its availability is further restricted by a tendency to precipitate in the presence of divalent and trivalent cations at neutral and alkaline pH. Microbes transform phosphorus in two main ways. In one, they mineralise organic P (occurring mainly as phosphate esters) to form inorganic phosphate in a process catalysed by phosphatase enzymes, which are produced by many bacteria and fungi. In the other, they transform insoluble, immobilised P to soluble or mobile P in a process normally mediated by the production of organic acids. Microbes release suffi cient P for their own use and that of plants and other soil organisms.

Mycorrhizal fungi produce oxalate to release phosphate from insoluble mineral P. This mobilisation of P by fungal symbionts is a major strategy that allows plants to overcome P-limitation. For example, several ectomycorrhizal basidiomycetous fungi have high-affi nity phosphate transporters that are expressed in extraradical hyphae in response to P defi ciency in their host. In New Zealand pasture soils, P-solubilising bacteria have been found in the Proteobacteria (in particular Pseudomonas), Actinobacteria, Firmicutes, and Bacteroidetes, but their numbers and diversity are affected by farm management strategies, with highest numbers in soils low in P. There is evidence that long-term application of P-rich fertiliser can alter the diversity of Actinobacteria and arbuscular mycorrhizal fungi in pasture soils.

Microbes as Agents for Recycling Wastes and Detoxification

Naturally occurring microorganisms – particularly bacteria and fungi – have evolved an impressive array of mechanisms to biodegrade or detoxify substances hazardous to human health or the environment. These microbial processes are being harnessed for bioremediation.

Biodegradation

Many years of laboratory studies have provided a wealth of information about how microbes biodegrade or detoxify organic contaminants. These studies describe the establishment of enrichment cultures for detection of biotransformation of contaminants under a range of environmental conditions: for example, pH, or nutrient or oxygen availability. The source of microbes for the enrichment cultures are typically soils contaminated with the compound of interest. Where possible, pure cultures that can degrade the contaminant are obtained and have been used for biochemical and molecular characterisation of the degradation pathways.

Heterotrophic bacteria in soil – for example Pseudomonas, Sphingomonas and Mycobacterium – have often been implicated in oil degradation. Pseudomonas, for example, has been well studied and the genes and enzymes responsible for degrading alkanes, monoaromatics, naphthalene, and phenanthrene as a sole carbon source under aerobic conditions are well understood. Knowledge of the mechanisms that microbes use to degrade oil has been applied in situ. For example, enhancing oil degradation in soil typically involves addition of nutrients (N and P) and sometimes oxygen and water. There is usually no need to add hydrocarbon-degrading bacteria to oil-contaminated sites because they are ubiquitous in soil and when oil is spilled they increase in numbers. However, high concentrations of hydrocarbons can deplete available nitrogen and phosphorus because these elements are assimilated during biodegradation; consequently, activity of the hydrocarbon degraders may become limited by these nutrients.

Bacteria and fungi also degrade pesticides. For example, the bacterium Arthrobacter nicotinovorans HIM, isolated from a New Zealand agricultural soil, degraded atrazine as a sole source of carbon and nitrogen. In addition to atrazine the bacterium also degraded the related triazine compounds simazine, terbuylazine, propazine, and cyanazine. Pesticides broken down rapidly in soil may not effectively control pests. Others like DDT, which was used extensively in New Zealand for the control of grass grub, are not readily degradable and persist in soil. Under aerobic conditions DDT is converted to DDE, which was considered a dead-end metabolite. However, Terrabacter sp. Strain DDE-1, isolated from soil from Winchmore Research Station, metabolised DDE when grown on biphenyl (i.e. when biphenyl was provided as an alternative for growth).

Ligninolytic fungi such as the white rot fungus Phanaerochaete chrysosporium can degrade a diverse range of environmental contaminants such as pentachlorophenol and dioxin under co-metabolic conditions (i.e. with alternatives for growth such as sawdust, straw or corn cobs). This impressive ability has been attributed to the mechanisms these fungi have evolved to degrade lignin. New Zealand strains of white rot fungi and also Zygomycetes degraded pentachlorophenol and selected dioxin and furan cogeners in soil samples from a former dip tank wood-treating operation in Whakatane.

Biodegradation in situ is a function of three independent but interrelated factors: the contaminant, the microbes, and the environment.

Both the chemical structure and the physical state of an organic contaminant affect the rate at which it is biodegraded. In general, microbes can degrade naturally occurring organic contaminants such as those associated with oil, whereas some synthetic molecules like DDT and aldrin persist in the environment and are not readily degraded. Synthetic molecules often contain novel arrangements rarely found in nature, which increases persistence. Resistance to degradation is linked with a decrease in water solubility, so larger molecules tend to be less soluble and harder to degrade. Many organic contaminants are hydrophobic or poorly soluble in water, and they bind to soil organic matter or clay surfaces. This may reduce their toxicity but it also reduces their biodegradability. Degradation may also be impeded when the contaminant concentration is too high because these organic contaminants, while serving as carbon and energy sources for microbes, may be toxic at high concentrations.

For biodegradation to proceed, microbes with the appropriate biodegradative ability must be present in suffi cient numbers. This will depend in part on how long they have been exposed to the contaminant. As some pollutants contain a mixture of compounds (e.g. oil) a mixture of microbes is required because no single microbe has the metabolic potential to degrade all contaminants. It is essential that the microbes and the contaminant are in contact for biodegradation to occur. Some bacteria are mobile and chemotactic, sensing the contaminant and moving towards it; other microbes such as fungi grow as filaments towards the contaminant.

The presence of the required microbial population is not enough: environmental conditions in situ must permit microbial growth or activity. Microbial growth and activity are sensitive to pH, temperature, moisture, nutrient availability and oxygen concentrations, with most microbes growing optimally over a narrow range of these conditions. Hydrocarbons are readily degraded aerobically whereas reactions involving dechlorination (e.g. degradation of trichloroethylene) often require anaerobic conditions. Other compounds, such as the alkylated benzenes, are degradable under aerobic and anaerobic conditions. Nutrients that may limit biodegradation in situ include nitrogen, phosphorus, potassium, and iron.

Detoxification of Heavy Metals

Environmental exposure of microbes to heavy metals has led to the evolution of Detoxification mechanisms. In soils, heavy metal contaminants include copper, mercury (Hg), zinc (Zn), lead, cobalt (Co) and cadmium (Cd); in New Zealand agricultural soils, cadmium accumulation is linked to the use of superphosphate fertiliser. Metals may be toxic to soil microbes due to their chemical affi nity for thiol groups on biomolecules such as proteins. To avoid cellular damage caused by these metals, bacteria have evolved three general mechanisms for metal tolerance. The fi rst is sequestration of the metals by binding to cell constituents, which reduces the concentration of free ions in the cytoplasm. The metals can be adsorbed by cell membranes, cell walls, and extracellular polymeric substances (EPS) such as polysaccharides. Several metals can be sequestered in EPS, including copper and lead. The second mechanism involves Detoxification through reduction of intracellular ions. For example, Hg^{+2} may be reduced to Hg^{o} by mercury reductase (encoded by the merA gene), and the Hg^{o} then diffuses from the cell because of its low evaporation point. The third mechanism involves extrusion of ions from the cell by effl ux systems. The cation/proton antiporter Czc, known for example in Alcaligenes eutrophus, mediates resistance to Cd^{2+}, $Zn^{2}+$ and Co^{2+} by expelling metals from the cytoplasm through the cell membrane to the environment. These microbial transformations of heavy metals are being harnessed for bioremediation of wastes containing heavy metals.

Reference

- Industrial-microbiology, chapter, boundless-microbiology: lumenlearning.com, Retrieved 6 August, 2019
- Diagnostic-microbiology, immunology-and-microbiology, topics, sciencedirect.com, Retrieved 5 July, 2019
- Pharmaceutical-microbiology-an-applied-branch-of-microbiology: pharmaceutical-microbiology.imedpub.com, Retrieved 4 June, 2019
- Role-microbes-microorganisms-used-wastewater-sewage-treatment: aosts.com, Retrieved 5 May, 2019

Permissions

All chapters in this book are published with permission under the Creative Commons Attribution Share Alike License or equivalent. Every chapter published in this book has been scrutinized by our experts. Their significance has been extensively debated. The topics covered herein carry significant information for a comprehensive understanding. They may even be implemented as practical applications or may be referred to as a beginning point for further studies.

We would like to thank the editorial team for lending their expertise to make the book truly unique. They have played a crucial role in the development of this book. Without their invaluable contributions this book wouldn't have been possible. They have made vital efforts to compile up to date information on the varied aspects of this subject to make this book a valuable addition to the collection of many professionals and students.

This book was conceptualized with the vision of imparting up-to-date and integrated information in this field. To ensure the same, a matchless editorial board was set up. Every individual on the board went through rigorous rounds of assessment to prove their worth. After which they invested a large part of their time researching and compiling the most relevant data for our readers.

The editorial board has been involved in producing this book since its inception. They have spent rigorous hours researching and exploring the diverse topics which have resulted in the successful publishing of this book. They have passed on their knowledge of decades through this book. To expedite this challenging task, the publisher supported the team at every step. A small team of assistant editors was also appointed to further simplify the editing procedure and attain best results for the readers.

Apart from the editorial board, the designing team has also invested a significant amount of their time in understanding the subject and creating the most relevant covers. They scrutinized every image to scout for the most suitable representation of the subject and create an appropriate cover for the book.

The publishing team has been an ardent support to the editorial, designing and production team. Their endless efforts to recruit the best for this project, has resulted in the accomplishment of this book. They are a veteran in the field of academics and their pool of knowledge is as vast as their experience in printing. Their expertise and guidance has proved useful at every step. Their uncompromising quality standards have made this book an exceptional effort. Their encouragement from time to time has been an inspiration for everyone.

The publisher and the editorial board hope that this book will prove to be a valuable piece of knowledge for students, practitioners and scholars across the globe.

Index